Mechanism
in
Organic Chemistry

MECHANISM

IN

ORGANIC CHEMISTRY

ROGER W. ALDER
University of Bristol

RAY BAKER
University of Southampton

JOHN M. BROWN
University of Warwick

WILEY – INTERSCIENCE

a division of John Wiley & Sons Ltd.

London New York Sydney Toronto

Library of Congress catalog card number 72–147196

ISBN 0 471 02050 8 Cloth bound

ISBN 0 471 02051 6 Paper bound

Printed in Great Britain by
William Clowes and Sons, Limited.
London, Colchester and Beccles

Preface

No student or qualified chemist needs to be reminded that organic chemistry is an extremely wide and diverse subject. It has, however, become increasingly possible over the years to rationalize the factors which govern the course of organic reactions, and to predict reactivity and reaction course in most situations. The quantitative study of organic reactions has become a subject in its own right, whose conclusions are relevant to the whole of organic chemistry

We believed that the current state of the mechanistic study of organic reactions was worth assessing, and in doing so we have considered the needs of third-year undergraduates and beginning postgraduate students, with the hope that the material presented and our approach to the subject will be of value to many more experienced research workers and teachers. Our basic assumption is that a student will have a working knowledge of a textbook of organic chemistry such as 'Morrison and Boyd' or 'Roberts and Caserio', and will have received instruction in the fundamentals of physical chemistry. The aim of the book is to provide the reader having this background with an appreciation of organic reaction mechanism and the attendant aspects of physical organic chemistry such that research papers can be read with a real appreciation of the chemical context and relevance. We also hope that it will assist considerably in the design and execution of mechanistic experiments, and in the choice of appropriate experimental conditions for most organic reactions.

There is a traditional approach to the structure of textbooks in organic reaction mechanism that relies on sub-division into the various accepted classes of reaction type. In considering how we might best present the subject, we felt that this style now had several shortcomings and would inhibit us from giving an accurate assessment of the subject in the way it is approached by modern research workers. We felt that any sub-division should as far as possible reflect not only the relationship between reactants and products but also the nature of the reaction transition-state, and that the latter presented the most fundamental feature of the process, since it related to the timing of bond-breaking and bond-making processes. We have therefore attempted to sub-divide reactions according to the degree of association or dissociation of bonds to reactant carbon atoms. Thus, after an introductory chapter relating to the factors governing reactivity, we devote successive chapters to processes which

v

proceed through dissociation of a bond to carbon, processes in which synchronous dissociation and bond formation are occurring, reactions in which several bonds are changing at once (multicentre), and associative reactions involving transition-states formed as a consequence of bond formation to carbon. It will be seen that this approach establishes the fundamental relationship between, for example, the chemistry of carbonium ions and carbanions, and between S_E2, S_N2, and $E2$ reaction or electrophilic additions to olefins and electrophilic substitution. It has the added advantage of comparability with the accepted approach to the much younger subject of inorganic reaction mechanisms.

In a book of this size it is evident that the treatment will not be comprehensive, and we have been selective in favour of heterolytic rather than homolytic reactions, and in favour of reactions in solution. Many fascinating fields such as mechanistic enzymology, photochemistry, and organometallic chemistry are scarcely mentioned. Nevertheless we hope that much of our approach will provide a valuable background to the reader in more detailed considerations of these fields. The text is very fully referenced, and for each major topic certain key papers have been distinguished by a few descriptive words in an attempt to assist those readers with limited time to explore the literature. Many interesting topics, which could not be fitted into the text, have been treated as exercises for the reader.

In writing this work the authors have been considerably helped by forthright and uncompromising criticisms from one another, and by the advice, comment, and suggestions provided by many friends and colleagues. We are particularly indebted to Drs. R. G. Coombes, B. R. Davis, J. B. Hobbs, B. T. Golding, T. J. Mason, and M. J. Perkins, and Professor M. C. Whiting, who either read and criticized whole chapters or made valuable specific contributions. It would certainly not be inappropriate to thank our wives for their forebearance, and Miss Janet Brown who typed most of the final manuscript.

July, 1970

R. W. A.
R. B.
J. M. B.

Contents

Chapter 1   MECHANISM AND REACTIVITY   1

Energies of Ground States   5
Energies of Transition States   11
Primary Kinetic Isotope Effects   14
Secondary Kinetic Isotope Effects   18

Factors Affecting Reactivity   20
Steric and Strain Effects   20
Stereoelectronic Effects   23
Electronic Effects   26
Linear Free Energy Relationships   28

Medium Effects   40
Intermolecular Forces and Types of Solvation   42
Correlation of Solvent Effects   43
*A Thermodynamic Dissection of Some Protic–Dipolar-aprotic Solvent
   Effects*   48

Catalysis   51
Electrophilic and Nucleophilic Catalysis   52
Catalysis of Non-ionic Reaction Mechanisms   54
'Physical' Catalysis   56
Acid–Base Catalysis   57
Intramolecular Catalysis   69
Enzymic Catalysis   70

Problems   72

Chapter 2   DISSOCIATIVE PROCESSES   78

CARBONIUM IONS   78
Stability of Carbonium Ions   78
Kinetic Evidence for S_N1 Reactions   82

Stereochemistry of S_N1 Reactions 84
Ion Pairs 85
Structural Factors Affecting Reactivity 92
Participation by Neighbouring Groups 97
Reactions of Carbonium Ions 99
Fragmentation of Carbonium Ions 102
Rearrangement and Formation of Bridged Carbonium Ions . . 105
σ-Bond Participation 111

CARBANIONS 120

Stabilization of Anionic Carbon 125

Hybridization Effects 125
Conjugation Effects; pπ-Conjugation 127
Conjugation Effects; dπ-Conjugation 130
Inductive Effects 133

Reactions Involving Carbanions 135

Isomerizations and Rearrangements 135
Reductions by Electron Transfer 138
Reactions of Ambident Anions. 139

Stable Carbanions in Solution 141

CARBENES 145

Structure of Bivalent Carbon Intermediates . . . 146
Reactions of Carbenes 148

FREE RADICALS 155

Generation and Stability of Free Radicals . . 158

Reactions of Free Radicals 161

Hydrogen Atom Abstraction 161
Electron Transfer 163
Intramolecular Reactions 165

Detection of Free Radicals 166
Detection of Free Radicals as Reaction Intermediates; CIDNP . 168

Problems 170

Chapter 3   SYNCHRONOUS REACTIONS   180

BIMOLECULAR ELECTROPHILIC SUBSTITUTION REACTIONS 181

Protons as Electrophiles   182
Metals as Electrophiles   190

S_N2 DISPLACEMENT REACTIONS   194

Electronic, Solvent, and Steric Effects   195
Nucleophilicity   202
Hard and Soft Acids and Bases   205
S_N2' Reactions   208

BIMOLECULAR ELIMINATION REACTIONS   .   .   209

E1cb Mechanism   210
Transition States in E2 Eliminations   213
Orientation in E2 Eliminations   214
Stereochemistry in E2 Eliminations   220
Competition between Elimination and Substitution   .   .   227
Thermal syn-*Eliminations*   230
γ-Elimination   231

Problems   233

Chapter 4   MULTICENTRE REACTIONS   238

The Orbital Symmetry Rules: Electrocyclic Reactions   .   243

Cycloaddition Reactions.   245
Sigmatropic Reactions   248

Mechanisms of Electrocyclic Reactions   .   .   .   251

Mechanisms of Cycloaddition Reactions   .   .   .   256

Mechanisms of Sigmatropic Reactions   .   .   .   266

Problems   274

Chapter 5   ASSOCIATIVE REACTIONS   279

ASSOCIATIVE REACTIONS OF NON-POLAR BONDS   .   279

Electrophilic Aromatic Substitution   280
Substituent Effects on Reactivity   283

Isotope Effects 286
ortho-Substitution. 288
'Anomalous' Substitution 289
Substitution in Fused and Heteroatom Systems 291

Homolytic Aromatic Substitution 293

Electrophilic Addition to Unsaturated Systems 295

Reagent Types 297
'Charge-controlled' Addition 298
Stereochemistry of 'Charge-controlled' Addition . . . 302
'Overlap-controlled' Addition 305
Strained Cyclic Olefins 306
Electrophilic Addition to Acetylenes 307

Free Radical Addition to Olefins 308

NUCLEOPHILIC ASSOCIATIVE REACTIONS . . . 310

Nucleophilic Addition 315

Nucleophilic Addition–Elimination Reactions with Aldehydes and Ketones 321

Acylation Reactions 328

Ester Hydrolysis 328
Hydrolysis of Some Acetates in Concentrated Solutions of Sulphuric Acid 333
Mechanism of Other Acyl Transfer Reactions 335
Comparison of the Hydrolysis of Phosphate and Carboxylate Esters . 337

Nucleophilic Vinyl Substitution 340

Nucleophilic Aromatic Substitution 342

Substitution via Benzyne Intermediates 345
Reactions via Phenyl Cations 347

Problems 348

Author Index 357

Subject Index 367

Chapter 1
Mechanism and Reactivity

The detailed study of the mechanisms of organic reactions dates back to around the turn of the century, and many famous chemists have been intimately associated with its development. The efforts of these pioneers, among whom we may mention Bartlett, Hughes, Ingold, Lapworth, Meerwein, and Winstein as outstanding contributors, have produced a situation where the factors governing the behaviour and reactivity of organic molecules are rather well understood, and broad principles of mechanism adequately defined. Notwithstanding this position, the complexities and subtleties of the subject are such that no chemist exploring new organic reactions will work for very long without encountering some entirely unexpected products, and be forced to resort to *ad hoc* explanations of his results. Only when theory and mechanistic expertise have advanced to the stage where the rates of uncharted reactions and the proportions of the various products can be predicted with accuracy will reaction mechanisms cease to play a vital role in organic chemical research. Indeed, if and when that situation ever arises all organic research will cease. Rest assured that it is unlikely to be the case during the lifetime of readers of this text! It might be thought that the advent of the high-speed digital computer would hasten the realization of Dirac's aphorism that chemistry is an exercise in applied mathematics. The present state of the theoretical chemist's art is that he may produce a rather accurate description of the reaction profile for the process $H—H + \cdot H^* \rightarrow H\cdot + H—H^*$ whereby a hydrogen molecule interacts with a hydrogen atom. When it is considered that this is a three-electron three-atom system, and that a typical organic reaction, the Diels–Alder addition of maleic anhydride to cyclopentadiene, is a twenty-atom eighty-six-electron system, it will be appreciated that the experimental organic chemist ought not to be made redundant by computers for some considerable time. However, the use of approximate methods in theoretical organic chemistry is becoming increasingly important in its predictive power. Whilst they have more frequently been applied in the past to the structure and energy of ground-states, applications to reactivity and transition-state structure are popular current themes. We shall refer to examples in the text, and it is to

1

be expected that many valuable conclusions will accrue from this kind of work.

The central problem of organic reaction mechanisms, therefore, is to be able to define the reactivity of a species or set of species in a given situation, and to predict which pathway is likely to be followed where a choice is available. It is thus necessary that we think carefully about what precisely is happening during the course of a chemical reaction. In the simplest case, only one species is involved and this is converted into product without involvement of external agents. A familiar example of such a unimolecular reaction is the Claisen rearrangement of allyl vinyl ether (1) to pent-4-enal (2) which occurs on heating in the vapour or liquid phase to around 150°C. Reaction involves breaking of a carbon–oxygen σ-bond and formation of a carbon–carbon σ-bond; we may anticipate later discussions by stating that these two processes are occurring simultaneously.

$$\text{(1)} \qquad\qquad \text{(3)} \qquad\qquad \text{(2)}$$

Energy is required both to orient the reactant in a product-like conformation and to induce bond-weakening, and an individual molecule acquires this energy through the population of successively higher vibrational quantum levels by intermolecular collision. We may illustrate this qualitatively by constructing a plot of free-energy against the distances between carbon and oxygen in the bond being broken (x) and between carbon and carbon in the bond being formed (y). There are an infinite number of pathways between 1 and 2 which may be expressed in terms of the three-dimensional energy surface (Figure 1a). At some point on any pathway a vibrational oscillation of (x) will result in a reduction of energy since it propels the vibrationally excited molecule in the direction of product. The surface shown in Figure 1a is saddle-shaped and the cross-section (i) represents the most favourable path; (ii) shows a path where approach between the atoms is unfavourably close during the reaction, and (iii) shows a situation where bond-lengths become too attenuated. Clearly, the reacting system will prefer to traverse path (i), and we define its highest point as the 'transition-state', which is the point at which reactions may be consummated without further input of energy. Alternatively, we may plot the curve corresponding to (i) in two dimensions with free-energy as ordinate and an arbitrary reaction coordinate as abscissa (Figure 1b). This can be a particularly valuable way of representing a multistep reaction.[1]

Two thermodynamic quantities may be expressed by means of our reaction profile: the free-energy of activation, $\Delta G^{\ddagger}$, which is a measure of the energy difference between reactants and transition-state, and the free-energy of reaction, ΔG_r, which measures the energy difference between reactants and products. Any chemical reaction must follow the 'principle of microscopic reversibility' so that the minimum-energy pathway between reactants and products must also be the minimum-energy pathway between products and

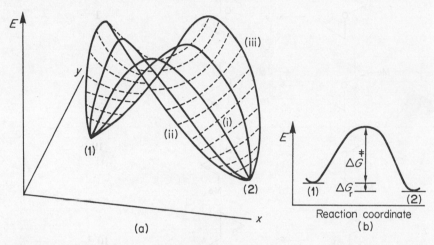

Figure 1. (a) Minimum-energy pathways between reactant and product, (b) Reaction free-energy profile.

reactants. ΔG_r therefore relates to the position of equilibrium in the reaction by means of the standard equation:

$$\Delta G_r = -RT \log_e K_r \qquad (4)$$

where K_r is the equilibrium constant.

The rate of reaction is totally independent of K_r, and may be completely expressed in terms of $\Delta G^{\ddagger}$. The equation derived by statistical thermodynamics requires inclusion of Boltzmann's constant, $\mathbf{k}$, and Planck's constant, h:

$$\Delta G^{\ddagger} = -RT \log_e k(h/\mathbf{k}T) \qquad (5)$$

Apart from emphasizing that the only factor determining reaction rate is the difference in free-energy between reactants and transition-state, this treatment illustrates an important limitation of kinetics. We cannot learn anything about the energy of any intermediate point on the reaction profile, or whether there are discrete steps in the reaction sequence before or after the transition-state. An important practical point may be introduced at this stage. The thermal rearrangement of allyl vinyl ether has a free-energy of activation of 32.7 kcal mole^{-1} at 150°C, and structural changes in the reactant can increase or decrease this value. The small differences in energy observed[2] (Figure 2) nevertheless correspond to quite substantial differences in reaction

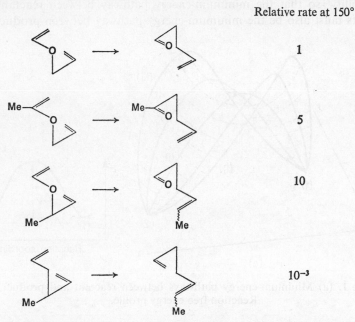

Relative rate at 150°

1

5

10

10^{-3}

Figure 2. Effect of structural changes on the rate of Claisen and Cope reactions; most of the alteration reflects changes in ΔH^{+} rather than ΔS^{+}.

rate, and demonstrate what is the most intriguing, and also perhaps the most frustrating, feature of mechanistic organic chemistry. Very minor changes in reactant may not only affect the rate markedly but they may cause reaction to proceed in an entirely new direction!

Free-energy of activation is a quantity which may be divided into enthalpy and entropy terms, according to the equation:

$$\Delta G^{+} = \Delta H^{+} - T\Delta S^{+} \tag{6}$$

Entropy normally plays the minor role in determining the free-energy of ground-states, but this is not the case when free-energies of activation are under consideration. $\Delta S^{\ddagger}$ is a measure of a change in the degree of ordering on going from reactants to transition-state, and it may provide valuable information on the structure of the latter. If the transition-state has a high degree of organization, as when several molecules must be brought together or when a single molecule may react only via a particular conformation, then negative values of $\Delta S^{\ddagger}$ result. Conversely, if the transition-state has a more random structure than the reactants, $\Delta S^{\ddagger}$ will be positive. Entropy of activation is directly related to the pre-exponential factor, A, of the Arrhenius equation (7) by the expression 8, and its determination may help us to make a decision between one reaction mechanism and another.

$$k = A\mathrm{e}^{-E/RT} \tag{7}$$

$$\Delta S^{\ddagger} = 4.576(\log_{10} A - 13.23) \qquad \text{(at 25°C)} \tag{8}$$

Typical values of entropy of activation are recorded in Table 1.

Table 1. Some typical values of entropy of activation (cal mole^{-1} deg^{-1}).

Reaction		Type		$\Delta S^{\ddagger}$
Cyclopropane $\xrightarrow{550°}$ Propene		Thermal	Dissociative	7
Bu^{t}Cl $\xrightarrow{H_2O}$ Bu^{t}OH		S_N1	Dissociative	14
Pr^{i}Cl $\xrightarrow{H_2O}$ Pr^{i}OH		S_N2	Synchronous	−8.1
MeCl $\xrightarrow{H_2O}$ MeOH		S_N2	Synchronous	−12.3
1-Me-cyclobutene $\xrightarrow{150°}$ 2-Me-buta-1,3-diene		Unimolecular	Multicentre	2.1
Cyclopentadiene + maleic anhydride (Diels–Alder)		Bimolecular	Multicentre	−36

Energies of Ground States

When considering reactions in the gas phase, the only ground-state terms required are the free-energies of formation of the reactants. A majority of reactions of interest, however, occur in solution where the solvent may interact with reactants and thereby alter their energy; we shall defer a study of such solvent effects until later in this chapter. If heats of formation, ΔH_f, and entropies of formation, ΔS_f, are known, then free-energies may be derived. The former quantity may be obtained calorimetrically from the heat of combustion, for example:

$$C_6H_8 + 8\,O_2 \rightarrow 6\,CO_2 + 4\,H_2O + \Delta H_c \tag{9}$$

Knowing the heat of combustion, ΔH_c, and the heats of formation of water and carbon dioxide, the heat of formation of C_6H_8 may be derived. Data are usually referred to a standard state which is the regular phase of the molecule at normal temperature (25°C) and pressure (760 mm Hg). Much less experimental data are available on entropies of formation although these may be calculated fairly accurately by consideration of the degrees of translational, vibrational, and rotational freedom available to the molecule. Consequently, ground-state thermodynamic functions are known for a large number of molecules to a high degree of accuracy.[3] In considering reactions, ground-state entropy normally makes the minor contribution towards free-energy terms, but it cannot always be ignored. A reaction which occurs at high temperatures in the gas phase is the conversion of one molecule of cyclobutane into two of ethylene (10). The reaction is strongly endothermic, that is, ΔH_r is positive despite the fact that reaction is favoured by a negative value of ΔG_r. The balance is swung by the favourable change in entropy when reaction occurs in the forward direction.

$$\square \xrightarrow[\quad]{800°K} \begin{array}{c} CH_2 \\ \| \\ CH_2 \end{array} + \begin{array}{c} CH_2 \\ \| \\ CH_2 \end{array} \tag{10}$$

$$
\begin{array}{lllll}
\Delta H & 6,400 & \xrightarrow{+18,600} & 12,500 + 12,500 & \text{cal mole}^{-1} \\
T\Delta S & 51,600 & \xrightarrow{-32,400} & 42,000 + 42,000 & \text{,, ,,} \\
\Delta G & -45,200 & \xrightarrow{-13,800} & 29,500 + 29,500 & \text{,, ,,}
\end{array}
$$

There are two rather distinct ways in which the energy of a molecule may be considered. From a wave-mechanical point of view, any configuration of the atoms in space may be associated with an electron configuration of minimum energy; it should therefore be possible to determine the ground-state energy (strictly enthalpy, since entropy is not taken into account in wave-mechanical calculations) by calculations in which molecular geometry is varied until an absolute minimum of electronic energy is reached. If calculations involving precise wave-functions are attempted, they become impossibly complex for all but the simplest molecules. Nevertheless, modern developments of the LCAO (linear combination of atomic orbitals) method such as CNDO (complete neglect of differential overlap), where approximate wave-functions are used to describe atom and bond integrals, may give a very successful simulation (see Table 2), particularly for systems containing only carbon and hydrogen.[4] Calculations of this type demand high-speed digital computing; as machine sophistication increases, the accuracy of these methods will improve and application to determination of ground-state enthalpy of increasingly complex molecules will ensue.

Table 2. Experimental heats of formation (kcal mole^{-1}) compared with values calculated by Dewar and Baird.[4]

Molecule	ΔH_f(obs.)	ΔH_f(calc.)	Molecule	ΔH_f(obs.)	ΔH_f(calc.)
Methane	−17.9	−17.9	Buta-1,3-diene	26.1	26.3
Ethylene	12.5	12.5	Cyclobutene	35.1	35.0
Cyclohexane	−30.0	−29.4	Cyclohexene	3.1	−1.7
Cyclopropane	12.1	12.7	Benzene	19.9	19.8
Cyclobutane	6.4	6.4			

An important point for our purposes is that the wave-mechanical approach does not factorize the total energy into components of individual bond-energies and does not separate any interactions between atoms or bonds which diminish the overall energy of the molecule.

It is frequently advantageous to the organic chemist to treat molecules as if the energy were localized in individual bonds. By a combination of results derived from heats of combustion, heats of hydrogenation, enthalpies of free-radical abstraction reactions (Chapter 2), and other techniques, a fairly comprehensive table of bond enthalpies may be produced (Table 3).

Table 3. Calculated values of bond enthalpy.

Molecule	Bond strength (kcal mole^{-1})		Comments
Ethane	C—H	98	
	C—C	82	
cis-But-2-ene	C=C	146	π-Bond energy $\approx$ 64 kcal mole^{-1}
But-2-yne	C≡C	200	π-Bond energy $\approx$ 2 × 59 kcal mole^{-1}
Cyclopropane	C—C	73	Strain energy $\approx$ 9 kcal mole^{-1} per C—C bond
Cyclopropene	C=C	111	π-Bond energy $\approx$ 38 kcal mole^{-1}
MeOH	C—O	85	
	O—H	110	
MeSH	C—S	65	
	S—H	80	
MeCOMe	C=O	180	ΔH_f(C=O) > 2ΔH_f(C—O)
Me$_3$N	C—N	73	
Me$_3$P	C—P	63	
MeF	C—F	118	Bond energy decreases sharply as the
MeCl	C—Cl	81	diffuseness of the outer-shell halogen
MeBr	C—Br	67	orbitals increases
MeI	C—I	54	

The most valuable aspect of this approach is that it permits an assessment of individual effects such as strain energy and interaction energies. A considerable effort has been made to determine ground-state energies by a purely empirical approach where the 'ideal' enthalpy of a molecule is estimated from bond-energy data, and the 'real' bond-energy is determined by summation of potential destabilizing factors. Since these factors are complex, the method has enjoyed most success in relatively simple systems, particularly in the calculation of the geometry of cyclic hydrocarbons which is associated with minimum energy.[5] As an example of this approach we may consider the structure of *cis*-1,3-dimethylcyclobutane (**11**). An intuitive possibility is that the carbon atoms lie at the corners of a square. In this conformation, however, the hydrogen atoms on each carbon atom are eclipsed by hydrogens on neighbouring carbon atoms, thereby introducing 'torsional strain' which is in addition to the 'angle strain' inherent in the structure, and a non-bonded interaction between the two methyl groups due to interpenetration of electron-clouds by their hydrogen atoms. If the molecule flexes to move the methyl groups apart, both non-bonded interaction and torsional strain will be diminished, although angle strain will be increased. A slight increase in bond-length will cause 'stretching strain' but may help to relieve non-bonded interactions. The empirical approach seeks to define the consequences of each of these types of strain, and to define a potential function for each one. These functions are often rather arbitrary, and derived from a 'because it works' standpoint, but torsional strain may be evaluated by reference to the ethane rotational barrier, and angle strain from spectroscopic data on vibrational bending force constants. Given values for the four potential functions, any geometry of *cis*-1,3-dimethylcyclobutane may be associated with a total strain energy, and, by stepwise variations in geometry, a minimum ground-state energy and geometry may be derived (Figure 3).

Most organic reactions occur at one or two localized sites in a molecule, and it is of advantage to assess specific stabilizing and destabilizing effects in the environment of the reaction centre. Reference to specific problems will indicate the kind of information which may be obtained.

It is a well established fact that solvolysis of tertiary esters may proceed by an S_N1 mechanism (discussed more fully in Chapter 2), and that the rate-determining stage then involves rehybridization at the reaction centre from sp^3 towards sp^2. Alkylation at the tertiary centre by bulky groups may produce a situation where the halide ground-state possesses a considerable amount of destabilization due to non-bonded H—H interactions between the alkyl groups. These interactions should be partly relieved as the carbon atom rehybridizes. We might therefore expect that increasing bulk in the alkyl group would lead to an increased solvolysis rate, and this expectation is

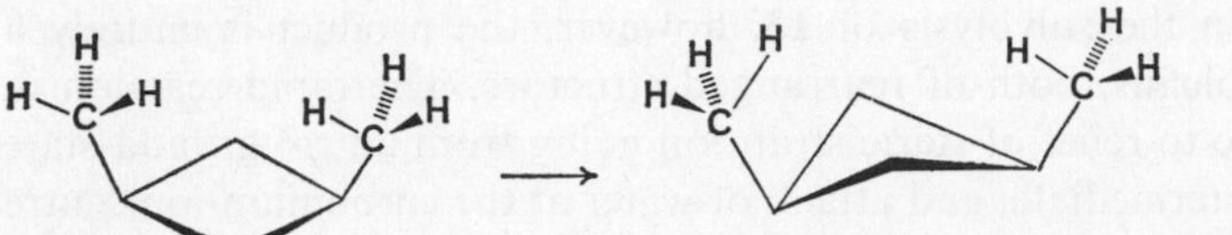

(11)

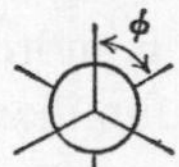

Torsional strain

$$E_t = \sum_\phi 0.165(1 + \cos 3\phi) \text{ over all}$$

angles at each C—C bond

Angle strain

$$E_\theta = 0.023(\theta - 112°) \text{ over all six angles}$$

at each tetrahedral C atom

C—C   *Bond-length strain* $E_r = \frac{1}{2}k(r - 1.54)^2$

H∴H   *Non-bonded interaction strain* $E_{H\cdots H} = 2.300\, e^{-3.6x} - 49.2/x^6$

Total enthalpy $= E_{\text{ideal bond energies}} - (E_t + E_\theta + E_r + E_{H\cdots H} + E_{H\cdots C})$

Figure 3. Typical potential functions used in strain-energy minimization calculations.

borne out in practice. t-Butyl *p*-nitrobenzoate (**12**) solvolyses very slowly in aqueous dioxan at 40° whereas tri-t-butylmethyl *p*-nitrobenzoate (**13**) reacts rapidly under identical conditions.[6] The ratio in reaction rate at 40° of 13,500/1 reflects a difference in free-energy of activation of 6 kcal mole^{-1}. As might be expected, on the basis of its intermediate steric effects, triisopropylmethyl *p*-nitrobenzoate (**14**) reacts at an intermediate rate 403 times greater than does t-butyl *p*-nitrobenzoate.

Solvolysis of 'normal' tertiary compounds in aqueous media usually produces a mixture of olefin (by an *E*1 mechanism) and tertiary alcohol (by

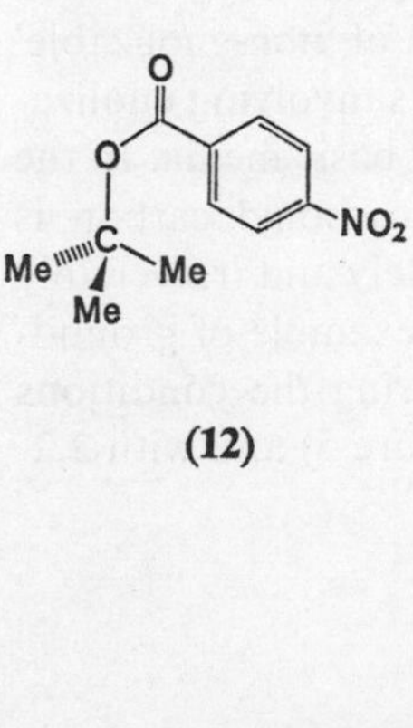

(12)

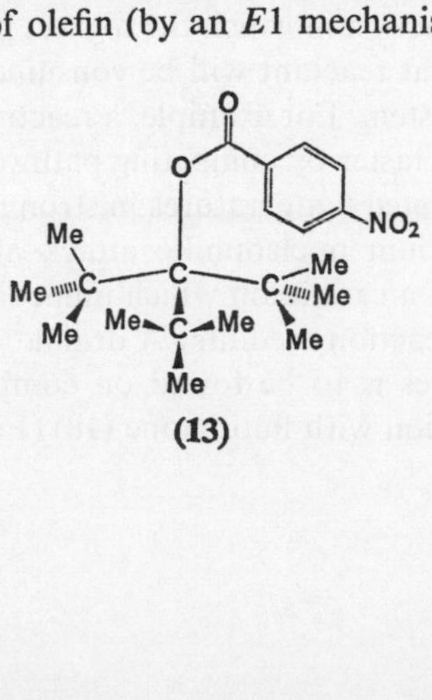

(13)

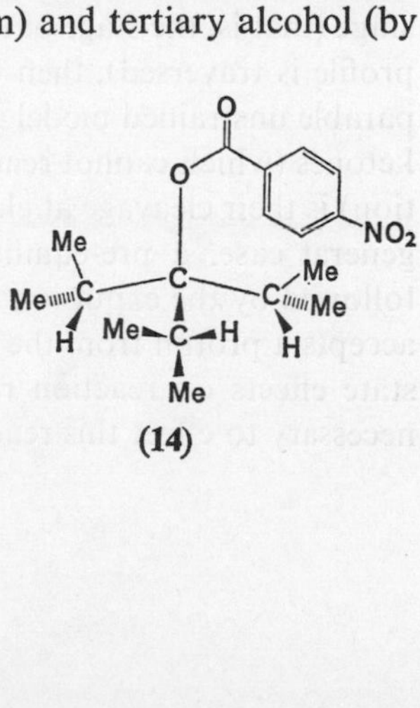

(14)

an S_N1 route). In the solvolysis of **13**, however, the product is entirely a mixture of two olefins, both of rearranged structure. The rapid reaction in this system is due to relief of steric strain on going from an sp^3 ground-state towards an sp^2 intermediate, and attack of water at the carbonium-ion centre reverses this favourable rehybridization. The carbonium ion therefore undergoes alkyl shift preferentially, leading to the observed products (Figure 4).

This example shows an increase of reaction rate due to steric crowding in the ground-state. When, as is commonly found, the transition-state is more crowded than the ground-state, the reverse effect is observed. In a familiar example, the rates of nucleophilic additions to acetone are invariably greater than corresponding additions to t-butyl methyl ketone since here the incoming

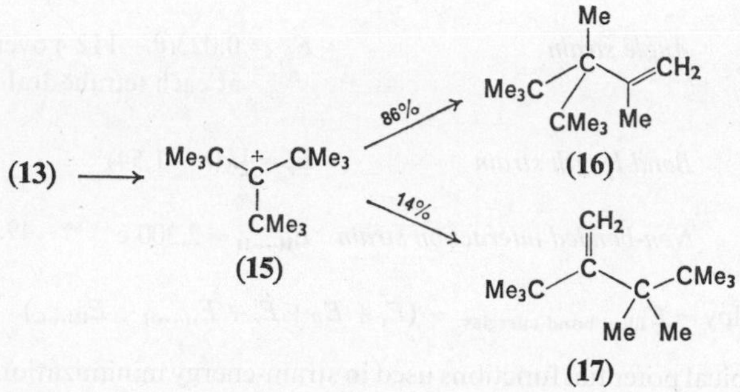

Figure 4. The solvolysis of **13**; **16** is formed by methyl shift followed by proton loss, and **17** by a sequence of methyl shift, t-butyl shift, and proton loss.

reagent experiences non-bonded repulsion by the hydrogens of the bulky t-butyl group.

The operation of ground-state steric effects in enhancing or decreasing reaction rates finds some parallel in situations where bond-angle strain is important. If angle strain in a given reactant is relieved at the rate-determining stage (that is, the stage of reaction in which the highest plateau of the reaction profile is traversed), then that reactant will be consumed faster than a comparable unstrained model system. For example, a reaction of 'non-enolizable' ketones (which cannot react faster by competing pathways involving enolization) is their cleavage at elevated temperatures in strongly basic media. In the general case, a pre-equilibrium nucleophilic attack at carbonyl carbon is followed by the expulsion of a carbanion which immediately and irreversibly accepts a proton from the reaction medium. A dramatic example of ground-state effects on reaction rates is to be found on comparing the conditions necessary to effect this reaction with fluorenone (**18**) (Figure 5) and with 2,2-

dimethylcyclopropanone[7] (19). The latter compound has about 60° of angle strain at carbonyl carbon, and this is partially relieved in the first stage of reaction which involves $sp^2 \rightarrow sp^3$ rehybridization, so that the total angle strain at this carbon atom is reduced to 49° (109° − 60°). In the rate-determining stage all the rest of this angle strain is being relieved. Both stages, and particularly the second one, may be expected to be enhanced by this relief of steric strain.

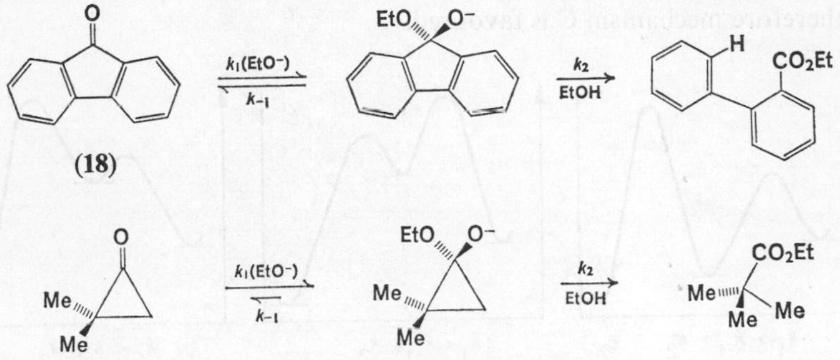

Figure 5. Base-promoted carbonyl cleavage of **18** at 200° and **19** at −25°.

These examples serve to illustrate the need for consideration of ground-state structure and energy in determining reactivity. Only when appropriate attention has been given to these may we go on to study the much more subtle and difficult questions associated with transition-state structure and energy.

Energies of Transition States

In our discussion of allyl vinyl ether thermolysis, we assumed that the energy profile for the reaction was a smooth continuous curve. This and most other reactions offer the possibility of a multistep mechanism. In many cases this is unambiguous, since it may be impossible to go from reactants to products in one single stage. Thus, in the oxidation of cyclohexanol by chromium trioxide in acetic acid, cyclohexanone is produced with concurrent reduction of the metal to chromium(III). Mechanistic evidence of the kind we shall explore later has led to the suggestion of a reaction pathway (Figure 6).

There are three possible situations which may be represented by the free-energy reaction profiles A, B, or C. In A, k_1 is a rapid step leading to the formation of a chromate ester which subsequently undergoes the oxidation step. It should be intrinsically possible to determine whether or not this is the route, by following the reaction spectrophotometrically; an intermediate having a visible spectrum different from that of either starting material or product should be observable at short reaction times. No intermediate is formed in appreciable concentrations in either B or C which, however, differ in the nature of the rate-determining step. Only in C is the carbon–hydrogen bond being broken at the transition-state, and reactions of this type are distinguished by a 'kinetic isotope effect' (page 14) whereby the reaction is slower if that hydrogen is replaced by deuterium or tritium. Experimentation showed the presence of a kinetic isotope effect and absence of build-up of intermediate, and therefore mechanism C is favoured.

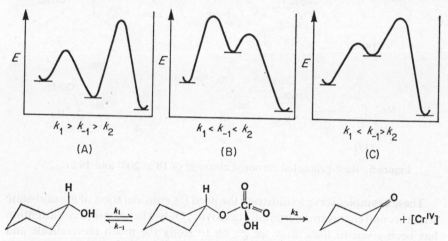

Figure 6. Possible mechanisms for the chromic acid oxidation of cyclohexanol.

In many reactions it is possible for a mixture of products to be formed from a common reactant or intermediate, and often the composition of this mixture varies with the reaction conditions. This situation arises when the transition-states of reactions determining product composition are accessible from the products themselves (which may be the case if $\Delta G_r \ll \Delta G^{\ddagger}$), and it illustrates the possibility of kinetic or thermodynamic control of reaction products. Thus, at low temperatures the reaction between butadiene and hydrogen bromide gives a mixture of 90% of **20** and 10% of **21** (Figure 7); if this mixture comes in contact with a trace of hydrogen bromide at 40° for 5 minutes the product proportions change to 15% of **20** and 85% of **21** which is the equilibrium mixture.[8] These observations are rationalized by considering the initial electrophilic addition of hydrogen bromide to be reversible, and to involve the *trans*-1-methylallyl cation (**22**) as intermediate. Bromide ion prefers to react with this cation at the site of highest charge-density, forming the thermodynamically less stable product (**20**), and at low temperatures the reaction is 'kinetically controlled'. At 40°, HBr-catalysed equilibrium between **20** and **22** is rapid, and this allows conversion into the more stable product (**21**) so that the reaction becomes 'thermodynamically controlled'.

In general, the excess energy of a transition-state over reactant ground-states will be largely produced by bond attenuation and steric interactions between reactants. Our main problem is to gain insight into the structure of a species which by its very nature is not amenable to direct examination. We would like to be able to answer questions such as "How much bond-making and bond-breaking has occurred at the transition-state?" or "Is any special spatial

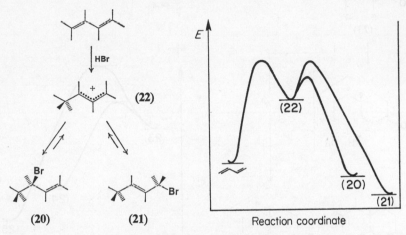

Figure 7. Kinetic and thermodynamic control in the addition of hydrogen bromide to butadiene.

arrangement of atoms required?", and so on. Many indirect techniques are available which will answer such questions in a relative manner, and some will be encountered later on. There is a broad, but very useful, generalization due to G. S. Hammond:[9] "If two states, as for example a transition state and an unstable intermediate, occur consecutively during a reaction process and have nearly the same energy content, their interconversion will involve only a small reorganization of the molecular structure". As an example of Hammond's postulate in practice, consider the protonation of the enolate anion of androst-4-ene-3,17-dione,[10] represented in the partial formula **23**. This could occur at the α-position, which is thought to have the higher charge-density owing to its closer proximity to the electron-withdrawing oxygen atom, giving **24**, or at the γ-position with production of **25**. Although **25** is more stable than **24**, this latter is more rapidly formed under conditions of kinetic control. The reaction should be highly exoenergetic (that is, involving a large negative free-energy change), and the transition-state (Figure 8) will be closer in structure to the carbanion than to the products. Consequently, the main factor influencing the direction of reaction will be the charge-distribution in this anion, and **24** will be preferred.

The enol **26** is an intermediate in the acid-catalysed equilibration of **24** and **25**, and if this reaction is carried out in deuteriated media the direction

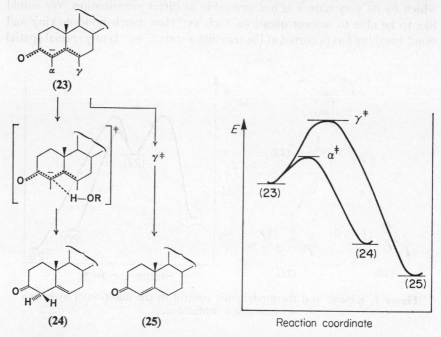

Figure 8. Hammond's postulate and the protonation of enolate anions.

of protonation of **26** may be discerned. The transition-state for this step will be close to the reactive cation **27a** or **27b**, and therefore the direction of reaction should be governed by the relative stabilities of these two species (Figure 9). **27a** would be expected to be the more stable owing to conjugation, and indeed this direction of reaction was shown to be greatly preferred.[11]

Primary Kinetic Isotope Effects[12]

A common reaction of free-radicals (Chapter 2) is hydrogen-atom abstraction from a neutral substrate, the reaction of chloroform with methyl radical being typical:

$$Cl_3C—H + \cdot CH_3 \rightarrow Cl_3C\cdot + CH_4$$

15

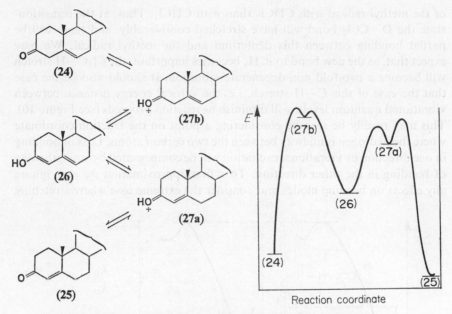

Figure 9. Hammond's postulate and the protonation of enols.

In assessing the activation energy for this process, it is relevant to consider the zero-point energy of the vibrational ground-state of the C—H bond. Infrared spectroscopic measurements show that there is a stretching vibration at 3019 cm^{-1} and a two-fold degenerate bending vibration at 1259 cm^{-1}, from which the total zero-point energy associated with the bond may be calculated:

$$\tfrac{1}{2}h\nu_{\text{C-H}} = \tfrac{1}{2}[8.63 + (2 \times 3.48)] = 7.80 \text{ kcal mole}^{-1} \qquad (28)$$

Now consider the situation in deuteriochloroform. It is known that infrared force constants and stretching frequencies depend on the reduced mass of the bonded atoms according to the equation:

$$\frac{\nu_{\text{D}}}{\nu_{\text{H}}} = \sqrt{\frac{m_{\text{H}}(m_{\text{C}} + m_{\text{D}})}{m_{\text{D}}(m_{\text{C}} + m_{\text{H}})}} = 0.734 \qquad (29)$$

and the observed infrared spectrum shows a C—D stretching vibration at 2256 cm^{-1} and a two-fold degenerate bending vibration at 908 cm^{-1}. This leads to a zero-point C—D energy of 5.83 kcal mole^{-1}.

$$\tfrac{1}{2}h\nu_{\text{C-D}} = \tfrac{1}{2}[6.45 + (2 \times 2.60)] = 5.83 \text{ kcal mole}^{-1} \qquad (30)$$

The lower zero-point energy of the C—D bond will lead to a slower reaction of the methyl radical with $CDCl_3$ than with $CHCl_3$. Thus, at the transition-state the D—CCl_3 bond will have stretched considerably, and there will be partial bonding between this deuterium and the methyl radical. We may expect that, as the new bond to $\cdot CH_3$ becomes important, the Cl_3C—H stretch will become a two-fold non-degenerate vibration. It should also be the case that the ease of this C—H stretch, i.e. the vertical energy distance between vibrational quantum levels, will diminish as reaction proceeds (see Figure 10). This may readily be seen by considering a point on the reaction coordinate where the hydrogen is midway between the two carbon atoms. Loss of bonding in one direction by vibrational excitation will be compensated by strengthening of bonding in the other direction. To a first approximation we may ignore any effects on bending modes and consider the extreme case where stretching

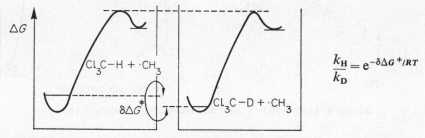

$$\frac{k_H}{k_D} = e^{-\delta \Delta G^{\ddagger}/RT}$$

Figure 10. The reaction of chloroform and deuteriochloroform with methyl radicals.

forces associated with the transferred atom are negligible at the transition-state. The difference in reaction rate between chloroform and deuteriochloroform will then be associated with their respective differences in C—H *stretching* zero-point energy, 1.09 kcal mole^{-1}. At 298°K:

$$k_H/k_D = e^{\Delta G^{\ddagger}/RT} = e^{(hc/2kT)(3019 - 2256)} = 6.3 \qquad (31)$$

With a more sophisticated treatment which took into account changes in the bending vibrational mode, a rather higher value of k_H/k_D would be obtained. The utility of primary kinetic isotope effects in reaction mechanisms is two-fold. Firstly, direct information may be obtained on whether a carbon–hydrogen bond is broken in the rate-determining stage of a given reaction. Secondly, it will be noted that our treatment assumed that the reaction transition-state was relatively symmetrical. If the carbon–hydrogen bond was only slightly perturbed at the transition-state and the stretching vibration was still strong, then the effect of isotopic substitution on the rate would be much smaller. A similar situation would exist in the event of a very late transition-state where the new bond to hydrogen is almost completely formed. We may therefore form some impression of the position of the transition-state on the reaction profile from a knowledge of the kinetic isotope effect and an appropriate comparison with model systems. Typical kinetic isotope effects are listed in Table 4.

The technique need not be confined to reactions in which bonds to hydrogen are being formed and broken, although it is most widely used for these. Owing to the relationship between isotope effect and reduced mass, rather small

Table 4. Examples of primary kinetic isotope effects.

Reaction			Isotope effect
$C_6H_5CH_2CH_2Br$	$\xrightarrow[\text{EtOH}]{\text{NaOEt}}$	$C_6H_5CH=CH_2$	$\dfrac{k_H}{k_D} = 7.1$
thiophene-H	$\xrightarrow[\text{Et}_2\text{O}]{\text{BuLi}}$	thiophene-Li	$\dfrac{k_H}{k_D} = 6.6$
decalone	$\xrightarrow{\text{H}_3\text{O}^+}$	enone	$\dfrac{k_H}{k_D} = 4.1$
^{18}O aryl benzoate	$\xrightarrow[\text{MeOH}]{\text{MeO}^-}$	methyl benzoate	$\dfrac{k_{16O}}{k_{18O}} = 1.018$
MeI	$\xrightarrow[\text{H}_2\text{O}]{\text{HO}^-}$	HOMe	$\dfrac{k_{12C}}{k_{14C}} = 1.09$

effects are observed when bonds between two heavier atoms are broken in the rate-determining stage of a reaction, but these are certainly measurable and may provide valuable insight into reaction mechanisms. The decomposition of **32** could take three possible courses in which either bond (a) or bond (b) broke first or these both broke simultaneously. Measurement of the ^{16}O/^{18}O isotope effect and the ^{12}C/^{13}C isotope effect at the atoms indicated (Figure 11) demonstrated that a two-step process operated where breaking of (a) only was important at the rate-determining transition-state.[13] The small rate-ratios involved were readily measured by mass-spectrometric analysis of the CO_2

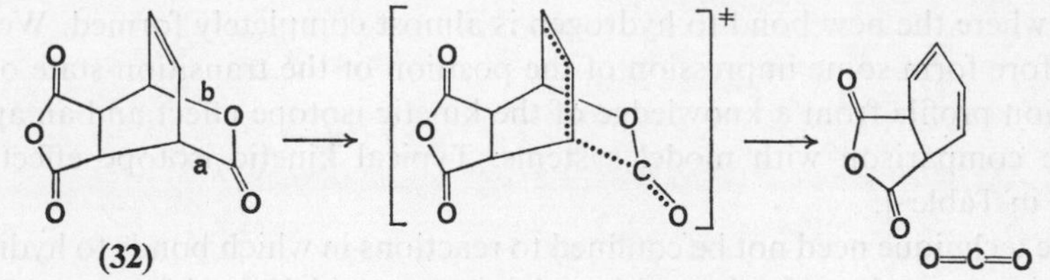

Figure 11. Isotope effects measured on evolved CO_2:

$$k_{12C}/k_{13C} = 1.032; \qquad k_{16O}/k_{18O} = 1.015$$

in agreement with extensive bond-breaking to acyl-carbon at the transition-state, but not with extensive O–Alkyl bond breaking.

produced, using a sample of **32** enriched in ^{18}O and containing ^{13}C at natural abundance (1.1 %).

Secondary Kinetic Isotope Effects[14]

There are many reactions in which a small, but measurable, rate effect is observed on substitution of a hydrogen atom by deuterium, even though that hydrogen atom is not being transferred at the rate-determining transition-state. These secondary isotope effects are commonly observed in two situations. Steric secondary isotope effects arise when there are severe repulsive H—H interactions arising at the rate-determining transition-state. The effective mean bond-length of a C—D bond is smaller than that of the progenitive C—H bond because of the lower C—D zero-point energy. Consequently the rate of a reaction involving severe interhydrogen repulsion is enhanced when these hydrogens are replaced by deuterium. In the racemization of **33** replacement of both methyls by CD_3 accelerates the reaction by a factor of 1.15 (Figure 12).

A more common situation where secondary isotope effects are encountered occurs when hydrogen is attached to a carbon atom undergoing rehybridization at the rate-determining transition-state. Detailed consideration has led to

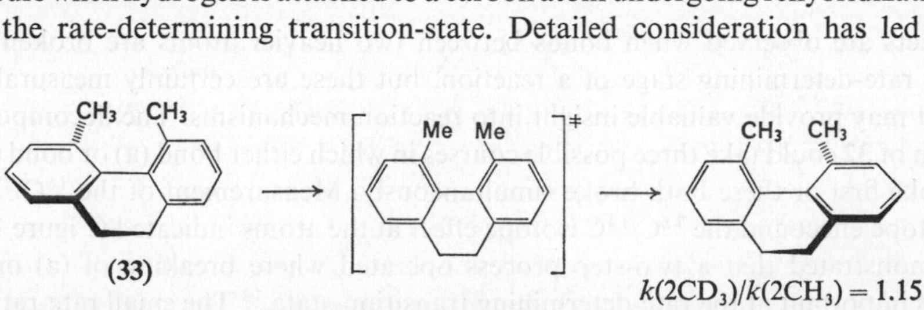

$$k(2CD_3)/k(2CH_3) = 1.15$$

Figure 12. A steric secondary isotope effect.

the conclusion that the most important factor here is the out-of-plane C—H bending vibration. For sp^3-hybridized C—H bonds this typically occurs at around 1400 cm^{-1}, and for sp^2-hybridized bonds at 800 cm^{-1}; deuterium substitution will lower these values to around 1040 and 600 cm^{-1} respectively. Consider a reaction in which rehybridization takes place, the Diels–Alder addition of tetracyanoethylene to anthracene (Figure 13).[15] If we assume that the transition-state is intermediate in structure between reactants and product, with out-of-plane C—H bending at 1100 cm^{-1}, then we will have to consider an increase in zero-point bending energy at C(9) and C(10) of $\frac{1}{2}hc(2 \times 300$ cm$^{-1})$, corresponding to 858 cal. Applying the same treatment to the reaction of 9,10-dideuterioanthracene with tetracyanoethylene, we may assume a transition-state with C—D out-of-plane bending at 820 cm^{-1} and an increase in zero-point bending energy at the transition-state of $\frac{1}{2}hc(2 \times 220$ cm$^{-1})$,

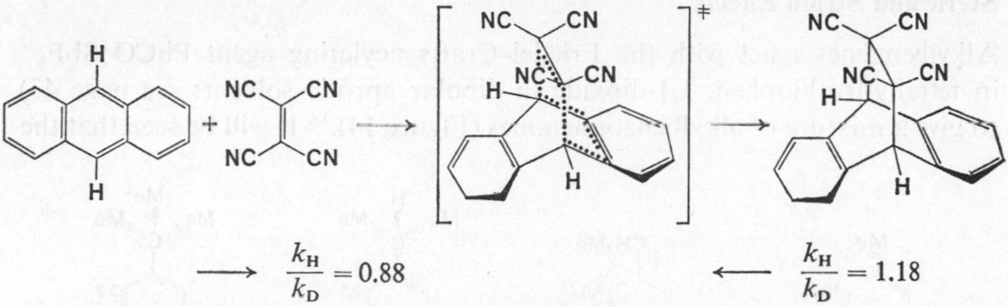

$$\xrightarrow{\hspace{1.5cm}} \quad \frac{k_H}{k_D} = 0.88 \qquad\qquad \frac{k_H}{k_D} = 1.18 \quad \xleftarrow{\hspace{1.5cm}}$$

Figure 13. Secondary isotope effects due to rehybridization.

corresponding to 629 cal. Thus, neglecting contributions from stretching zero-point energies, our treatment suggests that the deuterated compound, with a lower increase in zero-point bending energy on going from ground-state to transition-state, will have an activation energy 229 cal *lower* than that of anthracene. For reaction at 30°C, this corresponds to a rate ratio:

$$k_H/k_D = e^{-(229/303R)} = 0.68 \tag{34}$$

This simple treatment predicts the direction of the secondary isotope effect, with the deuterium compound reacting faster than its protium analogue, but exaggerates its magnitude—the actual experimental result is $k_H/k_D = 0.88$. For the reaction in the reverse direction then the protium compound should react the faster, and it has been shown experimentally that $k_H/k_D = 1.18$ for dissociation of this Diels–Alder adduct in ethanol at 50°C.

Secondary isotope effects of this kind may be generalized into a basic rule: *If a carbon atom undergoes rehybridization with increase of* p-*character in its* C—H *bonds at the rate-determining transition-state, then substitution of hydrogen by deuterium at that carbon atom will increase the reaction rate. If rehybridization occurs with decrease in* p-*character, deuterium substitution will decrease the rate.* The phenomenon has been useful in the determination of mechanism, but exceptions have been noted, and the technique must be evaluated with some caution.

Factors Affecting Reactivity

We earlier considered examples where specific features of reactant ground-states led to appreciable effects on reactivity. In general, it is important to consider *both* ground-state and transition-state, and any special factors which may operate to modify reaction rates.

Steric and Strain Effects

Alkylbenzenes react with the Friedel–Crafts acylating agent $PhCO^+SbF_6^-$ in tetrahydrothiophen, 1,1-dioxide (a dipolar aprotic solvent; see page 43) to give a mixture of alkylbenzophenones (Figure 14).[16] It will be seen that the

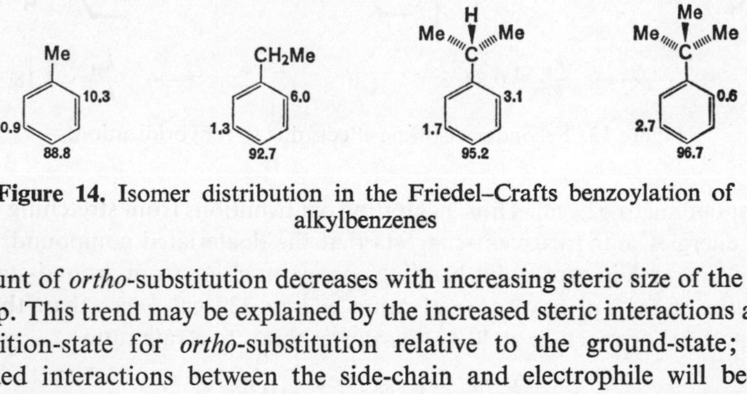

Figure 14. Isomer distribution in the Friedel–Crafts benzoylation of alkylbenzenes

amount of *ortho*-substitution decreases with increasing steric size of the alkyl group. This trend may be explained by the increased steric interactions at the transition-state for *ortho*-substitution relative to the ground-state; non-bonded interactions between the side-chain and electrophile will become increasingly serious as the former becomes larger. The data suggest a difference in free-energy for *ortho*-reaction between toluene and t-butylbenzene of about 2–3 kcal mole^{-1}.

Due regard for the sensitivity of transition-states to steric effects may influence the design of synthetic experiments. For instance, any attempts to prepare cyclopentadienone lead to the production of a dimer through Diels–Alder dimerization, and it is thought that this reaction is extremely fast, even at low temperatures. A t-butyl group at the 3-position appears to slow the dimerization sufficiently that the monomer (35) may be isolated, and a minimum value of 10^2 l mole^{-1} sec^{-1} at $-20°C$ for the rate constant of dimerization obtained. In this case two molecules of monomer may approach one another without any severe clash between the two t-butyl groups. 2,4-Di-t-butylcyclopentadienone (**36**), however, cannot dimerize without severe non-bonded interaction between the two t-butyl groups occurring at the transition-state; in consequence, this dimerizes about 10^8 times more slowly than **35**, is isolable near room-temperature, and may therefore be used as a model substance to demonstrate the spectral properties of cyclopentadienones.[17]

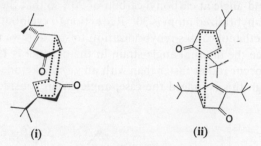

(i)   (ii)

Figure 15. Schematic transition-states in the dimerization of **35** (i) and **36** (ii).

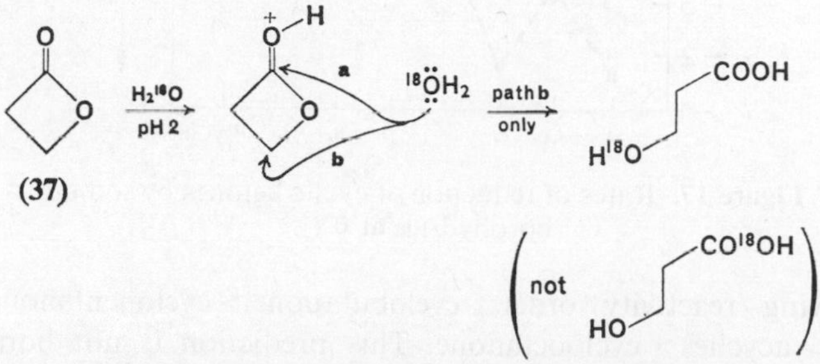

Figure 16. Lactone hydrolysis by O–Alkyl cleavage.

In many reactions of cyclic systems, factors other than bond-strength and repulsive non-bonded interactions must be taken into account in assessing the energies of transition-states. We have seen that severe angle-strain imposed on the ground-state may lower the energy difference between ground-state and transition-state, if that strain is partly relieved along the reaction co-ordinate. This factor may even change the direction of reaction, or allow the pursuit of another path between reactant and product. An example is afforded by acid hydrolysis of the β-lactone **37**, which provides an interesting case of the use of isotopic labelling in determining reaction mechanism. A general pathway for lactone hydrolysis is via cleavage of the O—Acyl bond, probably following a rate-determining nucleophilic attack at carbonyl carbon. In weakly acidic media, however, this particular β-lactone follows a quite distinct pathway where relief of angle-strain is important at the rate-determining transition-state, and O—Alkyl cleavage is conclusively demonstrated by the distribution of ^{18}O in the product[18] (see Chapter 5, page 329).

The above example illustrates the importance of angle-strain effects when ring-cleavage may occur. In some circumstances relief of angle-strain may be achieved without formal bond-breaking. Cyclobutanone may be expected to have a CCC bond-angle at carbonyl carbon of 90°, so that the internal angle-strain at this sp^2-hybridized atom is 30°. Reduction to cyclobutanol by sodium borohydride in ethanol causes rehybridization to sp^3, and we might therefore expect that, since the internal angle-strain in the product is 19° (109° − 90°), reaction might occur rather faster than with an acyclic ketone such as diethyl ketone. Although this is indeed the case, angle-strain considerations predict

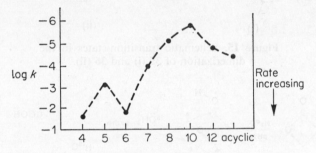

Figure 17. Rates of reduction of cyclic ketones by sodium borohydride at 0°C.

the following reactivity order: cyclobutanone > cyclopentanone > cyclohexanone > acyclics > cyclooctanone. This prediction is not borne out in practice, and the true picture is revealed in Figure 17.[19]

The data illustrate a general tendency, in that reactions where a single sp^2 carbon atom in a ring is rehybridized to sp^3 in the rate-determining stage are most rapid with the cyclobutane derivative, and slow for all the medium-ring (C_8—C_{12}) substrates. Conversely, reactions where an sp^3 centre is becoming sp^2 at the rate-determining transition-state show an inverse trend in reactivity for medium-ring compounds, exemplified by the solvolysis of cycloalkyl tosylates (Chapter 2, page 94). Part of the explanation lies in angle strain but an equally important factor is steric compression in the medium-ring compounds. These are known to be very flexible, with a number of conformers of comparable energy; the hydrocarbons have been subjected to a number of strain-energy minimization calculations, which suggest that around C_8—C_{10} there are no structures which avoid severe non-bonded or torsional interactions.[20] Consider the borohydride reduction of cyclooctanone, for which the best ground-state conformation is possibly **38**, and **39** a transition-state formed along the path of least steric hindrance. There are, however, considerable non-bonded interactions between the oxygen atom and hydrogens at C(4) and C(6) in **39**, and any twist in the molecule which relieves these only increases H—H interactions elsewhere, with consequent energy increase at this transition-state

(38)   (39)

Stereoelectronic Effects

The degree of overlap between two orbitals (other than s-type, which are spherically symmetrical) depends not only on the distance between their respective atomic centres, but also on the angular disposition of bonding lobes. For example, two p-orbitals in parallel planes may overlap to form a π-bond, but if their planes are orthogonal there is no net bonding. It therefore follows that, in any reaction, the relative spatial disposition of bonds being made and broken will be of crucial importance in determining the energy of activation. The basic principle of stereoelectronics is that, other things being equal, reactions will proceed via transition-states which have attained the most favourable orbital overlap possible. This principle governs the stereochemical course of familiar reactions such as bimolecular nucleophilic substitution, which occurs with inversion of configuration, and $E2$ elimination reactions, where a *trans*-orientation of breaking bonds is often preferred. Both of these reactions are discussed in more detail in Chapter 3. In the course of this book the reader will also discover in Chapter 2 that neighbouring-group participation and fragmentation reactions depend on specific stereoelectronic features; in Chapter 5 a variety of stereoelectronic facets in additions to olefins and carbonyl compounds; and in Chapter 4 a sophisticated modern development—the principle of orbital symmetry—which controls the stereochemistry of all reactions occurring via cyclic transition-states. The requirement for proper spatial orientation of bonds and orbitals is therefore of very general importance, and a simple example is to be found in the deprotonation of triptycene (**40**) and triphenylmethane (**41**) by strong bases, a reaction which may be followed by uptake of deuterium from the medium (Chapter 2). If we apply Hammond's rule to the deprotonation step in which a reactive carbanion is formed, it is clear that the transition-state will resemble that carbanion in structure. Triphenylmethylide anion (43) is considerably stabilized by delocalization of negative charge into the phenyl rings, but in the triptycenylide anion (42) delocalization is stereoelectronically impossible because the lone-pair of the carbanion and the π-electrons of the arene ring are orthogonal, with minimal interaction. Under standard conditions[21] the rate of deprotonation of 41 is about half-a-million times faster than that of 40.

(40)

(42)

(41)

(43)

cis- and trans-Divinylcyclobutane (44 and 45) are produced in the photodimerization of butadiene. Both isomers are converted into cycloocta-1,5-diene (46) on heating,[22] but the activation energy for the cis-compound is some 15 kcal lower, and other products, particularly 47, dominate the thermolysis of 45. The difference in activation energy reflects a difference in mechanism wherein the cis-compound undergoes a concerted Cope rearrangement, with breaking of the old σ-bond and formation of the new one occurring simultaneously at the transition-state. A concerted reaction of the trans-compound to give 46 is precluded on stereoelectronic grounds, since the termini of the vinyl groups are too far apart to allow effective bonding without introduction of excessive steric strain. In this case, therefore, the reaction pathway involves cleavage of the substituted cyclobutane bond with formation of a bis-allyl radical (48), which may only form 46 after a subsequent internal rotation.

A dramatic and clear-cut example[23] where reaction course is subject to stereoelectronic control is afforded by the deamination of the 2-aminocyclohexanols 49—52; the function of the t-butyl group is to hold the ring rigid and define the relative orientation of substituents. Deamination of amines by nitrous acid (Chapter 2) involves the production of diazonium ions, which are

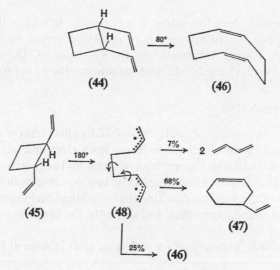

Figure 18. Other conformations of **45** may give rise to butadiene and **47** via geometrical isomers of **48**.

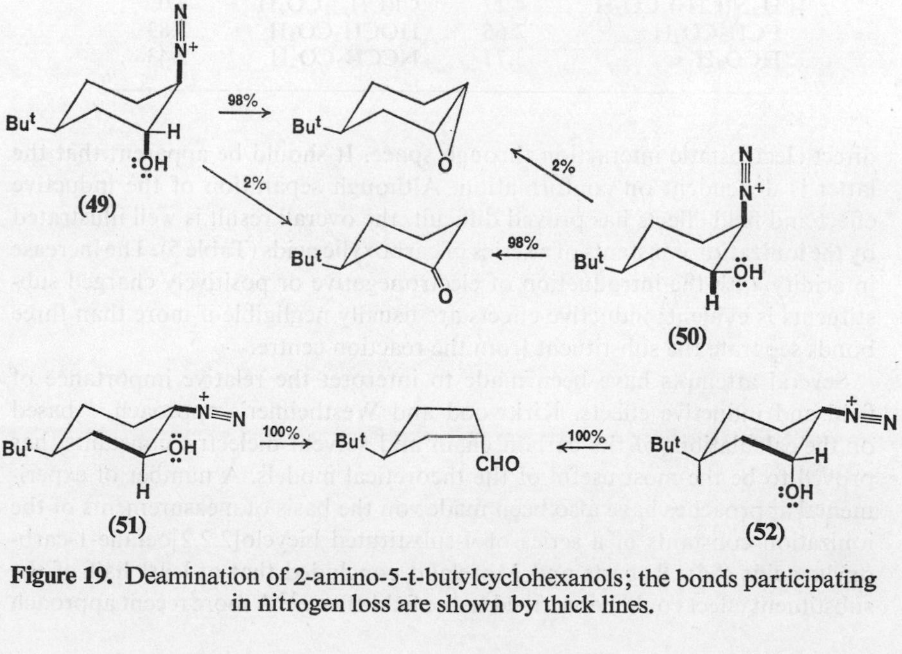

Figure 19. Deamination of 2-amino-5-t-butylcyclohexanols; the bonds participating in nitrogen loss are shown by thick lines.

highly reactive and which lose nitrogen very readily. In each of the cases under consideration this loss of nitrogen is probably concerted with migration of electrons from a bond antiparallel to the C—N bond. Depending on the nature of this bond, characteristic and specific products are formed.

Electronic Effects[1, 24-30]

Substituents in an organic substrate can markedly affect rates of reaction owing to their distinct electronic characteristics; such electronic effects are transmitted across molecules by the operation of polar and conjugative mechanisms. The polar or electrostatic effect consists of two separate components: firstly, a polarization of the bonds connecting the substituent and the reaction centre, usually termed an inductive effect, and secondly, the field effect which involves

Table 5. Values of pK for carboxylic acids in water at 25°.

Acid	pK	Acid	pK
$MeCO_2H$	4.80	$ClCH_2CO_2H$	2.86
$MeCH_2CO_2H$	4.88	Cl_2CHCO_2H	1.30
$Me_3\overset{+}{N}CH_2CO_2H$	1.83	Cl_3CCO_2H	0.65
$H_3\overset{+}{N}(CH_2)_4CO_2H$	4.27	$Cl(CH_2)_2CO_2H$	4.0
FCH_2CO_2H	2.66	$HOCH_2CO_2H$	3.83
HCO_2H	3.77	$NCCH_2CO_2H$	2.43

direct electrostatic interaction through space. It should be apparent that the latter is dependent on conformation. Although separation of the inductive effect and field effects has proved difficult, the overall result is well illustrated by the ionization constants of a series of carboxylic acids (Table 5). The increase in acidity with the introduction of electronegative or positively charged substituents is evident; inductive effects are usually negligible if more than three bonds separate the substituent from the reaction centre.

Several attempts have been made to interpret the relative importance of field and inductive effects. Kirkwood and Westheimer's approach,[31] based on the calculations of the carbon chain and solvent dielectric constants, has proved to be the most useful of the theoretical models. A number of experimental approaches have also been made; on the basis of measurements of the ionization constants of a series of 4-substituted bicyclo[2.2.2]octane-1-carboxylic acids (54), Roberts and Moreland concluded that at least half of the substituent effect could be ascribed to the field effect.[32] A more recent approach

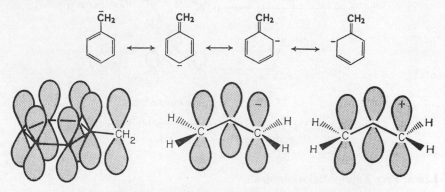

(53)　　　　　　　　　　　　　(54)

is that of Wilcox and Leung who compared the ionization constants of pairs of identically substituted acids (53 and 54) in which the field effects are similar but inductive effects are different.[33] On the basis of these results, the field effect was considered to make the dominant contribution to the total substituent effect.

Figure 20. π-Overlap in benzyl anion, and allyl anion and cation.

The second mechanism for the transmission of electronic effects is by a conjugative effect or delocalization. This is achieved by overlap of a p-orbital at the reaction centre with an adjacent π-electron system which in turn may overlap with p-electrons of substituents. Toluene is more acidic than methane, mainly because of conjugation in benzyl anion. Charge is distributed predominantly over four atoms by overlap of the π-orbitals, and generation of the anion is easier for toluene owing to the smaller difference in energy of the neutral and charged species. This overlap can also be drawn as in Figure 20, and other examples of conjugation are apparent in the ease of formation of the allyl anion and cation. A similar situation applies to the allyl radical. An indication of the marked effect the conjugative interaction produces in this case is that the bond-dissociation energy for formation of the n-propyl radical from propane is 98 kcal mole^{-1} compared to 88 kcal mole^{-1} for formation of the allyl radical from propene. The reader should consider the treatment of these cases by a linear combination of atomic p-orbitals (LCAO), and the molecular orbitals for the allyl radical are constructed in Figure 21.[34]

The most effective overlap of p-orbitals occurs when the orbitals are parallel to each other and in the same plane; the degree to which this may be achieved can affect the position of equilibrium and the course of many reactions. (See page 23 for discussion of stereoelectronic effects.)

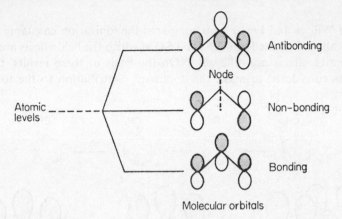

Figure 21. LCAO construction of molecular orbitals for the allyl radical $\cdot CH_2{-}CH{=}CH_2$.

Linear Free Energy Relationships

Some measure of the effect of substituents on the properties of molecules was seen in Table 5. Groups on benzene rings or other π-systems have large effects partly due to conjugative interactions. Two examples of electrophilic substitution reactions illustrate, qualitatively, the difference between two such groups, $-NO_2$ and $-OH$. Further nitration of nitrobenzene to m-dinitro-benzene occurs only under vigorous conditions because of the inductive and mesomeric conjugative effects of the nitro group which are both electron-withdrawing. In contrast, nitration of phenol in dilute nitric acid, to give o- and p-nitrophenols, takes place very readily as a consequence of the increase in electron-density in the ring due to the mesomeric effect of the $-OH$ group.

A number of correlations of rates and equilibrium constants have provided quantitative estimates of substituent effects. Hammett found that, for a large number of reactions of m- and p-substituted benzene derivatives, a plot of the logarithm of the rate constant (k) or equilibrium constant (K) for one reaction against $\log k$ or $\log K$ for another gave reasonably linear correlations.[35] A straight line was obtained when $\log K'$ for the ionization of a series of m- and p-benzoic acids (equation 55) was plotted against similar data for m- and p-phenylacetic acids (equation 56); each point on the graph referred to the ionization constant of a substituted benzoic acid derivative and an identically substituted phenylacetic acid derivative, e.g. p-$NO_2C_6H_4CO_2H$ and p-$NO_2C_6H_4CH_2CO_2H$.

$$m\text{- or } p\text{-}XC_6H_4CO_2H + H_2O \overset{K'}{\rightleftharpoons} m\text{- or } p\text{-}XC_6H_4CO_2^- + H_3O^+ \qquad (55)$$

$$m\text{- or } p\text{-}XC_6H_4CH_2CO_2H + H_2O \overset{K}{\rightleftharpoons} m\text{- and } p\text{-}XC_6H_4CH_2CO_2^- + H_3O^+ \qquad (56)$$

The equation for the linear correlation of ionization constants is:

$$\log K = \rho \log K' + C \qquad (57)$$

where ρ is the slope of the line and C the intercept. This equation refers to any substituent, so that when X is hydrogen:

$$\log K_0 = \rho \log K_0' + C \qquad (58)$$

where K_0' and K_0 are the ionization constants for benzoic acid and phenylacetic acid. Subtraction of equation **58** from equation **57** gives:

$$\log (K/K_0) = \rho \log (K'/K_0') \qquad (59)$$

Similar equations can also be written for other reaction series. Hammett defined a series of substituent constants, σ, which represent the effect of each substituent on the ionization of benzoic acid under standard conditions (water at 25°):

$$\log (K'/K_0') = \sigma$$

A numerical value was, in this way, assigned to each substituent (Table 6). Equation **59** then reduces to:

$$\log (K/K_0) = \rho\sigma \qquad (60)$$

and for rates of reaction:

$$\log (k/k_0) = \rho\sigma \qquad (61)$$

where k_0 is the rate of reaction for the unsubstituted derivative and k is the rate for any substituted derivative undergoing the same reaction, e.g. alkaline hydrolysis of m- and p-substituted methyl benzoates. In both equations, σ is the substituent constant which is a measure of the electron-donating or electron-withdrawing power of the substituent. ρ is the reaction constant, and is a measure of the sensitivity of the equilibrium or reaction to change in substituent.

Equations **60** and **61** are termed linear free-energy relationships, and the basis for this terminology is apparent from:

$$\Delta G_r = -2.303 \, RT \log K$$

$$\Delta G^{\ddagger} = -2.303 \, RT \log k(h/\mathbf{k}T) \qquad \text{(see page 3)}$$

where ΔG_r and $\Delta G^{\ddagger}$ are standard free-energy differences of reactants and product in an equilibrium process, and the free-energy of activation of a reaction, respectively.

Table 6. Substituent constants.[a]

Substituent	σ_I[b]	σ_m	σ_p	σ_p^+	σ_p^-
Me	0.00	−0.07	−0.17	−0.31	—
CF_3	0.41	0.43	0.54	0.52	—
CHO	0.31[c]	0.35	0.22	—	1.13
COMe	0.28	0.38	0.50	—	0.87
$CONH_2$	0.21	0.28	—	—	0.62
CO_2R	0.30	0.37	0.45	0.48	0.68
Ph	0.10	0.06	−0.01	−0.17	—
COO^-	0.05[c]	−0.10	−0.00	−0.03	—
CN	0.56	0.56	0.66	0.66	0.90
NMe_3^+	0.92	0.88	0.82	0.41	—
NO_2	0.63	0.71	0.78	0.79	1.24
NHCOMe	0.28	0.21	0.00	0.00	—
N=NPh	0.25[c]	—	0.64	—	—
$NHNH_2$	0.15	−0.02	−0.55	—	—
NMe_2	−0.10	−0.21	−0.83	−1.7	—
NH_2	0.10	−0.16	−0.66	−1.3	—
N_3	—	0.33	0.88	—	1.1
OCF_3	0.55[c]	0.47	0.28	—	—
OMe	0.25	0.12	−0.27	−0.78	−0.2
OH	0.25	0.12	−0.37	−0.92	—
OCOMe	0.39	0.39	0.31	—	—
O^-	−0.16	−0.17	−0.52	—	−0.81
N_2^+	—	1.76	1.91	—	—
F	0.52	0.34	0.06	−0.07	−0.02
Cl	0.47	0.37	0.23	0.11	—
Br	0.45	0.39	0.23	0.15	—
I	0.38	0.35	0.18	0.13	—
SOMe	0.52	0.52	0.49	—	—
SO_2Me	0.60	0.60	0.72	—	1.05
SMe	0.19	0.15	0.00	−0.60	—

[a] Ref. 27 except as noted.
[b] $\sigma = 0.45\sigma^*$ (XCH_2^-), from aliphatic derivatives.
[c] N.m.r. value.

Equation **61** can be written:

$$\Delta G^{\ddagger} - \Delta G_r^{\ddagger} = (2.303\ RT)\rho\sigma$$

and since σ is a measure of the free-energy difference in the ionization of benzoic and substituted benzoic acid derivatives, this relationship depends upon free-energy differences. These equations can be applied generally to a number of reaction series and equilibria by plotting either log K or log k against σ when a linear correlation is frequently obtained. Figure 22 displays a typical application of the Hammett equation; the alkaline hydrolysis of ethyl benzoates has a ρ-value, obtained from the slope of the line, of 2.54. Applications of linear free-energy relationships will be discussed later but it should already be apparent that they can be utilized for the correlation and prediction of rate constants. From a knowledge of a minimum of two rate constants in any reaction series, which enables the reaction constant (ρ) to

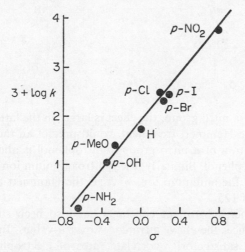

Figure 22. Hydrolysis of ethyl benzoates ($ArCO_2Et$) in NaOH–85% EtOH at 25°; correlation coefficient 0.974.

be evaluated, other rate constants in the same series can be predicted from the σ-values of the substituents.

Although the Hammett σ-values were the first to be assigned, alternative sets of substituent constants followed, derived from other model reactions. To appreciate the need for more than one set of substituent constants it is necessary to examine the nature of reactions and equilibria where the application of σ-constants would be expected to be productive. In Hammett's original considerations of benzoic acid ionizations, only *meta-* and *para*-substituted derivatives were considered since *ortho*-substituents introduce variable steric interactions. Steric effects must, therefore, be negligible for the successful application of σ-values in reaction series. Secondly, in the Hammett model reaction the O—H bond being broken is isolated from the benzene ring so that conjugative effects cannot be transmitted directly to the ionization

centre. Hammett correlations would, therefore, be expected to break down
in systems in which conjugative interactions are important. This can be illustrated by the effect which a *p*-nitro group has on the acidity of phenol and
benzoic acid. In *p*-nitrobenzoic acid the mesomeric effect of the nitro group
is transmitted through the ring of the *para*-carbon atom by a mesomeric
effect, but then only by induction to the –OH group. In *p*-nitrophenol, however, the nitro group is in direct conjugation with the reaction centre. Although,
therefore, both benzoic acid and phenol become substantially stronger acids

by substitution of a nitro group, the effect is larger in the latter case. Accordingly, whereas the Hammett treatment would predict an increase in acidity
of 36 on substitution of a nitro group, *p*-nitrophenol is about 600 times as
strong an acid as phenol. Similarly, the *p*-nitroanilinium ion is 4000 times as
strong an acid as the anilinium ion, whereas the Hammett treatment would
predict a factor of 125.

The use of other substituent constants has also been suggested, and in
certain circumstances they have advantages over σ-values. Brown found that
σ-values did not satisfactorily correlate rates of electrophilic substitution
reactions, owing to the direct conjugative effects between the substituent and
reaction centre. A set of substituent constants, termed σ^+-values, were obtained
from the measurement of the rates of solvolysis of a series of substituted
cumyl chlorides in 90% acetone–water (equation **62**).[36] Since no conjugative
effects are possible from *meta*-substituents, a reaction constant ($\rho = -4.54$)
was derived by a plot of $\log k$ (solvolysis rate) against σ_m. On this basis $\sigma_m = \sigma_m^+$
for the new scale. Measurement of the rates for *para*-substituted cumyl
chlorides, and the estimate of the substituent constant necessary for $\log k$
to fall on the line for *meta*-substituents, allowed a new set of σ_p^+-values to be
obtained ($\sigma_p^+ \neq \sigma_p$). This set of σ^+-constants can be used for the correlation
of the rates of electrophilic aromatic substitution and any reaction in which
direct conjugation between substituents and electron-deficient reaction

$$\text{(structure with Me, C–Cl)} \xrightarrow{\text{90\% acetone–water}} \text{(structure with Me, C}^+\text{)} \tag{62}$$

centres is possible. The effect of using σ^+ rather than σ for an example of this type of reaction is shown in Figure 23.

Although good correlations of rate constants and σ^+-values have been found for a large number of reactions, other linear free-energy relationships have been suggested involving additional parameters. As a consequence of the differing electron demand of reagents (see the discussion on selectivity

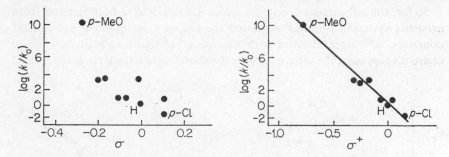

Figure 23. Bromination of substituted benzenes in acetic acid at 25°C.[36]

in Chapter 5, page 285), the extent of conjugative interactions in the transition-state varies with the reaction. Yukawa and Tsuno[37] suggested the use of the equation:

$$\log(k/k_0) = \rho[\sigma + r(\sigma^+ - \sigma)]$$

where r is a parameter dependent upon the importance of conjugative inter-actions at the transition-state. This equation inevitably yields better correla-tions of substituent effects in electrophilic substitution reactions but the parameter r has to be determined in each case. For example, a ρ-value of -8.82 is found for the bromination of benzene derivatives in acetic acid at 25°, and for the most effective correlation $r = 1.74$. Other approaches along similar lines to that of Yukawa and Tsuno have also been suggested.[38]

An alternative set of substituent constants, σ^--values, is available for the correlation of reactions in which conjugation between an electron-rich reaction centre and electron-withdrawing substituents can take place.[39] The standard reaction was the ionization of substituted phenols in water (equation **63**). The improvement in the correlation of the rates of reaction of *para*-substituted *o*-nitrochlorobenzenes with methoxide ion with σ^- rather than σ constants is evident in Figure 24.

$$NC-\langle\ \rangle-OH \underset{}{\overset{H_2O}{\rightleftharpoons}} NC-\langle\ \rangle-O^- \longleftrightarrow \bar{N}=C=\langle\ \rangle=O \qquad (63)$$

A scale of substituent constants has also been derived by the use of ^{19}F nuclear magnetic resonance of m- and p-XC_6H_4F compounds. The chemical shift of the fluorine was taken to be a measure of the electron-density around the fluorine. Since the inductive effect was assumed to be similar for the *meta*- and *para*-positions, the difference in chemical shifts was regarded as proportional to the conjugative effect of the substituent X. These substituent constants were termed σ_R-values.[41]

So far, the substituent constants scales discussed have been derived from aromatic systems. Taft has suggested the use of a set of polar substituent constants, σ^*-values, derived with the use of aliphatic substrates.[42] These constants represent the differences in activation free-energy changes produced

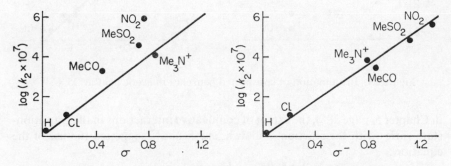

Figure 24. Substituent effects in the reactions of *para*-substituted *o*-nitrophenyl chlorides with methoxide ion in methanol at 25°C.[40]

by a substituent in base- and acid-catalysed hydrolysis of a series of substituted esters, RCO_2Me.

$$\sigma^* = \frac{1}{2.48}\left[\log\left(\frac{k}{k_0}\right)_{base} - \log\left(\frac{k}{k_0}\right)_{acid}\right]$$

where k_0 refers to the ester when R = Me, so that $\sigma^*_{Me} = 0$, and k is the rate constant for esters when R = XCH_2-, and X are different substituents. This approach is based on the assumption that steric effects are the same in both types of hydrolysis but that polar effects are significant only in the base-catalysed reactions. From a consideration of the two transition-states **64** and **65**, steric effects would be expected to be similar. The lack of substituent effects for acid hydrolysis arises since this is a multistep reaction with the effect of some steps cancelling those of others (equation **66**). For example, electron-releasing substituents will assist step 1 but hinder step 2. The factor $1/2.48$ was introduced to place the σ^*-values on the same scale as Hammett σ-values.

$$
\begin{array}{cc}
\overset{\delta-}{\underset{R-C----\overset{\delta-}{O}R}{\overset{\parallel}{O}}}\;
\underset{OH}{} &
\overset{\delta+}{\underset{R-C----\underset{\delta+}{O}R}{\overset{OH}{|}}}\;H\\[2mm]
(64) & (65)
\end{array}
$$

$$
\begin{array}{ccccc}
R-\overset{O}{\overset{\parallel}{C}}-OR + H^+ & \underset{\longleftarrow}{\overset{(1)}{\rightleftharpoons}} & R-\overset{OH}{\underset{+}{C}}-OR & \underset{\longleftarrow}{\overset{(2)}{\rightleftharpoons}} & R-\overset{OH}{\underset{\underset{+}{OH_2}}{C}}-OR
\end{array}
$$

$$ (66) $$

$$
R-\overset{O}{\overset{\parallel}{C}}-OH \rightleftharpoons R-\overset{OH}{\underset{\underset{+}{OH}}{C}} \rightleftharpoons R-\overset{OH}{\underset{H}{C}}-\overset{+}{O}R \rightleftharpoons R-\overset{OH}{\underset{OH}{C}}-OR
$$

These substituent constants can be used to correlate data for reactions by an equation analogous to the Hammett equation:

$$ \log(k/k_0) = \sigma^* \rho^* $$

Taft has also applied a similar treatment to the hydrolysis of *o*-, *m*-, and *p*-derivatives of benzoic esters. Since negligible polar effects were observed in the acid-catalysed hydrolysis, these reactions with a series of substituted carboxylic acids were used to define a set of steric substituent constants, E_s:

$$ \log(k/k_0)_{acid} = E_s $$

The use of linear free-energy relationships for the prediction of rates has already been discussed, but the application of substituent constants in this way appears to be much less important than the information that can be obtained on the electronic characteristics of transition-states by the consideration of reaction constants (ρ). Both the sign and size of ρ give specific information on the nature of transition-states. Consideration of a plot of log k against σ for any reaction leads to the conclusion that ρ will be negative if the reaction rate is increased by electron-supplying substituents, and positive if increased by electron-withdrawing substituents. It follows that:

(1) A negative ρ-value is indicative of a transition-state with an electron-deficient reacting centre relative to the initial state;

(2) A positive ρ-value is observed if the transition-state has a reaction centre which is electron-rich.

The size of ρ is also an important consideration and is a measure of the electron-demand or charge-development at the reaction centre. A number of the examples in Table 7 will be considered in the light of these conclusions; the appropriate substituent constants for the reactions are also stated.

Table 7. Values of ρ for some reactions.

Reaction	Solvent	Temp. (°C)	Subst. constant[a]	ρ
$ArH + Br_2 \rightarrow ArBr + HBr$	$MeCO_2H$	25	σ^+	−12.1
$ArH + Cl_2 \rightarrow ArCl + HCl$	$MeCO_2H$	25	σ^+	−8.06
$Ar(Ph)CHCl + EtOH \rightarrow Ar(Ph)CHOEt$	EtOH	25	σ^+	−4.1
$RCH(OEt)_2 + H^+ \rightarrow RCHO + EtOH$	50% dioxan	25	σ^*	−3.6
$ArNMe_2 + MeI \rightarrow ArNMe_3^+I^-$	90% acetone	35	σ^+	−3.3
$ArCH_2Cl + H_2O \rightarrow ArCH_2OH$	48% EtOH	30	σ^+	−2.18
$ArO^- + EtI \rightarrow ArOEt + NaI$	EtOH	43	σ	−0.99
$ArMe + Cl_2 \rightarrow ArCH_2Cl + HCl$	CCl_4	80	σ	−1.25
$ArSH + R\cdot \rightarrow ArS\cdot + RH$	PhMe		σ	−0.4
$ArCO_2H \rightleftharpoons ArCOO^- + H^+$	H_2O	25	σ	1.00
	MeOH	25	σ	1.76
$ArCH_2CO_2H \rightleftharpoons ArCH_2COO^- + H^+$	H_2O	25	σ	0.56
$ArCH_2CH_2CO_2H \rightarrow$				
$\qquad ArCH_2CH_2COO^- + H^+$	H_2O	25	σ	0.237
$PhCOCHR^1R^2$				
$\qquad + Br_2 \rightarrow PhCOC(Br)R^1R^2$	$NaOH-H_2O$		σ^*	1.59
$RCH_2OH \rightarrow RCH_2O^- + H^+$	PrOH		σ^*	1.36
$ArCO_2Me + HO^- \rightarrow ArCOO^-$	60% acetone	25		2.23
$ArCHO + HCN \rightarrow ArCH(OH)CN$	95% EtOH	20	σ	2.33
$ArCH_2ONO + {}^-OEt \rightarrow ArCHO + NO$	EtOH	30	σ	3.4
$ArCl^b + MeO^-Na^+ \rightarrow 4\text{-}X\text{-}2\text{-}NO_2\text{-}$				
$\qquad C_6H_3OMe$	MeOH	25	σ^-	3.94
$ArBr^b + C_6H_{11}NH_2 \rightarrow 4\text{-}X\text{-}2\text{-}NO_2\text{-}$				
$\qquad C_6H_3NHC_6H_{11}$		25	σ^-	4.9

[a] Constant for correlation (σ^* can also be used satisfactorily in those cases where σ is applied).
[b] 4-Substituted 2-NO_2-C_6H_3-.

The ρ-value (−4.1) for the solvolysis (**67**) of benzhydryl chloride in ethanol can be compared with the solvolysis of substituted benzyl chlorides in 48% aqueous ethanol which shows a ρ-value of −2.2. The much higher charge-development in the transition-state for the former carbonium ion reaction is apparent. The latter reaction has substantial S_N2 character in which the charge in the transition-state is more dispersed owing to bonding to incoming solvent. As expected, electrophilic substitution reactions display high negative values owing to the positively charged transition-states and direct conjugation from the substituents to the reaction centre. The influence of solvent on charge-dispersal in transition-states is made abundantly clear by the comparison of ρ-values of reactions in solution with that of -19 obtained by Lossing and coworkers for the ionization potentials of a series of substituted benzyl radicals (equation 68).[43] These measurements were made in a mass-spectrometer and the ionization potentials, converted from electron-volts to kcal mole^{-1}, plotted against σ^+-values. The large ρ-value in the gas phase compared to those in solution is indicative of the reduction in electron-demand of positive ions by solvation.

(67)

(68)

Reactions in which carbanions or electron-rich transition-states are produced are characterized by positive ρ-values. DePuy has suggested that the value of $+5$ found for the rate of polymerization of styrenes at 25° should be used as a guide for reactions involving carbanions.[44] This is of particular relevance to later discussions of elimination reactions (page 213) regarding the degree of carbanionic character in $E2$ transition-states. A value of $+4.9$ was obtained by Jaffé for the displacement of bromine from 4-substituted 2-nitrobromobenzenes by cyclohexylamine.[24]

A number of reactions have extremely small reaction constants. Radical reactions appear to depend much less on the stabilization of substituents than carbonium or carbanion reactions, and most have reaction constants of less than 1.6. A ρ-value of -1.37 for the reaction of substituted toluenes with bromine in carbon tetrachloride is indicative of a small amount of polar character in the transition-state. Reactions involving cyclic transition-states also display very small ρ-values; for the rearrangement of p-substituted cinnamyl p-tolyl ethers at 180°, ρ is -0.40 (equation 69).[45] On this basis little, if any, charge is developed in the transition-state. The fact that the rates of the vapour-phase pyrolysis of 1-arylethyl acetates ($\rho = -1.3$) are correlated by σ^+ better than by σ has been used as evidence for some ionic character in the transition-states.[46] This criterion has also been used for other reactions even though they have been associated with small ρ-values.

(69)

Two other general points about small ρ-values can be made. Small reaction constants are found if the reaction site is separated by a number of carbon atoms from the position of substituent change; for the ionization of $ArCO_2H$, $ArCH_2CO_2H$, and $ArCH_2CH_2CO_2H$ the reaction constants are 1.0, 0.56, and 0.24 respectively. Secondly, the value of ρ as an indicator of charge in the transition-state must be used with care if a multistep process, or prior equilibrium, is involved. Aqueous acid hydrolysis of benzamides involves a slow step following an equilibrium but the position of equilibrium is dependent upon the strength of acid employed.

$$RCONH_2 + H_3O^+ \overset{fast}{\rightleftharpoons} [RC(OH)NH_2]^+ + H_2O \qquad (70)$$

$$[RC(OH)NH_2]^+ + H_2O \overset{slow}{\longrightarrow} RCO_2H_2^+ + NH_3 \qquad (71)$$

In fairly dilute acid, where the equilibrium lies to the left-hand side of equation 70, the reaction was found to be almost independent of polar effects on the molecule.[47] Since polar effects would have opposite effects in equations 70 and 71, it was concluded that these are almost equal, with the resultant observation of negligible polar influences in the overall reaction. In strongly acidic conditions the prior equilibrium is complete and the large polar effects observed are those operating entirely in equation 71; the hydrolysis was accelerated by electron-attracting and retarded by electron-donating substituents. Considering the hydrolysis in dilute acid, if k_{obs} is the observed second-order rate constant, K_1 the equilibrium constant of the first step, and k_2 the first-order rate constant of the second step, then:

$$k_{obs} = K_1 k_2$$

If the constants for benzamide itself are denoted by suffix 'o' then:

$$(k/k_0)_{obs} = (K/K_0)_1 (k/k_0)_2$$
$$\log (k/k_0)_{obs} = \log (K/K_0)_1 + \log (k/k_0)_2$$
$$\rho_{obs} = \rho_1 + \rho_2$$

since $\rho_{obs} = +0.12$ in dilute acid, and $\rho_1 = -0.93$ from the basic ionization of amides, it was predicted that $\rho_2 = +1.05$. The reaction constant observed for the rate studies in strong aqueous acid agreed very closely with this value.

A number of examples exist where a non-linear correlation of rates and substituent constants has provided specific information on the mechanism of reactions. For the formation of semicarbazones from a series of aromatic aldehydes at pH 6.5, a plot of log k against σ rises steeply to a maximum for benzaldehyde but then decreases slowly as the substituent becomes more

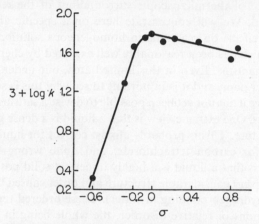

Figure 25. Plot of log k against σ for formation of semicarbazones of substituted benzaldehydes (ArCHO) at pH 6.5.

electron-withdrawing (Figure 25).[48] The failure to obtain a linear plot is probably caused by the change in rate-controlling step as a function of substituents (equation **72**). With electron-supplying substituents, the addition reaction is the slowest step and the overall rate decreases with increasing electron-supplying power of the substituent. For electron-withdrawing substituents, however, the dehydration step is rate-limiting and the overall rate decreases with increasing electron-withdrawing power of the substituent.

$$
\text{H}_2\text{N}-\overset{\overset{\text{O}}{\|}}{\text{C}}-\text{NH}-\overset{\underset{|}{\text{H}}}{\text{NH}} + \overset{\overset{\text{O}}{\|}}{\text{C}}-\text{H} \underset{k_{-1}}{\overset{k_1}{\rightleftharpoons}} \text{H}_2\text{N}-\overset{\overset{\text{O}}{\|}}{\text{C}}-\text{NH}-\overset{\overset{\text{H}}{|}}{\underset{|}{\text{N}}}-\overset{\overset{\text{OH}}{|}}{\underset{|}{\text{C}}}-\text{H}
$$

$$\Big\downarrow k_2 \qquad\qquad (72)$$

$$
\text{H}_2\text{N}-\overset{\overset{\text{O}}{\|}}{\text{C}}-\text{NH}-\text{N}=\text{CH}-\text{Ar} + \text{H}_2\text{O}
$$

From steady-state treatment: $k_{obs} = k_1 k_2/(k_{-1} + k_2)$

Medium Effects[49-54]

Organic reactions are normally written, and often discussed, as if they occurred between isolated molecules of the reactants. Although this situation is ideal for the purpose of theoretical discussion, it almost never occurs in practice and the influence of the microscopic surroundings of the reactants is very important indeed. We will concentrate here on a specific area of medium effects—solvent effects on reactions in homogeneous solution—since this is the only area which has been reasonably well explored by chemists interested in reaction mechanisms. Even in this limited area, our understanding of the situation is rather poor, and it is important to see why this is so.

In the first place it has not yet been possible to devise a satisfactory structural theory for liquids. One extreme view is that a liquid is a dense gas with almost no ordered structure. This is probably almost correct for liquid argon, a fair approximation for carbon tetrachloride, and quite wrong for water. An alternative view is that a liquid is a highly imperfect solid possessing regions in which there is very considerable structure and order caused by intermolecular forces (e.g. hydrogen-bonding in water). These ordered regions are interspersed with regions of relative disorder, the whole being in a state of flux, with the molecular motions continuously creating and destroying regions of order. The introduction of solute molecules or ions is certain to alter the arrangements of solvent molecules in their immediate vicinity. The predominant effect may be the destruction of the solvent structure on the one hand (known as structure-breaking) or the creation of relatively tightly organized solvation shells around the solute molecule or ion on the other. The basic difficulty in forming a clear picture of solvation is that (a) the physical situation is too complex and fluxional for a precise molecular model to be formulated, but (b) the effects, at least in the important region near a solute molecule, are too specific, and the numbers of solvent molecules involved too small, for a statistical approach or one based on the macroscopic properties of the solvent (e.g. dielectric constant) to be successful.

All this would matter little if solvation energies were small, but of course this is not so. The energy required to ionize Me_3CCl to $Me_3C^+Cl^-$ in the gas phase is 150 kcal mole^{-1}; in water this process is endothermic by only about 20 kcal mole^{-1}. The difference, 130 kcal mole^{-1}, is an approximate figure for the difference in solvation energy for the ions and for the neutral initial state. A large part of the ionic solvation energy is simply seen as a consequence of the reduced coulombic attraction ($e^2/r\epsilon$) between the ions in a medium of higher dielectric constant ($\epsilon = 78.4$ for water). To regard this as the only important effect is equivalent to regarding the solvent as a continuous medium. We know that this will not do; a variety of specific effects must be considered.

In this case these probably include hydrogen-bonding of H_2O to Cl^-, loose coordination of the lone-pair electrons of the oxygen atoms with Me_3C^+,

Table 8. Solvent parameters.

Solvent	ϵ at 25°	Y	Z	E_T	Solvent	ϵ	Z	E_T
H_2O	78.4	3.49	94.6	63.1	$HCONMe_2$	37	68.5	43.8
NH_3	23 (−33°)				$MeCONMe_2$	37.8	66.9	
HF	83.6 (0°)				MeCN	36.2	71.3	46.0
H_2SO_4	100				PhCN	25.2		42.0
SO_2	15.4				$MeNO_2$	35.9		46.3
MeOH	32.6	−1.09	83.6	55.5	$PhNO_2$	34.8		42.0
MeOH (50%[a])		1.97			Me_2SO	46.6	71.1	45.0
EtOH	24.3	−2.03	79.6	51.9	$(Me_2N)_3PO$	30 (20°)	62.8	
EtOH (80%[a])	67.0	0.00	84.8	53.6	Sulpholan[b]	44 (30°)	77.5	44.0
EtOH (70%[a])		0.59	86.4		Me_2CO	20.7	65.7	42.2
Me_2CHOH	18.3	−2.73	76.3	48.6	Pyridine	12.3	64.0	40.2
Me_3COH	12.2	−3.26	71.3	43.9 (30°)	CH_2Cl_2	8.9	64.2	41.1
HCO_2H	57.9 (20°)	2.05			$CHCl_3$	4.64	63.2	
$MeCO_2H$	6.17 (20°)	−1.68	79.2		CCl_4	2.22		32.5
CF_3CO_2H	8.32	~4.5	~88		C_6H_6	2.27	54	34.5
Et_2O	4.2			34.6	PhMe	2.37		33.9
Dioxan	2.21			36.0	PhF	5.42		
					PhCl	5.62		37.5
Tetrahydro-furan	7.4			37.4	PhBr	5.40		37.5
					PhI	4.60		
$HCONH_2$	109.5	0.60	83.2	56.6	Ph_2O	2.53 (75°)		35.3 (30°)
MeCONHMe	165.5 (40°)		77.9		$n-C_6H_{14}$	1.89		30.9

[a] v/v in H_2O.
[b] Tetramethylene sulphone or tetrahydrothiophen 1,1-dioxide.

disruptive effects of the hydrophobic Me_3CCl on the structure of the surrounding water, etc. It is apparent that taking account of the appropriate specific effects when comparing the rate of a reaction in one solvent with that in another is likely to be a highly complicated task. In the face of these difficulties we will adopt the following approach. First we will briefly discuss our general ideas of the forces responsible for solvation. Secondly, we will consider two rather different approaches which have been made to the quantitative study of solvent effects on reaction rate. In one, correlations are sought between solvent effects on the rate of a reaction and on some model process, or property, which may be something as simple as the dielectric constant or may be the solvent effect on another (model) reaction or on, say, the position of a spectroscopic absorption. The second approach is a more circumspect (and time-consuming) one which seeks to dissect the solvent effect on a rate of reaction into effects on the initial and transition states and in turn to dissect these into enthalpy and entropy contributions.

Intermolecular Forces and Types of Solvation

The forces responsible for solvation are, of course, those responsible for intermolecular interactions in general. They are basically of two types derived from the exchange and coulombic contributions to the potential-energy in the wave equation for the total system, and they dominate the situation in the non-polar and polar case respectively.

The non-polar exchange interactions, also known as van der Waals forces, London forces, or dispersive interactions, arise from weak coupling or correlation of electronic motions in the molecules concerned. The coupling is strongest for the least strongly bound electrons, and so the attractive force is greatest between molecules containing weakly bound electrons in π-orbitals (e.g. in aromatic systems) or in lone-pair orbitals, especially in the heavier elements (e.g. iodine). The attraction is generally not very specific geometrically although, in the most strongly interacting cases, well defined 'π-complexes' result (e.g. the iodine–benzene complex).

Polar solvation is, by definition, dominated by coulombic or electrostatic interactions, giving rise to dipole–dipole, dipole–ion, and ion–ion forces. The solute molecules or ions we shall be considering will be nucleophiles (Lewis bases) or electrophiles (Lewis acids), or at least will have nucleophilic or electrophilic centres requiring solvation. Any dipolar solvent molecule possesses a centre capable of providing solvation to a nucleophile (the positive end of its dipole) or an electrophile (the negative end). In general, however, with one notable exception, the positive ends of the dipoles of most solvent molecules are well buried inside the molecule, so that solvation of a nucleophile is relatively ineffective or at least requires severe restriction of molecular motions (entropy of solvation negative). The one exception is in molecules with X—H bonds where X is an electronegative atom. The positive end of the H—X dipole is exceptionally exposed and hydrogen-bonding results. This picture of a hydrogen bond Y$\cdots$H—X as an electrostatic interaction is an approximation but not a very serious one from our point of view. In what follows, molecules with H—X groups are described as hydrogen-bond donors; molecules containing Y: lone-pair dipoles form the major class of hydrogen-bond acceptors. This unfamiliar approach into familiar territory is intended to show the reader the special position occupied by protic solvents (those capable of acting as hydrogen-bond donors). The distinction is so important that we characteristically consider three categories of solvents: non-polar, polar but aprotic (with the dielectric constant greater than about 20, but incapable of acting as hydrogen-bond donors), and protic (moderate to high dielectric constant, capable of acting as both hydrogen-bond donors and acceptors).

These ideas can be illustrated by considering the solvation of some common cations and anions. Typical metal cations are solvated by Y: groups; there is a difference in degree but not in kind between weak solvation in, for example, ether solutions containing potassium ions, $Et_2O\cdots K^+$, and the formation of stable metal complexes such as $Cr(OH_2)_6{}^{3+}$. In this type of solvation there is no distinction between protic and aprotic solvents, each exerting its effect according to its Lewis basicity or nucleophilicity towards the ion in question. The solvation of anions, however, is rather different because the hydrogen-bonding from protic solvents is generally considerably more efficient than any other form of electrophilic solvation. Anions are therefore rather poorly solvated in dipolar aprotic solvents and are therefore rather reactive. Thus, the rate of the reaction of azide ion with methyl iodide at 0° increases by a factor of 4.5×10^4 on changing from the protic solvent methanol ($\epsilon = 32.6$) to the aprotic solvent dimethylformamide ($\epsilon = 37$).[55] The rate of racemization of 2-methyl-3-phenylpropionitrile by methoxide ion [via the carbanion $PhCH_2C(Me)CN^-$] is increased by a factor of no less than 5×10^7 on changing solvent from methanol to 98.5% dimethyl sulphoxide–1.5% methanol.[56] Although the solvation of the transition-state must be considered (see page 48), it is likely that less efficient solvation of the reactant anions in the aprotic solvents is an important factor in producing these large rate enhancements.

Correlation of Solvent Effects

If electrostatic interactions governing the separation and combination of charges really were the dominant influence in the effects of solvent on reaction rate, we should be able to correlate these with the dielectric constant of the solvent.[49] The rates of reactions which go from non-polar reactants to a non-polar transition-state should be insensitive to solvent change, and other charge types should respond as shown in Table 9. Typical results of a dielectric constant–reaction rate correlation[57] are shown in Figure 26. (Dielectric constants and other parameters of many common solvents are in Table 8.) It can be seen from Figure 26 that the expected effect of increasing dielectric

Table 9. Solvent effects according to an electrostatic model.

Charge type	Change in charge distribution on going from initial state to transition-state	Effect of a solvent with increased dielectric constant on reaction rate
$R{-}X \rightarrow R^{\delta+}\cdots X^{\delta-}$	Increased separation	Large increase
$R{-}X^+ \rightarrow R^{\delta+}\cdots X^{\delta+}$	Dispersed	Small decrease
$Y^- + RX \rightarrow Y^{\delta-}\cdots R\cdots X^{\delta-}$	Dispersed	Small decrease
$Y + RX \rightarrow Y^{\delta+}\cdots R\cdots X^{\delta-}$	Increased separation	Large increase
$Y^- + RX^+ \rightarrow Y^{\delta-}\cdots R\cdots X^{\delta+}$	Reduced separation	Large decrease
$Y^- + R_2CO \rightarrow Y^{\delta-}\cdots(R_2)C\cdots O^{\delta-}$	Dispersed	Small decrease

constant (Table 9) is occurring but that the correlation is only approximate. Other properties of the solvent are clearly playing a part, as expected. For example, the increase in rate on going through the halogenobenzenes from fluoro- to iodo-benzene suggests that one factor of importance could be dispersive interactions between the transition-state and the solvent.

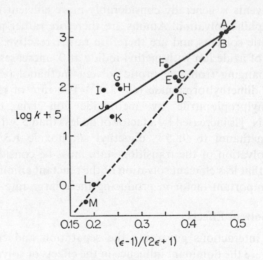

Figure 26. Effect of dielectric constant on the rate of reaction between triethylamine and ethyl iodide; the function of dielectric constant (ϵ) shown is proportional to log k on simple electrostatic theory.

A PhNO$_2$; B PhCN; C PhCl; D PhF; E PhBr; F PhI; G PhOPh; H PhCH$_2$Ph; I dioxan; J benzene; K PhMe; L C$_6$H$_{12}$; M C$_6$H$_{14}$.

To get a more precise correlation, a more perfect model process must be sought. The best model for one chemical reaction should be another of similar mechanism. Grunwald and Winstein,[58] studying solvolysis mechanisms, compared solvent effects on various solvolysis reactions with those on solvolysis of Bu^tCl. Solvolysis of Bu^tCl was chosen as the model reaction in setting up the Y-scale of solvent polarity because its solvolysis mechanism was considered to be well established as being S_N1 (see Chapter 2, page 82). If k is the rate constant for some reaction in some chosen solvent, and k_0 is the rate in 80% ethanol–water, then:

$$\log(k/k_0) = mY$$

where Y is the measure of ionizing power of the solvent ($Y = 0$ for 80% ethanol–water, the standard solvent) and m is a measure of the sensitivity of the reaction to solvent ionizing power ($m = 1$ for the standard reaction, solvolysis of Bu^tCl, so that Y-values for solvents are evaluated using this reaction). Excellent correlations with m near unity are obtained for compounds similar to Bu^tCl solvolysing by an S_N1 process (1-phenylethyl chloride, 1.195; Bu^tBr, 0.90). As a further example, the rates of solvolysis[59] of 2-methyl-2-phenylpropyl toluene-p-sulphonate (neophyl tosylate; see Chapter 2, page 106) at 75° are: EtOH 5.23×10^{-6}; $MeCO_2H$ 2.0×10^{-5}; HCO_2H 3.0×10^{-3}; CF_3CO_2H 1.12×10^{-1} sec^{-1}. For the first three solvents there is a fair correlation with Y-values. The Y-value of CF_3CO_2H is unknown, but these results suggest that it is very high, about +4.5. It can be seen that ionizing power as measured by Y has very little to do with dielectric constant (see the entries in Table 8 for $MeCO_2H$, H_2O, and CF_3CO_2H). A very important factor must be acid catalysis of the separation of the leaving group. For a tosylate which solvolyses by S_N2 reaction with the solvent, the order of reactivity is quite different.[59] Thus, for ethyl tosylate reaction rates at 75° are: EtOH 2.98×10^{-5}; $MeCO_2H$ 7.72×10^{-7}; HCO_2H 1.89×10^{-5}; CF_3CO_2H 2.26×10^{-7} sec^{-1}. There is no correlation with Y; solvent nucleophilicity is now the most important factor. The high rate for formic acid does, however, suggest that acid-catalysis (hydrogen-bonding to the leaving group) is still important. It is interesting that this solvent has a high autoprotolysis constant: $2HCO_2H \rightleftharpoons HCO_2H_2^+ + HCO_2^-$, $\log K = -6.2$, and so should be good at providing simultaneous acidic and nucleophilic catalysis.

While the Y-scale is, therefore, a useful scale of solvent polarity for ionization reactions, Y cannot be measured easily in a wide range of solvents, and so several workers have set up related scales by using certain spectroscopic processes as models for ionization. The two most successful scales are the Z-scale of Kosower[60] and the E_T scale of Dimroth, Reichardt, and coworkers.[61] The Z-scale is based on the charge-transfer band in the visible spectrum of

(73)

(75)

(74)

1-ethyl-4-methoxycarbonylpyridinium iodide (see **73**), whose position varies from 314.6 nm in methanol to 450.8 nm in *cis*-1,2-dichloroethylene. The Z-value is simply the energy (in kcal mole^{-1}) associated with the transition ($Z = h\nu$, where ν is the frequency of the light-wave at the absorption band). The conversion factor is $Z = 2.859 \times 10^4/\lambda$, where λ is the wavelength in nanometres. The E_T values are derived in a similar way from the visible spectrum of the pyridinium betaine **74**. A plot of E_T-values against Z-values gives quite good correlation (Figure 27), and plots of Y-values against Z-values are quite accurately linear (Figure 28). The Z/E_T scale may, therefore, be a quite useful extension of the Y-scale but, as has been stressed above, these scales must not be regarded as a generally valid measure of solvent polarity. For example, correlation of the rates of the reaction $Et_3N + EtI$ with E_T (Figure 29) is no better than with $(\epsilon - 1)/(2\epsilon + 1)$ (Figure 26). Obviously, the model process is an inappropriate one. The best model process for one amine quaternization reaction might be another similar reaction; this is illustrated by Figure 30.

Before leaving the subject of correlation of solvent effects we should remember that the quite crude correlations which are obtained are sometimes all that is required. Thus the fact that the rate of dimerization of cyclopentadiene varies by only a factor of 2.5 between the gas phase and ethanol solution[63] surely means that the transition-state is about as non-polar as the starting diene and rules out the zwitterion (**75**) as an intermediate. A more serious worry about the use of this correlation approach is that it can easily conceal interesting data about mechanism. A striking illustration of this is to be found in Bu^{t}Cl solvolysis, the very reaction used to construct the Y-scale.

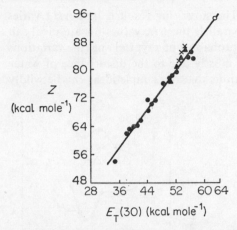

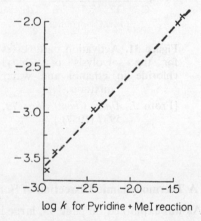

Figure 27. A plot of Z-values against $E_T(30)$ values in various solvents.

[After Dimroth et al, *Ann.* **661**, 1, (1963).]

Figure 28. The correlation of Y with Z for methanol–water (M), ethanol–water (E), and acetone–water (A) mixtures.

[From *J. Amer. Chem. Soc.*, **80**, 3253 (1958).]

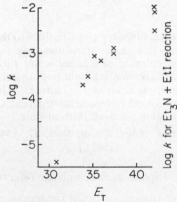

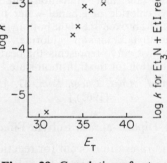

Figure 29. Correlation of rates for $Et_3N + EtI \rightarrow Et_4N^+I^-$ with E_T (cf. Figure 26).

Figure 30. Plots of rates for the $Et_3N + EtI$ reaction[57] against those for the Pyridine + MeI reaction.[62]

The fairly smooth variation of $\Delta G^{\ddagger}$ (and thus log k) with composition of ethanol–water mixtures is the result of compensation between the $\Delta H^{\ddagger}$ and $-T\Delta S^{\ddagger}$ values which undergo quite wild fluctuations (Figure 31).[64] By measuring the heat of solution of Bu^tCl in ethanol–water mixtures, Arnett and coworkers[65] were able to deduce how the heat of solution of the transition-state varied (Figure 32). The surprising result is that ΔH_s^g varies more than ΔH_s^{ts} and more closely parallels the ΔH_s values for a typical salt ($Me_4N^+Cl^-$). Although this result is probably not a typical one (the variations in the ΔH_s terms are thought to be mostly due to the disturbance of water structure by the solutes), it does warn us that our simple ideas *can* be wildly wrong.

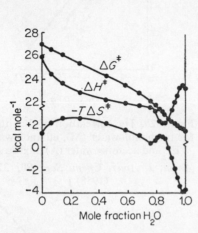

Figure 31. Activation parameters for the solvolysis of t-butyl chloride in ethanol and water mixtures.

[From *J. Amer. Chem. Soc.*, **79**, 5937 (1957).]

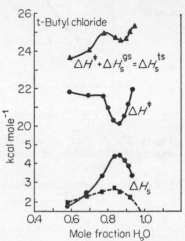

Figure 32. Enthalpy contributions to the ground-state and transition-state of t-butyl chloride in ethanol–water mixtures. The broken line at the bottom is for tetramethylammonium chloride. The uppermost curve represents the heat of solution for the transition-state.

[From *J. Amer. Chem. Soc.*, **87**, 1541 (1965).]

A Thermodynamic Dissection of Some Protic–Dipolar-aprotic Solvent Effects

As was explained on page 43, large increases in rates often occur for reactions involving anionic nucleophiles when we change from a protic to a dipolar aprotic solvent of similar dielectric constant. In this section we will take a closer look at three reactions (**76—78**) and try to analyse the solvent effect on the activation parameters ΔH^+, ΔG^+, and ΔS^+ in terms of initial-state and transition-state effects.

Changes (δH) in the enthalpy of the reactants on going from one solvent to another can be found if the heats of solution of the reactants in each solvent are measured. δH for some reactants on transfer from methanol to dimethyl-formamide (DMF) is equal to its heat of solution in DMF minus its heat of

$$\text{MeI} + \text{SCN}^- \longrightarrow \text{MeSCN} + \text{I}^- \tag{76}$$

k_2 at $0° = 3 \times 10^{-5}$ in MeOH, 6.9×10^{-3} in DMF

$$O_2N\!-\!\!\left\langle\!\!\bigcirc\!\!\right\rangle\!\!-\!I + N_3^- \longrightarrow O_2N\!-\!\!\left\langle\!\!\bigcirc\!\!\right\rangle\!\!-\!N_3 + I^- \tag{77}$$

k_2 at $100° = 3.4 \times 10^{-6}$ in MeOH, 1.17×10^{-2} in DMF

$$\left\langle\!\!\bigcirc\!\!\right\rangle\!\!-\!CH_2Cl + \left\langle\!\!\bigcirc\!\!\atop N\right\rangle \longrightarrow \left\langle\!\!\bigcirc\!\!\right\rangle\!\!-\!CH_2\!\!-\!\overset{+}{N}\!\!\left\langle\!\!\bigcirc\!\!\right\rangle Cl^- \tag{78}$$

k_2 at $25° = 3.46 \times 10^{-6}$ in MeOH, 3.74×10^{-6} in DMF

solution in MeOH. This has been done for the reactants in equations 76—78 by Haberfield.[66] It is, of course, impossible to measure the heat of solution of a single ion; the measurement must be made with an electrically neutral salt. The value obtained for the salt can be split into values for the separate ions only by making an arbitrary assumption. Haberfield assumed that the contributions of the anion and cation to the heat of solution of $\text{Bu}^n_4\text{N}^+\text{Bu}^n_4\text{B}^-$ are equal, and he could then subtract the value obtained for Bu^n_4N^+ from the heat of solution of say $\text{Bu}^n_4\text{N}^+\text{SCN}^-$ to obtain the heat of solution of SCN^-. In a similar way the free-energy changes (δG) for the reactants on going from one solvent to another can be measured by measuring solubilities in each solvent or distribution coefficients between each solvent and a third immiscible solvent. This is so because the reactant in the solvent is in equilibrium with the reactant as solid or as a solution in the immiscible solvent. Parker[67] has done this for all the reactants in question except pyridine. [In his review, the results are expressed in terms of activity coefficients, $^M\gamma^S$, for transfer from methanol to the dipolar aprotic solvent, in our case DMF. The relationship between our δG and $^M\gamma^S$ is $(\delta G)_{\text{RX from MeOH to DMF}} = 2.303\ RT\log_{10} {}^{\text{MeOH}}\gamma_{\text{RX}}^{\text{DMF}}$.] As with the δH quantities, δG values for single ions require a similar arbitrary assumption.

As a result of Haberfield's and Parker's measurements, the sums (Σ) of δH, δG, and therefore δS ($=\delta H - \delta G/T$) for the reactants together can be found. These may then be combined with the values of $\Delta H^{\ddagger}$, $\Delta G^{\ddagger}$, and $\Delta S^{\ddagger}$ for the activation process (determined from the kinetics of the reaction) to get δH etc. values for the transition-state. This is illustrated in Figure 33 and the results are shown in Table 10.

If we consider the results in Table 10 in conjunction with what was said on page 43, we see that SCN^- is actually solvated better in DMF than in MeOH (δH negative) but at a tremendous cost in entropy ($\delta S = -23.6$ cal deg^{-1} mole^{-1}). Azide ion is rather better solvated in methanol; again solvation in

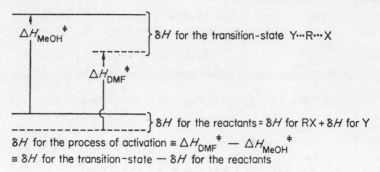

$$\delta H \text{ for the process of activation} \equiv \Delta H_{\text{DMF}}^{\ddagger} - \Delta H_{\text{MeOH}}^{\ddagger}$$
$$\equiv \delta H \text{ for the transition-state} - \delta H \text{ for the reactants}$$

Figure 33. The δH value for the transfer of the transition-state from MeOH to DMF is found from δH for the reactants and δH for the activation process (given by $\Delta H_{\text{DMF}}^{\ddagger} - \Delta H_{\text{MeOH}}^{\ddagger}$, the difference in the enthalpies of activation in the two solvents). Similar relationships hold amongst the free-energy (δG) and entropy (δS) terms. Also, $\delta S_{\text{RX}} = (\delta H_{\text{RX}} - \delta G_{\text{RX}})/T$ etc.

Table 10. Transfer of reactions **76—78** from MeOH to DMF at 25°.

	δH (kcal mole^{-1})	δG (kcal mole^{-1})	δS (cal mole^{-1} deg^{-1})
76 (MeI + SCN$^-$)			
MeI	−0.65	−0.7	+0.2
SCN$^-$	−3.35	+3.7	−23.6
Σ Reactants	−4.00	+3.0	−23.4
Activation	−4.3	−3.0	−4.4
Transition-state	−8.3	0.0	−27.8
77 (p-nitroiodobenzene + N$_3^-$)			
p-Nitroiodobenzene	−1.80	−1.6	−0.5
N$_3^-$	+0.44	+6.7	−21.0
Σ Reactants	−1.36	+5.1	−21.5
Activation	−5.6	−5.8	+0.3
Transition-state	−7.0	−0.7	−21.2
78 (benzyl chloride + pyridine)			
PhCH$_2$Cl	−0.83	+0.20	−3.5
Pyridine	+0.86	+0.3[a]	+1.9
Σ Reactants	+0.03	+0.5	−1.6
Activation	−5.3	−0.05	−17.6
Transition-state	−5.3	+0.45	−19.2

[a] An estimate based on distribution coefficients by R. W. Alder. Other data from refs. 66 and 67.

DMF is expensive in terms of entropy. Solvation of the neutral nucleophile pyridine and of the organic halides does not lead to any outstanding trends. Possibly one might suggest that the anion solvation in DMF is only achieved by ordering solvent molecules where previously little order had been present. In methanol the prevailing ordered state is perhaps not much disturbed. All the transition-states are quite large and polarizable, and therefore susceptible to (non-polar) dispersive interactions. These are provided better by DMF than by methanol. In the case of the anionic nucleophiles there is little further cost in entropy on going to the transition-state, but the energetic advantage in reaction **78** is only gained at considerable cost in entropy, so the rates at room temperature in the two solvents are almost equal (at $-60°$, k_{DMF}/k_{MeOH} would be about 30; at $+100°$ it would be about 0.1).

Few reactions have yet been studied from this point of view, so it is unwise to draw any general conclusions from these results. However, this type of analysis should ultimately be very helpful in understanding solvent effects, and we can hope for considerable developments along these lines.

Catalysis[68-76]

Catalysts in general operate by providing an alternative pathway for a reaction in which $\Delta G^{\ddagger}$, the free-energy of activation, is lower than for the uncatalysed reaction. The position of equilibrium, determined by the overall free-energy change, ΔG_r, is however unaffected by the catalyst (Figure 34). In accomplishing this, catalysts play a wide variety of roles in the mechanisms of organic reactions. In ionic (heterolytic) reactions, the catalyst frequently enters the mechanism as an electrophile or nucleophile. Catalysis also occurs in reactions involving non-ionic and concerted mechanisms, an important

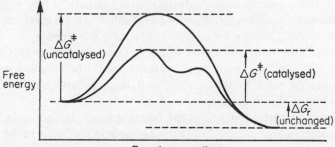

Figure 34. Free-energy relationships for catalysed and uncatalysed reactions. The reaction coordinate will normally have a different physical significance for the catalysed and the uncatalysed path.

example being the hydrogenation of olefins. In bimolecular reactions part of the entropy of activation is involved in assembling the two reactants, and catalysts sometimes function by providing some special way of bringing about this assembly. This type of catalysis might be called 'physical', as opposed to 'chemical' where the catalyst is directly involved in the mechanism. These types of catalysis will be surveyed and then one important special case of electrophilic–nucleophilic catalysis, i.e. acid–base catalysis, will be examined in more detail. Intramolecular catalyses, that is, cases where the catalyst is part of one of the reactant molecules, will then be discussed. These cases are intriguing in their own right but are also interesting in relation to the question of how enzymes perform their remarkable feats of catalysis.

Almost all the examples will be of homogeneous catalysis. Heterogeneous catalysis, while very important, is so involved with physical effects at surfaces that it has become the province of physical chemists. It seems likely too that a mechanistic understanding of heterogeneous catalysis will follow only when homogeneous catalysis is well understood.

Electrophilic and Nucleophilic Catalysis

Some examples of electrophilic catalysis are shown in Figure 35. Friedel–Crafts alkylation of aromatic compounds (Figure 35a) normally requires catalysis by metal halides which play an electrophilic role in assisting the breaking of the carbon–halogen bond in the alkyl halide. Friedel–Crafts acylation (Figure 35b) is similar but here the 'catalyst' is consumed in the reaction, remaining coordinated to the product ketone. If we adopt the common definition of a catalyst as 'an agent which affects the velocity of a chemical reaction without appearing in the final products of the reaction', we cannot call the metal halide a catalyst in this case, but we might call it a promoter. Many metal ions promote the hydrolysis of alkyl halides (see Figure 35c). The effectiveness of a given metal ion or metal halide in these reactions depends on the nucleophilic atom it attacks and is broadly governed by the Principle of Hard and Soft Acids and Bases which is discussed further in Chapter 3. Briefly we can say that small 'hard' cations should be effective with small 'hard' anions or nucleophilic centres, and large polarizable 'soft' cations should go best with soft anions.

An interesting metal-ion promoted rearrangement is shown in **79**. 2,2'-Pyridil undergoes the benzilic acid rearrangement in methanol in the presence of nickel acetate or cobalt acetate, under conditions where benzil is quite unchanged.[78] The sequence shown suggests that the metal ion assists in two steps of the reaction: the addition of hydroxide and the rearrangement step.

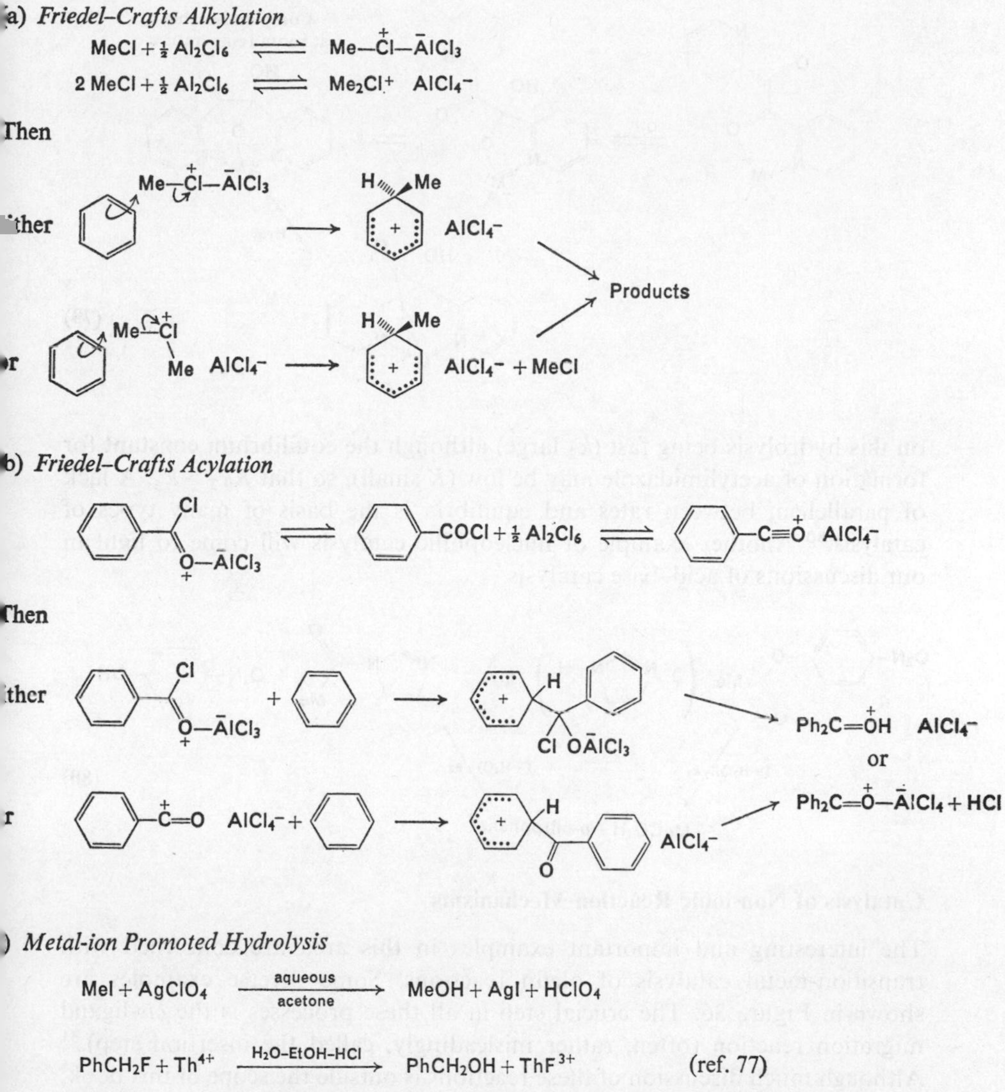

(a) *Friedel–Crafts Alkylation*

$$MeCl + \tfrac{1}{2} Al_2Cl_6 \rightleftharpoons Me\overset{+}{-}Cl-\bar{A}lCl_3$$

$$2\,MeCl + \tfrac{1}{2} Al_2Cl_6 \rightleftharpoons Me_2Cl^+\ AlCl_4^-$$

Then

either

or

Products

(b) *Friedel–Crafts Acylation*

Then

either

or

$$Ph_2C=\overset{+}{O}H\quad AlCl_4^-$$

or

$$Ph_2C=\overset{+}{O}-\bar{A}lCl_4 + HCl$$

(c) *Metal-ion Promoted Hydrolysis*

$$MeI + AgClO_4 \xrightarrow[\text{acetone}]{\text{aqueous}} MeOH + AgI + HClO_4$$

$$PhCH_2F + Th^{4+} \xrightarrow{H_2O-EtOH-HCl} PhCH_2OH + ThF^{3+} \qquad \text{(ref. 77)}$$

Figure 35. Electrophilic assistance by metal ions. The active electrophile in the Friedel–Crafts reactions is not known with certainty.

An example of nucleophilic catalysis is shown in **80**. Hydrolysis of *p*-nitrophenyl acetate is accelerated by imidazole.[79] When the imidazole concentration is high, the intermediate acetylimidazole can be detected spectroscopically. The rate-determining step is hydrolysis of this intermediate. Catalysis depends

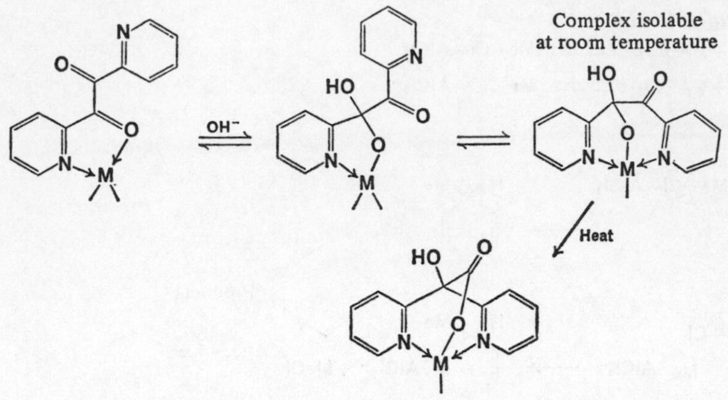

(79)

on this hydrolysis being fast (k_2 large) although the equilibrium constant for formation of acetylimidazole may be low (K small), so that $Kk_2 > k_3$. A lack of parallelism between rates and equilibria is the basis of many types of catalysis.[80] Another example of nucleophilic catalysis will come to light in our discussions of acid–base catalysis.

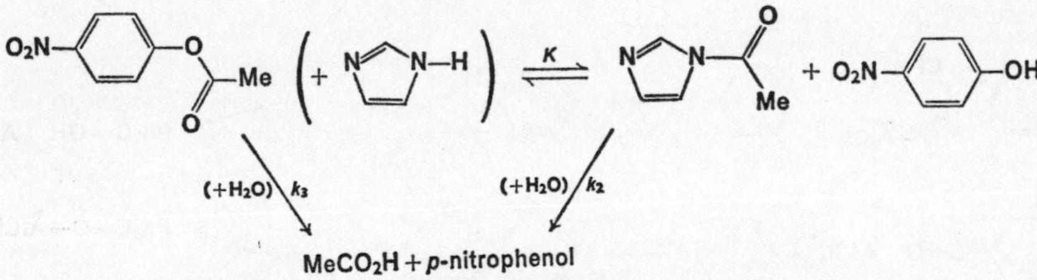

(80)

Catalysis of Non-ionic Reaction Mechanisms

The interesting and important examples in this area are concerned with transition-metal catalysis of olefin reactions. Some diverse examples are shown in Figure 36. The crucial step in all these processes is the *cis*-ligand migration reaction (often, rather misleadingly, called the insertion step).[81] Although much discussion of these reactions is outside the scope of this book, we may mention two properties of transition-metals which make them unique catalysts for this type of rearrangement. (*i*) They change their coordination number relatively easily; notice that this occurs in the *cis*-ligand migration reaction. (*ii*) The *d*-orbitals on the metal are probably intimately concerned in maintaining bonding throughout the transitional stages of this reaction.

The *cis*-ligand migration reaction is closely related to the multicentre reactions discussed in Chapter 4, and we may expect it to be allowed as a

Homogeneous Hydrogenation

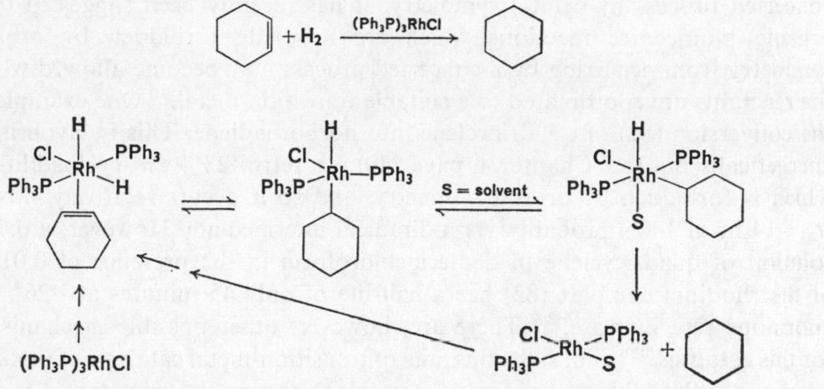

Oxidation of Ethylene (Wacker Process)

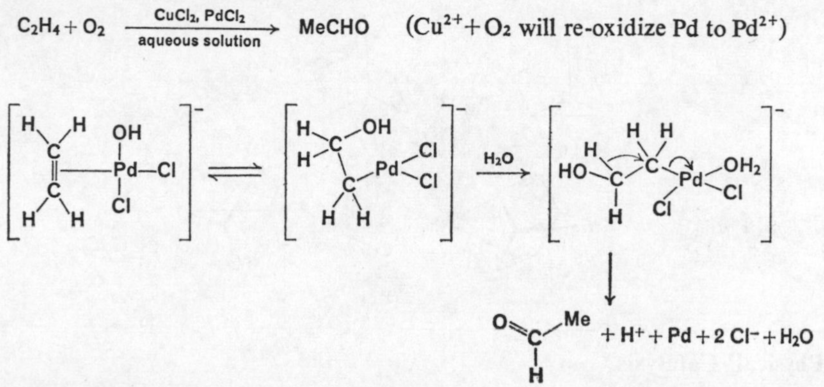

$$C_2H_4 + O_2 \xrightarrow[\text{aqueous solution}]{\text{CuCl}_2,\ \text{PdCl}_2} \text{MeCHO} \quad (Cu^{2+} + O_2 \text{ will re-oxidize Pd to Pd}^{2+})$$

Polymerization of Olefins

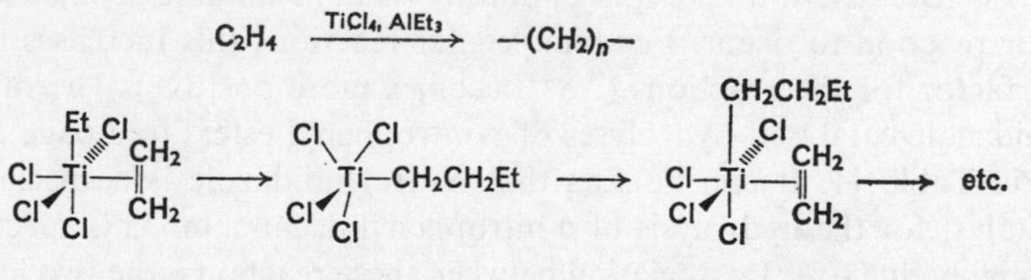

$$C_2H_4 \xrightarrow{\text{TiCl}_4,\ \text{AlEt}_3} -(CH_2)_n-$$

cis-Ligand Migration Reaction

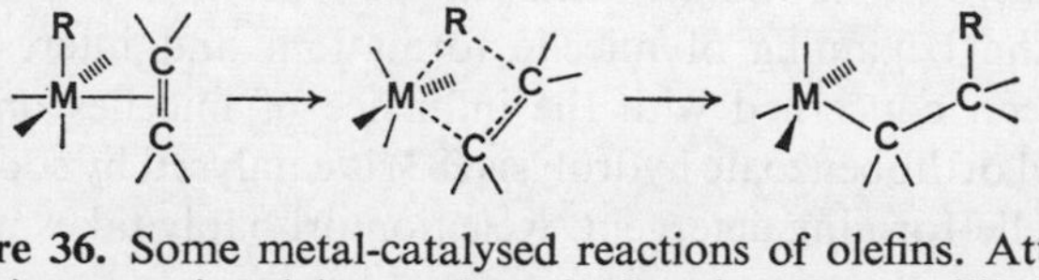

Figure 36. Some metal-catalysed reactions of olefins. Attention is focused on the addition steps involving *cis*-ligand migration. Pre-equilibria involving the coordination of the olefin and the other reactant are, of course, essential to the overall reaction.

3

concerted process by orbital symmetry. It has recently been suggested that organic multicentre reactions which are normally forbidden by orbital symmetry from occurring by a concerted process may become allowed when the reactants are coordinated to a suitable transition-metal.[70] One example is the conversion (81) of quadricyclene into norbornadiene. This is favourable energetically but (see Chapter 4, page 240) is a retro-$(2\pi + 2\pi)$-cycloaddition which is forbidden by orbital symmetry, and so it occurs relatively slowly $(t_{\frac{1}{2}} > 14$ hr at $140°)$ probably via a diradical intermediate. However, a 0.7M-solution of quadricyclene in deuteriochloroform in the presence of 0.014M of the rhodium complex (82) has a half-life of only 45 minutes at $-26°$, an enormous rate increase.[82a] There are, however, other possible mechanisms for this catalysis,[82b] and, since this area of transition-metal catalysis of organic reactions is the subject of vigorous research at present, we may expect to see our present ideas considerably refined in the near future.

(81)

(82)

'Physical' Catalysis

Some catalytic effects are the result of the catalytic agent's ability to assemble the reactants before reaction takes place, putting them in suitable geometrical positions for reaction to occur. For bimolecular reactions this increases the probability factor for the reactions ($\Delta S^{\ddagger}$ becomes more positive). The rates of some imidazole-catalysed hydrolyses of p-nitrophenyl esters (see page 53) are shown in Table 11. It can be seen that N-decylimidazole is a specially effective catalyst for the hydrolysis of p-nitrophenyl decanoate. This specific effect is probably due to 1:1 association between these reactants, the two long hydrocarbon chains being forced together to provide minimum disturbance of the water structure and favourable van der Waals attractions.[83]

Such association is the beginning of micelle formation, and interesting studies have recently been concerned with the influence of micelles on reaction rates. Thus methyl orthobenzoate hydrolysis (83) is catalysed by sodium dodecyl sulphate, a micelle-forming detergent. Reaction probably takes place

Table 11. Rate constants (1 mole^{-1} sec^{-1}) of catalysed hydrolysis of p-nitrophenyl esters at 25°C.

Catalyst	Acetate ester	Decanoate ester	$\dfrac{k_{decanoate}}{k_{acetate}}$
Hydroxide ion	11.6	0.477	0.041
N-Ethylimidazole	0.38	0.014	0.036
N-Decylimidazole	0.55	10.2	18.6
$\dfrac{k_{decylimidazole}}{k_{ethylimidazole}}$	1.43	740	515

in the outer regions of the micelle where there is a high concentration of $-OSO_3^-$ groups. This leads to an enhanced concentration of H_3O^+ ions and also probably to electrostatic stabilization of the developing oxonium cation.[84] Micellar catalysis may provide a model for some aspects of enzymic catalysis.

$$PhC(OMe)_3 + H_3O^+ \xrightarrow[\text{step}]{\text{slow}} Ph-C \overset{OMe}{\underset{OMe}{\overset{\cdot\cdot}{\cdot\cdot}}}{}^+ + MeOH + H_2O \xrightarrow{\text{fast}}$$

$$PhCO_2Me + MeOH + MeOH_2^+ \qquad (83)$$

Acid–Base Catalysis

This section is concerned with catalysis by proton-transfer. Acid–base catalysis was historically very important to the development of physical organic chemistry, and much interesting research continues to be done in this important area. A brief resumé of some basic ideas concerning acidity and basicity seems appropriate at the outset. We adopt Brønsted's definition of an acid as a proton-donor and a base as a proton-acceptor for the purpose of this section. Acid HA_1 is said to be stronger than acid HA_2 if K in equation **84** is greater

$$HA_1 + A_2^- \overset{K}{\rightleftharpoons} A_1^- + HA_2; \qquad K = \frac{[A_1^-][HA_2]}{[HA_1][A_2^-]} \qquad (84)$$

than 1. A_1^- is the conjugate base of acid HA_1 and HA_2 is the conjugate acid of A_2^-. Frequently the solvent is itself an acid and/or a base. This puts a limit on the strengths of acids and bases which can be studied in that solvent, since we cannot expect to have present in dilute solution appreciable concentrations of acids stronger than the protonated solvent (H_3O^+ in water) or bases stronger than the deprotonated solvent (OH^- in water). We assume that the reader is familiar with the pH scale of acidity for dilute aqueous solution. In that situation we have a precise and thermodynamically rigorous measure of acidity. Outside the area of dilute aqueous solution (other solvents, concentrated solutions) the quantitative measurement of acidity is much more difficult and uncertain. It is usually possible to compare acids HA_1 and HA_2 as in equation **84** and obtain a value for K. However, if HA_1 is compared with a third acid HA_3, and then HA_2 with HA_3, the values obtained for K do not usually form an internally consistent set. Hammett[85] originally thought that internal consistency might be achieved by restricting the acids and bases to one charge type only. Thus one can set up an acidity function, H_0, for a given acid system by measuring the degree of protonation of a series of uncharged bases of similar type as indicators (how this is done will be explained shortly). H_0 would, it was then hoped, provide a measure of acidity appropriate to the study of the protonation of any other uncharged base. Other acidity functions H_- and H_+ would be needed for anionic and cationic bases. Concentrated solutions of mineral acids, such as H_2SO_4, in water are of great importance since they provide the common reaction medium for carrying out hydrolyses of esters, amides, etc., and much research has been done on the H_0 acidity function and on trying to relate acid catalysis to it. As a result we now know that there is no unique H_0 function which covers all neutral bases but that individual acidity functions are needed for bases of different structural type. Thus the original H_0 function, based on primary nitroanilines as indicators, is different from one based on amides and even from one based on tertiary nitroaniline indicators. In spite of their restricted scope, however, acidity functions have proved useful in studies of acid–base catalysis, and a brief description of how one is set up will now be given.

For a base, B_1, used as an indicator, we can determine, for some acidic solution which causes partial protonation, the ratio of the concentration of the protonated form to that of the free base, $[B_1H^+]/[B_1]$, by a spectroscopic method, usually in the visible or ultraviolet region, provided that we know the spectra of the free base and of B_1H^+. When we change the acid concentration, $[B_1H^+]/[B_1]$ will alter and a graph can be plotted of $\log([B_1H^+]/[B_1])$ against acid concentration (see Figure 37a). A second indicator base, B_2, of a similar type will produce a parallel curve, the vertical separation between the curves being $pK_{B_1H^+} - pK_{B_2H^+}$, the basicity difference between B_1 and B_2. Our acidity function, H, is simply constructed by shifting the curves as shown in Figure 37b. This provides a logarithmic scale of acidity like pH, extending over several logarithmic units. In general the choice of zero for H is quite arbitrary but if we are working with concentrated solutions in water it is normal to choose the scale so that H merges with pH as we pass to dilute solutions.

The equation for H is:

$$H = pK_{BH^+} - \log([BH^+]/[B])$$

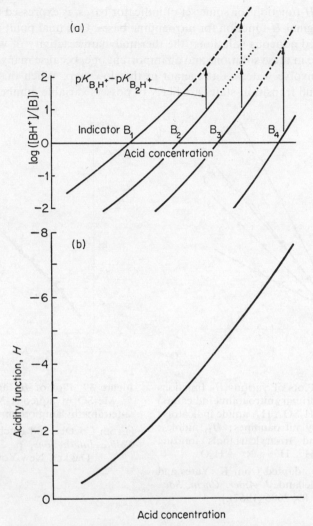

Figure 37. The construction of an acidity function. The separate curves for each indicator in (a) are shifted to create the continuous curve in (b).

As we have mentioned, different H-scales are obtained using indicator bases of different structural types. For the sulphuric acid–water system, however, one H-function varies linearly with another when neutral indicators are used (see Figure 38), so we may write:

$$H_B = mH_0 + \text{constant}$$

H_B, the H-function for some set of indicator bases, is expressed in terms of H_0, the original H-function for nitroaniline bases. One final point concerning concentrated aqueous solutions: the thermodynamic activity of water, a_{H_2O}, is a variable in these solutions and an important one, because many hydrolytic reactions involve water as a reactant in the slow step. When this is not so, reactants and transition-states are likely to possess variable numbers of water

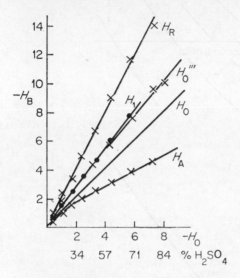

Figure 38. Plots of various H_B functions against H_0 (primary nitroaniline indicators) in aqueous H_2SO_4. H_A, amide indicators; H_0''', tertiary nitroanilines; H_1, indoles; H_R, di- and tri-arylcarbinols ionizing
$$ROH + H^+ \rightarrow R^+ + H_2O.$$

[This figure adapted from K. Yates and R. A. McClelland, *J. Amer. Chem. Soc.*, **89**, 2686 (1967).]

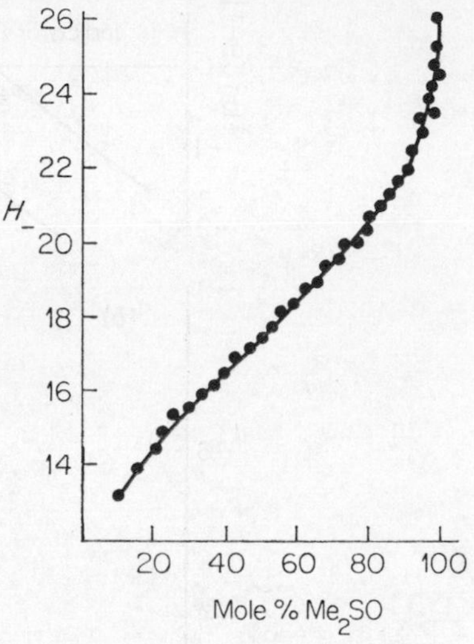

Figure 39. Plot of H_- against mole % Me_2SO in aqueous Me_2SO–0.01 M-tetramethylammonium hydroxide.

[From Coetzee and Ritchie, *Solvent-Solute Interactions*, p. 234, Marcel Dekker, New York.]

molecules bound to them by hydrogen-bonds so that, in general, reaction rates are a function of a_{H_2O} in these solutions.[86] Values of a_{H_2O} are known for mixtures of sulphuric acid and water, and in general it is found that:

$$\log k + H = r \log a_{H_2O} + \text{constant}$$

where H is an appropriate acidity function and r is a constant which is a rough measure of the number of extra water molecules involved in the transition-state compared with those hydrating the reactants. The value of r is dependent on the mechanism in an understandable way. We make use of these ideas of acidity function and hydration in a discussion of ester hydrolysis in Chapter 5, page 333. An alternative, possibly preferable, way to treat acidity-function deviations by linear free-energy relationships has been discussed by Bunnett and Olsen.[87]

The H_- acidity function is mainly of interest in basic solutions where the ionization of many uncharged acids (neutralization of anionic bases) occurs:

$$BH + OH^- \rightleftharpoons B^- + H_2O$$

and is potentially useful for comparing the acidity of many C—H acidic compounds (see Chapter 2, page 121), but, just as with H_0, it seems that there is a dependence on indicator structure. An interesting H_- plot is shown in Figure 39.[88] The deprotonating power of OH^- increases dramatically as the proportion of dimethyl sulphoxide in the solvent mixture increases; this is probably due to the stripping away of its hydrogen-bonded solvation by water (see the section on Medium Effects).

So far in this section we have not gone beyond the topic of acid–base equilibria, but of course an understanding of the *rates* of proton-transfer reactions is important for acid–base catalysis. We know from simple experience that many proton-transfers are very fast, and progress in this field required the development of techniques for studying very rapid reactions. Eigen,[74] one of the pioneers of these techniques, has provided much of the data on rates of proton-transfers (see Table 12). A proton-transfer is shown in Figure 40 as occurring in three simple steps. The rates will be at a maximum of the energy profile as is shown in (a), since then the rate of proton-transfer from the stronger acid HA^+ to the weaker acid HB^+ will be simply the rate of diffusion together of HA^+ and B ($k_{diffusion}$ is about 10^9—10^{10} l mole^{-1} sec^{-1} in aqueous solution at room temperature). The rate of proton-transfer in the reverse, unfavourable, direction will be $k_{diffusion}/\Delta K_a$, where $\Delta K_a = K_{a(BH^+)} - K_{a(AH^+)}$. A plot of log k (where k is the rate of proton-transfer) against ΔpK_a for reaction of a series of bases with a single acid under these conditions would give the curve (r) shown in Figure 40. It seems that in aqueous solution many proton-transfers from oxygen atoms to oxygen atoms (e.g. $H_2O + OH^-$) and between oxygen and nitrogen atoms ($NH_4^+ + H_2O$) are of this type. Proton-transfers to carbanions, and transfers from carbon acids, are often considerably slower processes, at least in aqueous solution. This can arise if formation of the necessary $(AH\cdots B)$ complex is an unfavourable equilibrium [see Figure 40(b)], or if there is an activation barrier for the actual proton-transfer step [see Figure 40(c)] (or, of course, if both of these occur). Both of these possibilities are quite reasonable. If AH is a carbon acid and B is, say, an oxygen base, AH will not be involved in the solvent hydrogen-bonding network, but

B will. It will therefore be necessary to supply energy to break B out of the network and associate it with AH (solvent reorganization needed). The second possibility (barrier to actual proton-transfer) may arise if there is considerable

Table 12. Rates of proton-transfer reactions in water at 25°.

$$HA + H_2O \underset{k_2}{\overset{k_1}{\rightleftharpoons}} A^- + H_3O^+$$

$$HA + OH^- \underset{k_4}{\overset{k_3}{\rightleftharpoons}} A^- + H_2O$$

The figures in the table are the logarithms of the rate constants in $l\ mole^{-1}\ sec^{-1}$.

Reactions with $\log k \geqslant +9$ are close to the diffusion-controlled limit $[\Delta G^+ \leqslant (3\ kcal\ mole^{-1})]$.

Acid, HA	pK_a	$\log k_1$	$\log k_2$	$\log k_3$	$\log k_4$
HOH	15.75	−6.3	11.2		
$PhOH$	9.98			10.1	4.4
$MeCO_2H$	4.74	4.1	10.6		
$H_3\overset{+}{N}H$	9.25	−0.4	10.6	10.5	4.0
$Me_3\overset{+}{N}H$	9.79	−1.1	10.4	10.3	4.4
$MeC(OH){=}CHCOMe^a$	8.24	0.5	10.5	7.3	−0.2
$MeCOCH_2COMe$	9.0	−3.7	7.1	4.6	−2.1
$HCH(CN)_2$	11.2	−3.5	9.4		
HCH_2COMe	~20	−11.0	~10.7	−0.6	~3.7
HCH_2NO_2	10.2	−9.1	2.8		

From the definitions of K_a and K_b:

$$pK_b = 14.0 - pK_a$$

$$\log k_1 - \log k_2 = -pK_a - 1.75 \quad (1.75 \equiv \log[H_2O]\ in\ H_2O)$$

$$\log k_3 - \log k_4 = pK_b + 1.75 = 15.75 - pK_a \quad (14.0 \equiv -\log K_w)$$

^{a} Internal hydrogen-bond in enol form of acetylacetone.

structural reorganization on going from the carbon acid to its anion $[CH_3NO_2$ to $H_2C{=}N^+(O^-)_2$ is a case in point]. This structural reorganization may also require further solvent reorganization. The result of energy profile (b) or (c) will be the $\log k/\Delta pK_a$ curve (s) shown in Figure 40. Some indication that solvent reorganization is an important factor has come from recent studies in dipolar aprotic solvents such as dimethyl sulphoxide where the difference between carbon acids and oxygen or nitrogen acids seems to be much less marked.[89]

The very rapid proton-transfers between e.g. an oxygen acid and an oxygen base are unlikely to be the slow steps in proton-catalysed organic reactions

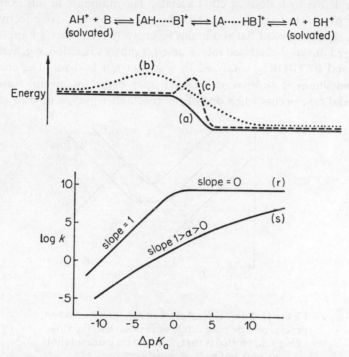

$$AH^+ + B \rightleftharpoons [AH\cdots B]^+ \rightleftharpoons [A\cdots HB]^+ \rightleftharpoons A + BH^+$$
(solvated)   (solvated)

Figure 40. Proton transfer as a sequence of three steps, with three possible energy profiles. Profile (a) leads to maximum possible rates with a $\log k - \Delta pK_a$ curve like (r). Energy profile (b) or (c) leads to a $\log k - \Delta pK_a$ curve like (s). The quantity α can be identified with the Brønsted coefficient (see p. 67).

unless, for example, (i) they are concerted with some other more energy-demanding process (e.g. concerted fission of another bond), or (ii) they are associated with a very unfavourable pre-equilibrium:

$$A + B \rightleftharpoons C + D$$

$$D + H^+ \rightarrow \text{products (rate-determining)}$$

The slower proton-transfers to and from carbon are, of course, proportionally more likely to be encountered as rate-determining steps in catalysed reactions.

With all these preliminary considerations in mind we can at last turn to acid–base catalysis itself. One convenient way to display the sensitivity of a reaction to acid–base catalysis in aqueous solution is the pH–log k_{rate} profile (Figure 41). Curve (a) is that of a reaction catalysed by both acid and base, for example the hydrolysis of ethyl acetate. The minimum in this example is at about pH 4, showing that catalysis by base is rather more effective than that by acid. The broad flat minimum in curve (b) indicates an appreciable uncatalysed or water-catalysed rate. Curve (c) shows a reaction, e.g. hydrolysis of an acetal $RCH(OR)_2$, catalysed by acid but not by base. The curve (d), with a maximum at an intermediate pH, may occur for several reasons. A fairly trivial case occurs when one of the reactants is, say, a dibasic acid and

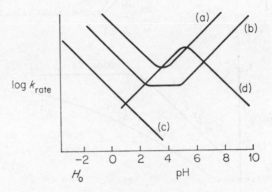

Figure 41. Some typical pH–log k_{rate} profiles. When general acid or base catalysis is observed, the value of log k_{rate} plotted is that obtained on extrapolation to zero buffer concentration (see p. 67).

only the mono-anion is reactive, so that the rate is at a maximum when pH = $(pK_2 - pK_1)/2$. A more important cause of curve (d) may be change of mechanism with pH, or, more often, change of rate-determining step in the same mechanism. For example, some step in the mechanism which is rate-determining in alkaline solution may become so rapid in acid solution that another step in the mechanism becomes the slowest. pH–log k_{rate} profiles of this type are, therefore, especially interesting.

For many reactions the observed rate does not simply depend on pH but is also a function of the *total* concentration of the buffer solution normally used to control pH. The buffer solution may be a weak acid (HA)–salt (M^+A^-) combination; pH depends on the $[HA]/[M^+A^-]$ ratio. When the reaction rate depends on pH *and* on the concentration of the undissociated acids present in the buffer, we speak of *general* acid catalysis. If the reaction rate depends only on pH, we have *specific* acid catalysis. Figure 42 shows how the distinction is made in practice. What does the distinction mean in terms of the mechanism of the catalysed reaction?

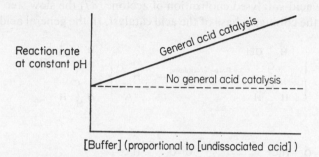

Figure 42. The test for general acid catalysis.

Specific acid catalysis occurs when the reactant is converted in a pre-equilibrium step into its conjugate acid:

$$R + H^+ \rightleftharpoons RH^+$$

and the subsequent rate-determining step involves RH^+ but not the anions A^- of any weak acids which may be present. The concentration of RH^+, however small, depends only on the pH of the solution, so the reaction rate is independent of buffer concentration. This mechanism commonly occurs for the hydrolysis of acetals,[90] the rate-determining step being unimolecular breakdown of the conjugate acid of the acetal, as in equation **85**.

$$
\begin{array}{c}
\text{R}\!\!-\!\!\underset{\text{H}}{\overset{\text{OR}}{\text{C}}}\!\!-\!\!\text{OR} + \text{H}_3\text{O}^+ \rightleftharpoons \text{R}\!\!-\!\!\underset{\text{H}}{\overset{\text{O}^+\!-\!\text{R}}{\text{C}}}\!\!-\!\!\underset{\text{R}}{\text{O}} + \text{H}_2\text{O}
\end{array}
$$

$\Big\downarrow$ Rate-determining   (85)

$$
\text{R}\!\!-\!\!\underset{\text{H}}{\text{C}}\!\!=\!\!\text{O} + \text{ROH} \quad \overset{\text{fast}}{\leftarrow\ \leftarrow} \quad \text{R}\!\!-\!\!\underset{\text{H}}{\text{C}}\!\!=\!\!\overset{+}{\text{O}}\!\!-\!\!\text{R} + \text{H}\!\!-\!\!\text{O}\!\!-\!\!\text{R}
$$

General acid catalysis involves rate-determining proton-transfer. This must be so because at equilibrium the concentration of RH^+, the protonated reactant, like that of any other acid, is determined solely by the pH. In rate-determining proton-transfer, however, each individual acid or base present will contribute individually to the total rate. We have already seen on page 61 that the high rates of proton-transfer reactions between simple oxygen- and nitrogen-bases impose some limitations on when this type of mechanism can occur. Many cases of simple rate-determining proton-transfer to carbon bases are known, a good example[91] being that of the hydrolysis of ethyl vinyl ether (86). In the acid-catalysed enolization of acetone (87) the slow step is proton-transfer to the conjugate base of the acid catalyst. In the general acid catalysed

$$(86)$$

$$(87)$$

hydrolysis of orthoesters (88) proton-transfer in the slow step is concerted with C—O bond-breaking. The E2 elimination reaction (89) (discussed in Chapter 3) is general base promoted; proton-transfer and C—X bond-breaking are concerted. Here we have a convenient parallel in drawing the reader's attention to a problem. We cannot distinguish an E2 from an S_N2 reaction by kinetics alone since the rates of both are proportional to [Base] [RCH_2CH_2X]. We also cannot distinguish kinetically general acid or base catalysis as for example in 87 where A^- acts as a base from nucleophilic catalysis. Thus acetate

$$(88)$$

$$RCO_2R + ROH$$

$$(89)$$

anion catalyses the hydrolysis of aryl acetates; does it act as base or nucleophile? The probable alternative mechanisms are set out in Figure 43. Gold[92] was able to distinguish between these by trapping any acetic anhydride formed, by its rapid reaction with aniline to form acetanilide. Although acetanilide is also formed by direct reaction of aniline with the aryl acetate, allowance could be made for this complication, and it was found that acetic anhydride was formed (nucleophilic catalysis) from aryl acetates derived from fairly acidic phenols, such as p-nitrophenol, but not from phenyl acetate itself. The weak catalysis observed in this case is probably general base catalysis. We may point out, in passing, that this is reasonable since the tetrahedral

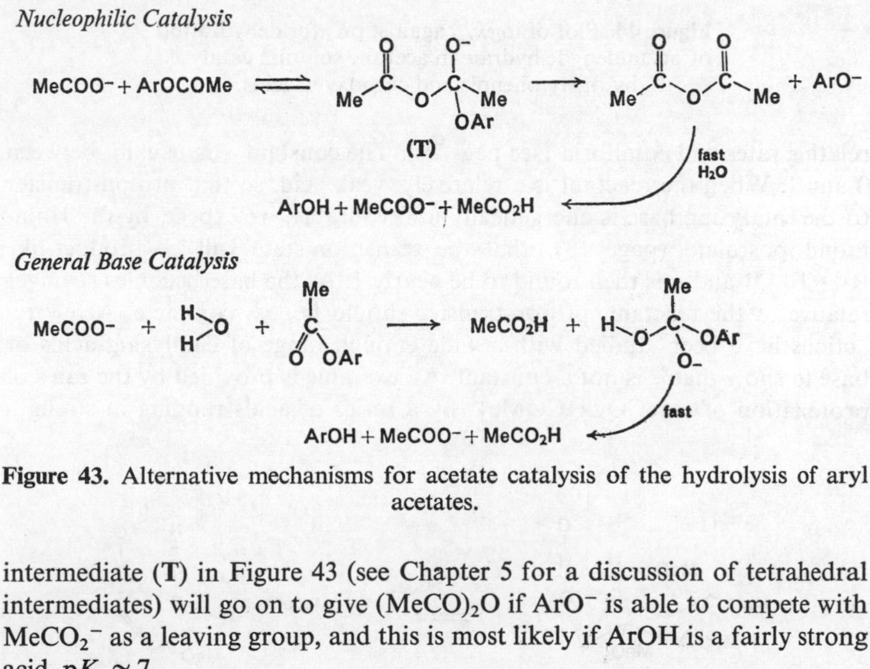

Nucleophilic Catalysis

General Base Catalysis

Figure 43. Alternative mechanisms for acetate catalysis of the hydrolysis of aryl acetates.

intermediate **(T)** in Figure 43 (see Chapter 5 for a discussion of tetrahedral intermediates) will go on to give $(MeCO)_2O$ if ArO^- is able to compete with $MeCO_2^-$ as a leaving group, and this is most likely if ArOH is a fairly strong acid, $pK_a \approx 7$.

When general acid or base catalysis is observed, we may ask if there is any relationship between the catalytic constant for the acid or base and its pK. A relationship of the type:

$$\log k_{cat} = -\alpha pK_a + \text{constant}$$

is often observed especially when all the catalysing acids and bases are of similar structure. A good example[93] is of catalysis of dehydration of acetaldehyde hydrate in acetone solution (Figure 44). This relationship, the Brønsted equation, is another example of a linear free-energy relationship

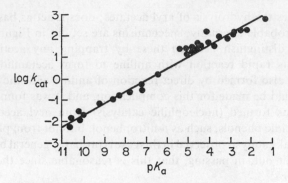

Figure 44. Plot of $\log k_{cat}$ against pK_a for dehydration
of acetaldehyde hydrate in acetone solution catalysed
by thirty phenols and carboxylic acids.

relating rates and equilibria (see page 29). The constant α is usually between
0 and 1. When the reactant is a relatively weak acid, so that proton-transfer
to the catalysing base is energetically unfavourable, we expect, by the Hammond postulate (page 13), that the transition-state will be product-like
R····H—B and α is then found to be nearly 1. As the base becomes stronger
relative to the reactant, proton-transfer should be less complete. A few reactions have been studied with a wide enough range of catalysing acids or
base to show that α is not a constant. An example is provided by the rates of
protonation of [MeCOCHCOMe]⁻ by a series of acids ranging in strength

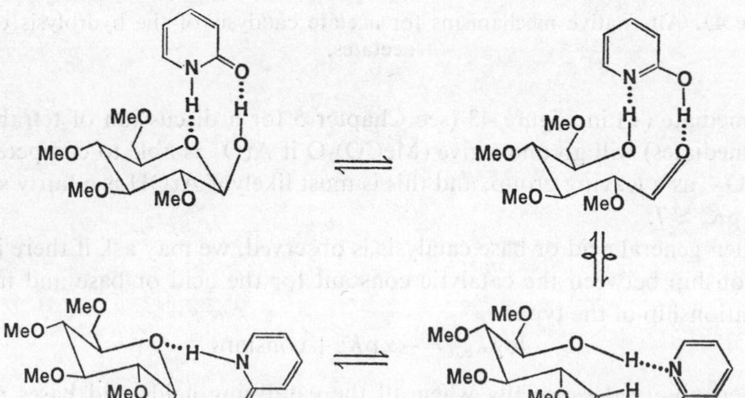

Figure 45. One possible scheme for bifunctional catalysis by 2-pyridone
of the mutarotation of 2,3,4,6-tetramethylglucose.

from H_3O^+ to H_2O. A curve like (s) in Figure 40 is obtained (ΔpK_a is pK_a for the catalysing acid minus pK_a for acetylacetone). It is often suggested that α be used as a measure of the degree of proton-transfer at the transition-state, a quantity which can also be approached by measuring isotope effects (see page 17).

Finally in this section the attractive possibility of push-pull catalysis is mentioned. In benzene solution, mutarotation of 2,3,4,6-tetramethylglucose, which involves ring-opening and reclosure, is catalysed by 2-pyridone, which is much more effective than an equivalent concentration of pyridine and phenol although it is a much weaker base than pyridine and a weaker acid than phenol. Swain and Brown[94] suggested that 2-pyridone is a bifunctional catalyst, acting as both acid and base in a concerted process (Figure 45). It is also possible that it is the unstable tautomer of 2-pyridone (2-hydroxy-pyridine) which is responsible for the catalysis.[95] Bifunctional catalysis or push-pull catalysis might play its part in enzymic catalysis (see page 70).

Intramolecular Catalysis

Chemists have long been intrigued by unusually rapid reaction at one site in a molecule caused by catalysis by a group elsewhere in the same molecule. Intramolecular catalysis is one possible outcome of neighbouring-group participation, the general phenomenon of involvement of one functional group in the reactions of a neighbouring functional group (see Figure 46). More will be said about neighbouring-group participation as it affects nucleophilic substitution at saturated carbon in Chapter 2. Our example[96] of intramolecular catalysis is concerned with substitution at acyl carbon: hydrolysis of p-nitrophenyl γ-(N,N-dimethylamino)butyrate (Figure 47).

Not only is hydrolysis considerably accelerated when compared with the p-nitrophenyl butyrate itself, but in this example we can make a direct comparison of this *intra*molecular catalysis with an *inter*molecular analogue, trimethylamine-catalysed hydrolysis of p-nitrophenyl acetate. Since the latter is a bimolecular reaction with a rate constant in litre mole^{-1} sec^{-1} while the intramolecular reaction is unimolecular with a rate constant in sec^{-1}, the rate ratio is a molarity, in fact 5370 molar at 25°C. This might be regarded as the fictional concentration of trimethylamine which would be necessary to achieve comparable catalysis to that in the intramolecular case. The enthalpy of activation is 11.9 kcal mole^{-1} in the intramolecular case and 12.3 kcal mole^{-1} in the trimethylamine catalysis, and for series of p-substituted phenyl esters the ρ-values are 2.5 and 2.2 respectively, so there is little doubt that the two catalyses operate by similar mechanisms as shown in Figure 47. The greater effectiveness of the intramolecular catalysis is mostly due to the

Formation of the normal product but at an unusually high rate. E.g., alkaline

hydrolysis of is 10^5 times faster than that of

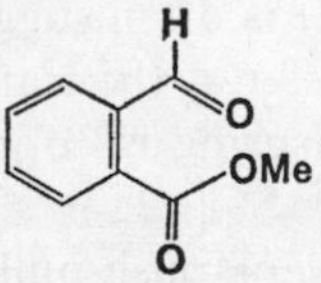

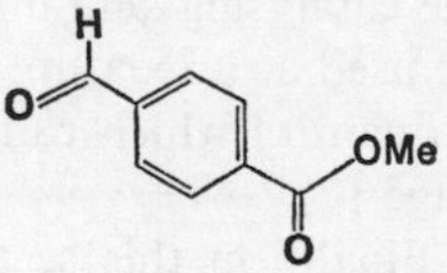

This is Intramolecular Catalysis

Formation of abnormal products (also possibly at enhanced rates):

(ring formation)

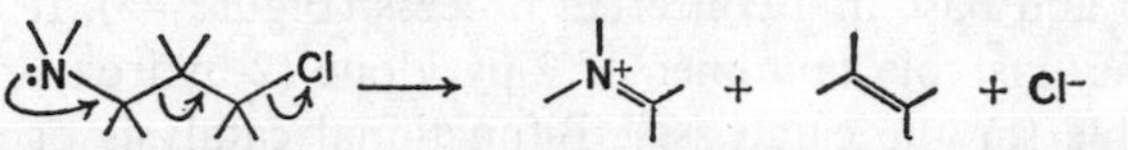

(fragmentation)

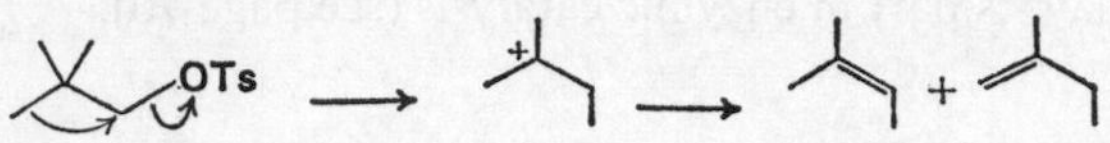

(rearrangement)

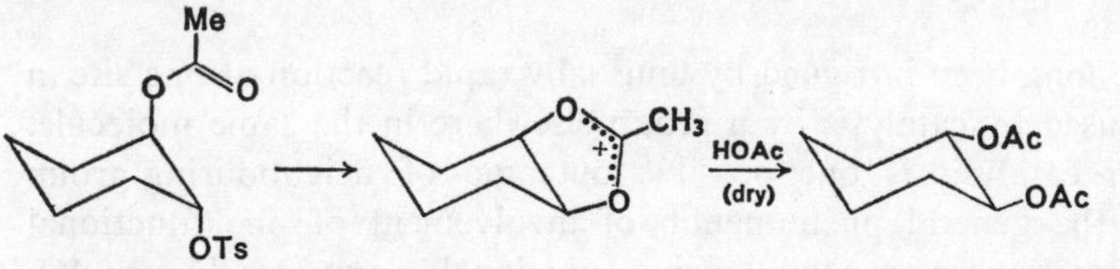

(abnormal
                                                               stereochemistry)

trans-Isomer   *trans*-Isomer

Figure 46. Intramolecular catalysis as one possible outcome of neighbouring-group participation. For the mechanism of the hydrolysis of methyl *o*-formylbenzoate see Chapter 5, p. 332.

continuous availability of the dimethylamino group for participation whenever the molecule acquires enough energy for reaction to occur.

Enzymic Catalysis

The striking efficiency of enzyme systems as catalysts has long been the marvel of organic chemists. However, we are probably rapidly approaching the time when the mechanism of action of some enzymes will be reasonably comprehensible. This field of study is probably the most exciting in mechanistic research at present, and we close this section with a list of factors thought to be important in enzymic catalysis. While this can only be in the nature of an appetizer, we hope it will stimulate the reader to go to one of the excellent discussions now available on this topic.[75, 76]

Intramolecular Catalysis

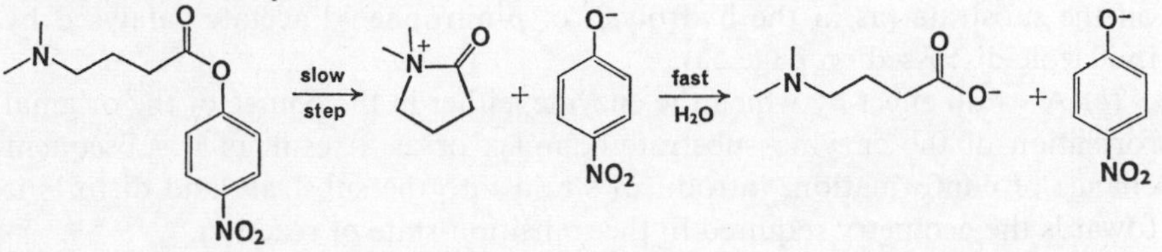

Intermolecular Analogue

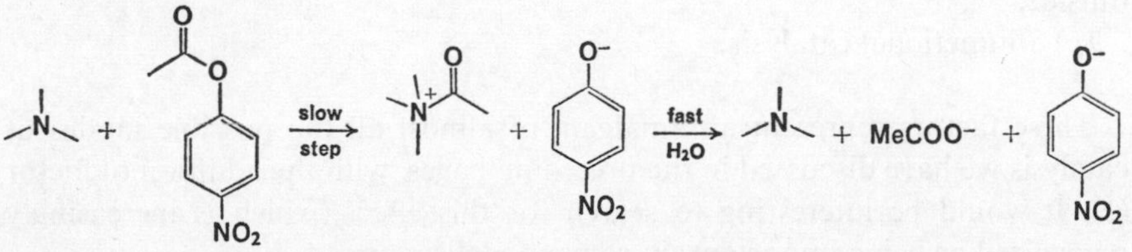

Figure 47. Intramolecular catalysis of the hydrolysis of *p*-nitrophenyl γ-(*N*,*N*-dimethylamino)butyrate and an intermolecular analogue, trimethylamine-catalysed hydrolysis of *p*-nitrophenyl acetate.

Enzymes are proteins with their attendant primary, secondary, and tertiary structures; some enzymes have prosthetic (non-protein derived) groups attached and they may require the presence of various cofactors (small molecules and ions). Enzymes, of course, only operate over a certain pH and temperature range and essentially only in aqueous solution. Catalytic activity is always lost when the secondary and tertiary structures are destroyed by extremes of pH or by heat. It seems that a preliminary step in enzymic catalysis is formation of an enzyme–substrate complex, and subsequent reactions take place on this complex before the transformed substrate is released to the solution. Not surprisingly, therefore, chemists have sought analogies with enzymic catalysis in intramolecular catalysis. The area of the enzyme surrounding the substrate is known as the active site and contains those functional groups on the enzyme which perform the chemical transformation. Some or all of the following effects may then contribute to the catalysis.

(*a*) A proximity effect due to the continuous proximity of reactive groups in the substrate and on the enzyme.

(*b*) An orientation effect related to (*a*) by which the groups involved are sterically disposed in optimum positions for reactions to occur.

(*c*) General base or general acid catalysis including bound metal ions acting as Lewis acids.

(*d*) Nucleophilic catalysis whereby a group is both a better nucleophile

than the final acceptor and also a better leaving group than that originally
on the substrate (as in the hydrolysis of p-nitrophenyl acetate catalysed by
imidazole discussed on page 53).

(e) A strain effect by which the enzyme, either in the course of the original
formation of the enzyme–substrate complex or as a result of a subsequent
change of conformation, introduces strain into the substrate and distorts it
towards the geometry required in the transition-state of reaction.

(f) A microenvironment effect, in which the active site perhaps acts as an
environment quite different from that prevailing in the aqueous solution
outside.

(g) Bifunctional catalysis.

These factors represent an amalgam of almost all the possible modes of
catalysis we have discussed in the preceding pages, with the addition of factor
(e). It would be interesting to search for this effect (which is increasingly
recognized as being important) in some model systems.

PROBLEMS

1. In the deprotonation of nitroethane by a series of bases, the following values of
entropy of activation, ΔS^+, have been determined in aqueous solution:

$$B: + MeCH_2NO_2 \underset{k_{-1}}{\overset{k_1}{\rightleftharpoons}} \overset{+}{B}H + Me\overset{-}{C}HNO_2$$

B	NH$_3$	MeNH$_2$	Me$_2$NH	OH$^-$
ΔS_1^+	−17.9	−12.0	−6.6	−15.5

Offer a reasonable explanation for the divergence of these values; do you consider
ground-state or transition-state effects to be responsible?

[See L. L. Shaleger and F. A. Long, Adv. Phys. Org. Chem., 1, 1 (1963)]

2. The following isotope effects have been observed in the acetolysis of deuteriated
cyclohexyl tosylates:

| k_H/k_D | 1.436 | 1.202 | 2.565 | 1.784 |

Can any mechanistic information be derived from these data?

[V. J. Shiner and J. G. Jewett, J. Amer. Chem. Soc., 86, 945 (1964)]

3. Explain why quinuclidinone (formula below) is chemically more akin to an amino-ketone than to an amide, e.g. in its formation of an oxime.

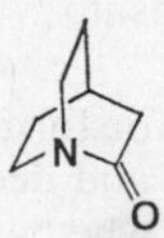

4. A ρ-value of $+2.65$ was obtained for the reaction of sodium borohydride with a series of substituted fluorenones. Deviations from the Hammett correlation were observed for the 2-OMe-, 3-OMe-, 2-NH_2-, and 3-NH_2-fluorenones.

What conclusions do you draw from these results?

[See G. G. Smith and R. P. Bayer, *Tetrahedron*, **18**, 323 (1962)]

5. Rate studies on the isomerization of phenyl-substituted *cis*-cinnamic acids in 30% aqueous sulphuric acid indicate that the rate rises gradually over the first 20% of reaction, followed by a constant rate of reaction (k). A plot of log k against σ^+ gives a good straight line of slope -4.3.

Suggest a mechanism for these reactions.

[See D. S. Noyce and H. S. Avarbeck, *J. Amer. Chem. Soc.*, **84**, 1644 (1962)]

6. The equilibrium constant for the following reaction is strongly solvent-dependent.

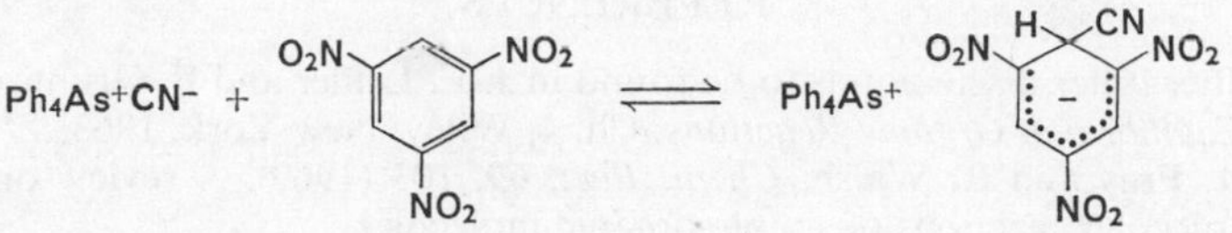

Make a plot of ΔG^0 for this reaction against E_T, and comment on, and interpret, the thermodynamic data as far as you can along the lines suggested on pages 48–51.

	ΔG^0 (kcal mole^{-1})	ΔH^0 (kcal mole^{-1})	ΔS^0 (cal mole^{-1} deg^{-1})
MeOH	-2.17	0 ± 1.5	$+7.2 \pm 5$
EtOH	-4.23	-6.5 ± 0.5	-8.7 ± 3
Pr^{i}OH	-5.46	-7.1 ± 0.6	-5.5 ± 3
Bu^{t}OH	-7.78	-15.5 ± 1.0	-25 ± 3
Me$_2$CO	-7.05	—	—

[See E. Buncel, A. R. Norris, W. Proudlock, and K. E. Russell, *Can. J. Chem.*, **47**, 4129 (1969)]

7. It is possible that the differences between the various H-functions in Figure 38 are mainly a result of different hydration changes on protonation for the various bases involved.

Consider whether this yields a chemically reasonable picture in view of the structure and hydrogen-bonding capabilities of the various indicator bases and their cations. (A plot of log a_{H_2O} is in Figure 42 of Chapter 5.) Is the relationship $H_B = (mH_0 + \text{constant})$ a linear free-energy relationship?

8. Consider the full sequence of steps for formation of an acetal $RCH(OR')_2$ from RCHO and R'OH with acid catalysis, and decide for each step you write whether acid or base catalysis is involved, and whether it is likely to be general or specific (see also Chapter 5, page 316).

9. (*a*) The reaction:

$$cis\text{-}[Co(\text{triethylenetetramine})(NH_2CH_2CO_2Et)Cl]^{2+} + NH_2CH_2CO_2Et \rightarrow$$
$$cis\text{-}[Co(\text{triethylenetetramine})(\text{glycylglycine OEt})]^{3+}$$

occurs rapidly in DMF at 25°.

(*b*) Hydration of 2-cyano-1,10-phenanthroline to the corresponding amide is accelerated by a factor of 10^7 by Ni^{2+} ions.

For both (*a*) and (*b*) write mechanisms where (i) the metal acts as an electrophilic catalyst, (ii) a coordinated ligand acts as a nucleophile. How might these be distinguished?

[See D. A. Buckingham, L. G. Marzilli, and A. M. Sargeson, *J. Amer. Chem. Soc.* **89**, 2772 (1967); and R. Breslow, R. Fairweather, and J. Keana, *ibid.*, p. 2135]

REFERENCES

1. A rather fuller discussion is to be found in J. E. Leffler and E. Grunwald, *Rates and Equilibria of Organic Reactions*, Ch. 4, Wiley, New York, 1963.
2. H. M. Frey and R. Walsh, *Chem. Rev.*, **69**, 103 (1969). A review on thermal unimolecular reactions of simple organic molecules.
3. S. W. Benson et al., *Chem. Rev.*, **69**, 279 (1969). A very useful collection of thermochemical data for organic compounds.
4. Examples of the sort of accuracy attainable are to be found in: M. J. S. Dewar and G. Klopman, *J. Amer. Chem. Soc.*, **89**, 3089 (1967); and in M. J. S. Dewar, *The Molecular Orbital Theory of Organic Chemistry*, McGraw-Hill, New York, 1969.
5. See F. Westheimer in *Steric Effects in Organic Chemistry*, Ch. 12 (ed. M. S. Newman), Wiley, New York, 1956; and J. E. Williams, P. J. Stang, and P. von R. Schleyer, *Ann. Rev. Phys. Chem.*, **19**, 531 (1968).
6. P. D. Bartlett and T. T. Tidwell, *J. Amer. Chem. Soc.*, **90**, 4421 (1968).
7. Cf. the review by N. J. Turro in *Accounts Chem. Res.* **2**, 25 (1969).
8. M. S. Kharasch, E. T. Margolis, and F. A. Mayo, *J. Org. Chem.*, **1**, 393 (1936).
9. G. S. Hammond, *J. Amer. Chem. Soc.*, **77**, 334 (1955).
10. S. K. Malhotra and H. J. Ringold, *J. Amer. Chem. Soc.*, **86**, 1997 (1964).
11. S. K. Malhotra and H. J. Ringold, *J. Amer. Chem. Soc.*, **87**, 3228 (1965).
12. A good book on the fundamental aspects of this subject is by Lars Melander, *Isotope Effects on Reaction Rates*, Roland Press, New York, 1960.
13. M. J. Goldstein and G. L. Thayer, Jr., *J. Amer. Chem. Soc.*, **85**, 2673 (1963).

14. Reviewed by E. Halevi in *Progr. Phys. Org. Chem.*, **1**, 109 (1963).
15. P. Brown and R. C. Cookson, *Tetrahedron*, **21**, 1993 (1965).
16. G. A. Olah, J. Lukas, and E. Lukas, *J. Amer. Chem. Soc.*, **91**, 5319 (1969).
17. E. W. Garbisch, Jr., and R. F. Sprecher, *J. Amer. Chem. Soc.*, **91**, 6785 (1969).
18. A. R. Olsen and J. L. Hyde, *J. Amer. Chem. Soc.*, **63**, 2459 (1941).
19. Data are taken from H. C. Brown and K. Ichikawa, *Tetrahedron*, **1**, 221 (1957).
20. J. B. Hendrickson, *J. Amer. Chem. Soc.*, **89**, 7036, 7047 (1967).
21. A. Streitwieser, Jr., and G. R. Ziegler, *J. Amer. Chem. Soc.*, **91**, 5081 (1969).
22. G. S. Hammond and C. D. DeBoer, *J. Amer. Chem. Soc.*, **86**, 899 (1964).
23. M. Chérest, H. Felkin, J. Sicher, F. Šipoš, and M. Tichy, *J. Chem. Soc.*, **1965**, 2513.
24. H. H. Jaffé, *Chem. Rev.*, **53**, 191 (1953).
25. P. R. Wells, *Chem. Rev.*, **63**, 171 (1963). Reviews of linear free-energy relationships.
26. L. M. Stock and H. C. Brown, *Adv. Phys. Org. Chem.*, **1**, 35 (1963). Quantitative treatment of substituent effects in aromatic substitution reactions.
27. C. D. Ritchie and W. F. Sager, *Progr. Phys. Org. Chem.*, **2**, 323 (1964). Critical analysis of application of structure–reactivity relationships.
28. S. Ehrensen, *Progr. Phys. Org. Chem.*, **2**, 195 (1964). Theoretical discussion of linear free-energy relationships.
29. P. R. Wells, S. Ehrensen, and R. W. Taft, *Progr. Phys. Org. Chem.*, **6**, 147 (1968). Substituent effects in naphthalene series.
30. P. R. Wells, *Linear Free Energy Relationships*, Academic Press, London, 1968. Summary of quantitative relationships.
31. J. G. Kirkwood and F. H. Westheimer, *J. Chem. Phys.*, **6**, 506, 513 (1938).
32. J. D. Roberts and W. T. Moreland, *J. Amer. Chem. Soc.*, **75**, 2167 (1953).
33. C. F. Wilcox and C. Leung, *J. Amer. Chem. Soc.*, **90**, 33 (1968).
34. J. N. Murrell, S. F. A. Kettle, and J. M. Tedder, *Valence Theory*, Wiley, London, p.336, 1965; and also ref. 4.
35. L. P. Hammett, *Physical Organic Chemistry*, McGraw-Hill, New York, 2nd Edn., 1970.
36. H. C. Brown and Y. Okamoto, *J. Amer. Chem. Soc.*, **80**, 4979 (1958).
37. Y. Yukawa and Y. Tsuno, *J. Amer. Chem. Soc.*, **81**, 2007 (1959).
38. C. G. Swain and E. C. Lupton, Jr., *J. Amer. Chem. Soc.*, **90**, 4328 (1968).
39. A. I. Biggs and R. A. Robinson, *J. Chem. Soc.*, **1961**, 388.
40. J. F. Bunnett, F. Draper, Jr., P. R. Ryason, P. Noble, Jr., R. G. Tonkyn, and R. E. Zahler, *J. Amer. Chem. Soc.*, **75**, 642 (1953).
41. R. W. Taft, Jr., E. Price, I. R. Fox, I. C. Lewis, K. K. Andersen, and G. T. Davis, *J. Amer. Chem. Soc.*, **85**, 3146 (1963).
42. R. W. Taft, Jr., *Steric Effects in Organic Chemistry* (ed. M. S. Newman), Wiley, New York, 1956, Ch. 13; R. W. Taft, Jr., *J. Amer. Chem. Soc.*, **79**, 1045 (1957).
43. A. G. Harrison, P. Kebarle, and F. P. Lossing, *J. Amer. Chem. Soc.*, **83**, 777 (1961).
44. C. H. DePuy, G. F. Morris, J. S. Smith, and R. J. Smat, *J. Amer. Chem. Soc.*, **87**, 2421 (1965).
45. W. White and W. K. Fife, *J. Amer. Chem. Soc.*, **83**, 3846 (1961).
46. R. Taylor and G. G. Smith, *Tetrahedron*, **19**, 937 (1963).
47. J. A. Leisten, *J. Chem. Soc.*, **1959**, 765.
48. D. S. Noyce, A. T. Bottini, and S. Smith, *J. Org. Chem.*, **23**, 752 (1958).

49. C. K. Ingold, *Structure and Mechanism in Organic Chemistry*, 2nd Edn., Bell, London, 1969, pp. 457–471. Discusses the electrostatic theory.
50. E. M. Kosower, *An Introduction to Physical Organic Chemistry*, Wiley, New York, 1968. Discusses correlations with model processes and several aspects of solvent effects not discussed here.
51. *Solute-Solvent Interactions* (ed. J. F. Coetzee and C. D. Ritchie), Dekker, New York, 1969. Many aspects of solvation discussed; see especially the chapter by C. D. Ritchie, p. 219, which discusses dipolar aprotic solvents and is also of interest in connection with proton-transfer rates (see catalysis section).
52. C. Reichardt, *Angew. Chem. Internat. Edn. Engl.*, **4**, 29 (1965). Discusses correlations with model processes.
53. A. J. Parker, *Adv. Phys. Org. Chem.*, **5**, 173 (1967); *Chem. Rev.*, **69**, 1 (1969). Dipolar aprotic solvents and the thermodynamic analysis of solvent effects.
54. E. M. Arnett and D. R. McKelvey, *Rec. Chem. Progr.* **26**, 185 (1965). Thermodynamic analysis of solvent effects.
55. A. J. Parker, *J. Chem. Soc.*, **1961**, 1328.
56. D. J. Cram, B. Rickborn, C. A. Kingsbury, and P. Haberfield, *J. Amer. Chem. Soc.*, **83**, 3678 (1961).
57. H. G. Grimm, H. Ruf, and H. Wolff, *Z. Phys. Chem.*, **B13**, 301 (1931); E. Tommila and P. Kauranen, *Acta Chem. Scand.*, **8**, 1152 (1954).
58. E. Grunwald and S. Winstein, *J. Amer. Chem. Soc.*, **70**, 846 (1948); A. H. Fainberg and S. Winstein, *ibid.*, **78**, 2770 (1956).
59. A. Diaz, I. Lazdins, and S. Winstein, *J. Amer. Chem. Soc.*, **90**, 6546 (1969).
60. E. M. Kosower, *J. Amer. Chem. Soc.*, **80**, 3253 (1958); see also p. 301 of ref. 50.
61. K. Dimroth, C. Reichardt, T. Siepmann, and F. Bohlmann, *Ann. Chem.*, **661**, 1 (1963); see also ref. 52.
62. N. J. T. Pickles and C. N. Hinshelwood, *J. Chem. Soc.*, **1936**, 1353.
63. H. Kaufmann and A. Wassermann, *J. Chem. Soc.*, **1939**, 870.
64. S. Winstein and A. H. Fainberg, *J. Amer. Chem. Soc.*, **79**, 5937 (1957).
65. E. M. Arnett, W. G. Bentrude, J. J. Burke, and P. McC. Duggleby, *J. Amer. Chem. Soc.*, **87**, 1541 (1965); see also ref. 54.
66. P. Haberfield, L. Clayman, and J. S. Cooper, *J. Amer. Chem. Soc.*, **91**, 787 (1969); see also R. Fuchs, J. L. Bear, and R. F. Rodewald, *ibid.*, p. 5797.
67. Ref. 53, and R. Alexander, E. C. F. Ko, A. J. Parker, and T. J. Broxton, *J. Amer. Chem. Soc.*, **90**, 5049 (1968).
68. *Friedel-Crafts and Related Reactions* (ed. G. Olah), Wiley-Interscience, New York. Electrophilic catalysis by metal ions and metal halides.
69. J. P. Candlin, K. A. Taylor, and D. T. Thompson, *Reactions of Transition Metal Complexes*, Elsevier, Amsterdam, 1968; J. P. Collman, *Accounts Chem. Res.*, **1**, 136 (1968); R. Cramer, *ibid.*, **1**, 186 (1968); L. Vaska, *ibid.*, **1**, 335 (1968); R. F. Heck, *ibid.*, **2**, 10 (1969); J. Tsuji, *ibid.*, **2**, 144 (1969). Various aspects of transition metal catalysis of olefin reactions.
70. F. D. Mango, *Adv. Catalysis*, **20**, 291 (1969). Transition metals and orbital symmetry effects. Previous volumes in this series contain many interesting reviews.
71. R. B. Dunlap and E. H. Cordes, *Accounts Chem. Res.*, **2**, 329 (1969). Micellar catalysis; E. J. Fendler and J. H. Fendler, *Adv. Phys. Org. Chem.*, **8**, 271 (1970).
72. R. P. Bell, *The Proton in Chemistry*, Methuen, London, 1959. Acidity and basicity and acid-base catalysis.

73. M. A. Paul and F. A. Long, *Chem. Rev.*, **57**, 1 (1957); R. H. Boyd and C. D. Ritchie, articles in ref. 51; C. H. Rochester, *Quart. Rev.*, **20**, 511 (1966). Acidity functions, C. H. Rochester, *Acidity Functions*, Academic Press, London, 1970.
74. M. Eigen, *Angew. Chem. Internat. Edn. Engl.*, **3**, 1 (1964); C. D. Ritchie, article in ref. 51. Rates of proton transfer reactions and acid–base catalysis.
75. T. C. Bruice and S. J. Benkovic, *Bio-organic Mechanisms*, Benjamin, New York 1966. Acid–base, nucleophilic, including intramolecular, catalysis, and enzymic catalysis; W. P. Jencks, *Catalysis in Chemistry and Enzymology*. McGraw-Hill, New York, 1969.
76. H. Gutfreund and J. R. Knowles in *Essays in Biochemistry*, Vol. 3 (ed. R. N. Campbell and G. D. Greville), Academic Press, New York, 1967, p. 25; L. Cunningham in *Comprehensive Biochemistry* (ed. M. Florkin and E. H. Stotz), Vol. 16, p. 85; M. R. Hollaway, *Ann. Reports* (Chem. Soc.), **68**, 601 (1968). Enzyme catalysis.
77. H. R. Clark and M. M. Jones, *J. Amer. Chem. Soc.*, **91**, 4302 (1969).
78. D. St. C. Black and R. C. Srivastava, *Austral. J. Chem.*, **22**, 1439 (1969).
79. T. C. Bruice and G. L. Schmir, *J. Amer. Chem. Soc.*, **80**, 148 (1958); M. L. Bender and B. W. Turnquest, *ibid.*, **79**, 1656 (1957).
80. D. S. Kemp, *J. Amer. Chem. Soc.*, **90**, 7153 (1968).
81. P. Cossee, *Rec. Trav. Chim.*, **85**, 1151 (1966).
82a. H. Hogeveen and H. C. Volger, *J. Amer. Chem. Soc.*, **89**, 2486 (1967).
82b. L. Cassar and J. Halpern, *Chem. Comm.*, **1970**, 1082.
83. J. R. Knowles and C. A. Parsons, *Nature*, **221**, 53 (1969).
84. R. B. Dunlap and E. H. Cordes, *J. Amer. Chem. Soc.*, **90**, 4395 (1968).
85. L. P. Hammett and A. J. Deyrup, *J. Amer. Chem. Soc.*, **54**, 2721 (1932).
86. J. F. Bunnett, *J. Amer. Chem. Soc.*, **83**, 4956, 4968, 4973, 4978 (1961).
87. J. F. Bunnett and F. P. Olsen, *Can. J. Chem.*, **44**, 1899, 1917 (1966).
88. D. Dolman and R. Stewart, *Can. J. Chem.*, **45**, 911 (1967).
89. C. D. Ritchie, *J. Amer. Chem. Soc.*, **91**, 6749 (1969).
90. E. H. Cordes, *Progr. Phys. Org. Chem.*, **4**, 1 (1967).
91. A. J. Kresge and Y. Chiang, *J. Chem. Soc.* (B), **1967**, 53, 58; M. M. Kreevoy and R. Eliason, *J. Phys. Chem.*, **72**, 1313 (1968).
92. V. Gold, D. G. Oakenfull, and T. Riley, *J. Chem. Soc.* (B), **1968**, 515.
93. R. P. Bell and W. C. E. Higginson, *Proc. Roy. Soc.*, **A197**, 141 (1949).
94. C. G. Swain and J. F. Brown, Jr., *J. Amer. Chem. Soc.*, **74**, 2534, 2538 (1952).
95. P. R. Rony, *J. Amer. Chem. Soc.*, **91**, 6090 (1969).
96. T. C. Bruice and S. J. Benkovic, *J. Amer. Chem. Soc.*, **85**, 1 (1963).

Chapter 2

Dissociative Processes

A large class of reactions occur where intermediates are formed by cleavage of a bond, followed in some cases by rehybridization at the reaction centre. These intermediates can be considered to be formed by dissociative processes. Bonds may be broken heterolytically when carbonium ions, carbanions, or carbenes are intermediates, or homolytically to form free-radicals, and, in this chapter, reactions involving these intermediates are discussed.

CARBONIUM IONS[1-15]

Substitution of saturated carbon by nucleophilic reagents has been one of the most extensively studied of all chemical processes. Although these reactions are themselves of wide utility, an important by-product has been that the ideas, experience, and research methods developed in the investigations of these reactions have been applied generally to many more complex ionic reactions. These reactions were classified by Ingold into two types of mechanism, S_N1 and S_N2.[16] Kinetic considerations distinguish between these two mechanisms but, in solvolysis reactions, where the solvent is in large excess over the substrate, this definitive criterion is lost. For these reactions the two mechanisms can be regarded as the extremes for which the character of the reaction is dependent upon the substrate, solvent, and leaving group, and many nucleophilic substitutions occur which have characteristics of both types of mechanism. In this section we concentrate on the formation, stability, and reactions of carbonium ions.

Stability of Carbonium Ions

Carbonium ions can be generated in a large number of ways, and the main types of dissociative processes are summarized in Table 1; associative processes for carbonium ion formation are discussed in Chapter 5. Numerous methods have been developed for the examination of the factors governing the stability of carbonium ions. Stability is, of course, only a relative term since in chemical

78

Table 1. Formation of carbonium ions by dissociative processes.

(a) Heterolysis of a C—X bond so that the carbon atom loses both bonding electrons:
$$RX \rightarrow R^+ + X^-$$
(X = halogen, OTs, OBs, $OCOC_6H_4NO_2$, etc.)

(b) Cleavage of 'onium ions:
$$R_2S^+\!\!-\!R \rightarrow R_2S + R^+$$

(c) Deamination:
$$RNH_2 \xrightarrow{HNO_2} RN_2{}^+ \rightarrow R^+ + N_2$$

(d) Decomposition of oxonium ions formed by protonation of an alcohol:
$$ROH \xrightarrow{H^+} \overset{+}{R}OH_2 \rightarrow R^+ + H_2O$$

(e) Reaction of ethers and Lewis acids:
$$3Ph_3COEt + 4BF_3 \rightarrow 3Ph_3C^+ + 3BF_4{}^- + B(OEt)_3$$

(f) Reaction of halides and Lewis acids:
$$Bu^tF + SbF_5 \rightarrow t\text{-}Bu^+ + SbF_6{}^-$$

(g) Hydride abstraction by oxidizing agents:
$$Ph_3CH + Cr(VI) \rightarrow Ph_3C^+ + Cr(IV)$$

systems carbonium ions are frequently associated with anions and solvent. Intermolecular ion–solvent forces can be discounted if studies are undertaken in the gas phase, and scales of stabilities have been obtained by the measurement of ionization potentials (Table 2).

$$R\cdot \rightarrow R^+ + e^-$$

Table 2. Ionization potentials ($kcal\ mole^{-1}$) of radicals derived from electron-impact studies.[17]

Radical	Ionization potential	Radical	Ionization potential
Me	229.4 ± 0.7	$p\text{-}NCC_6H_4CH_2$	197.9 ± 2.3
$MeCH_2$	202.5 ± 1.2	$p\text{-}MeOC_6H_4CH_2$	157.7 ± 2.3
$MeCH_2CH_2$	200.4 ± 1.2	$CH_2{=}CH$	217.9 ± 1.2
Me_2CH	182.2 ± 1.2	$CH_2{=}CHCH_2$	188.2 ± 0.7
Me_3C	171.8 ± 1.2	$MeCH{=}CHCH_2$	177.8 ± 1.2
$PhCH_2$	178.9 ± 1.8	$HC{\equiv}CCH_2$	190.2 ± 1.8
Ph_2CH	168.8 ± 2.3		

A number of carbonium ions can be formed which are extremely stable and can be kept as salts almost indefinitely (1—3); 1 and 2 owe their stability to conjugative interaction whilst 3 is an example of a 6 π-electron system which is predicted to be very stable by molecular-orbital theory. Other positive ions

$$Ph_3C^+X^-$$

(1)   (2)   2/1   (3)

may be formed in strong acid solution such as sulphuric acid (equation 4) and characterized by their u.v. spectra. As the degree of stabilization of the carbonium ion decreases, generation in sulphuric acid becomes increasingly difficult. The use of highly acidic solvents (termed super-acids) such as SbF_5 and FSO_3H–SbF_5 has made possible the generation of an almost infinite variety of carbonium ions. Coupled with this approach, n.m.r. spectrometry has been developed as a highly discriminating technique for assignment of carbonium ion structure (equations 4—7). Thus the cation generated as shown in equation 7 exists for only a short time before cyclization and a further rearrangement takes place[1, 2] (see page 105).

$$\xrightarrow{\text{H}_2\text{SO}_4} \quad \longleftrightarrow \quad \tag{4}$$

$$\text{OH} \qquad \text{HSO}_4^- \qquad\qquad \text{HSO}_4^-$$

$$Me_2CHF \xrightarrow{\text{SbF}_5} \overset{+}{Me}CHMe \tag{5}$$

$$R \diagup\diagdown \xrightarrow{\text{FSO}_3\text{H–SbF}_5} R \diagup\overset{+}{\diagdown} \tag{6}$$

$$\xrightarrow{\text{FSO}_3\text{H–SbF}_5} \quad \longrightarrow \tag{7}$$

$$\text{OH}$$

Further evidence has been obtained by these techniques for the comparative stability of primary, secondary, and tertiary carbonium ions; n-butyl, s-butyl, isobutyl, and t-butyl fluorides all lead to the formation of the same t-alkyl-carbonium ion in SbF_5. The n.m.r. spectrum of the isopropyl cation has been studied in SO_2ClF–SbF_5. Below 0° a doublet and septet were observed, ascribed to the methyl and methine protons, but between 0° and 40° spectrum changes indicated a rapid interchange between these protons.[18] This was assumed to occur via primary propyl cations, and, by matching calculated curves with the observed spectra, the rates of hydrogen-exchange could be estimated. The energy of activation of this process, E_A (16.4 kcal mole^{-1}), was then calculated and gives some estimate of the difference between a primary and a secondary propyl cation. In a similar way, by studying the rearrangements in the t-pentyl cation, the difference in stability of secondary and tertiary ions was found to be similar to that for primary and secondary ions.[19] It should be appreciated that the estimates of carbonium ion stability made by this approach cannot be precise since the energy barriers for the

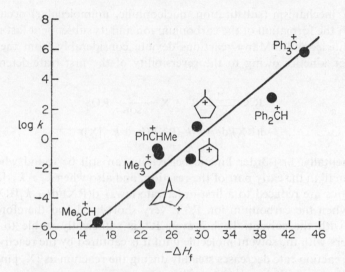

Figure 1. Plot of ΔH_f for various cations against log k for the solvolysis of the corresponding chloride in ethanol at 25°.

rearrangement processes are not known (see below) but reasonable agreement is found between these measurements in solution and those in the gas phase.

Measurements of the rates of carbonium ion formation are widely used for estimating the extent to which a positive charge can be accommodated by an organic structure. This method depends on the thesis that the greater the stability of the ion the more rapidly it will be formed; since the rate comparison is a measure of the free-energy difference between the parent molecule and the transition-state it is a measure of stability only as far as the energies of the transition-state reflect the respective energies of carbonium ions. Support for this generalization is found in the good correlation observed in Figure 1. Here we see that the heats of formation, ΔH_f (obtained by solution calorimetry), of a number of ions from their respective alcohols in HSO_3F–SbF_5 bear a close relationship to the rates of ethanolysis of the corresponding chlorides.[19]

Kinetic information is, then, only an approximation but has provided knowledge of the effect of structure on the stability of carbonium ions, and from this knowledge has come the ability to predict relative reaction rates and product compositions.

Kinetic Evidence for S_N1 Reactions

The S_N1 mechanism (substitution, nucleophilic, unimolecular) occurs step-wise with the formation of the carbonium ion and its subsequent fast reaction with a nucleophile. Many reactions deviate considerably from the simple first-order kinetics owing to the reversibility of the first (rate-determining) step:

$$RX \underset{k_{-1}}{\overset{k_1}{\rightleftharpoons}} R^+ + X^- \xrightarrow[\text{SOH}]{k_2} ROS \qquad (8)$$

$$-d(RX)/dt = k_1 k_2 [RX]/(k_2 + k_{-1}[X])$$

Experimentally, first-order kinetic behaviour can still be found when $[X^-]$ is very small in the early part of the reaction, and also when $k_2 \gg k_{-1}[X^-]$, i.e. the kinetics are reduced to a first-order rate law, $-d(RX)/dt = k_1[RX]$. This occurs when the carbonium ion R^+ is very short-lived and therefore reacts rapidly with the solvent. If, however, R^+ is sufficiently stable to survive encounters with the solvent molecule until it is captured by the reactive anion X^-, the reaction rate decreases steadily during the reaction as $[X^-]$ increases. This is the case with diaryl cations.[20]

$$PhCH(Cl)Ph + H_2O \underset{k_{-1}}{\overset{k_1}{\rightleftharpoons}} Ph\overset{+}{C}HPh + Cl^- \xrightarrow{k_2} PhCH(OH)Ph \qquad (9)$$

This 'mass-law' effect or 'common-ion' effect produces kinetic deviations of a distinctive and calculable form. The common-ion effect can be augmented by the deliberate addition of salts containing the anion X^-, thus providing a simple probe for the presence of carbonium ions in solution (discussed below). In fact, the formation of carbonium ions does not proceed as simply as described in equation **8**. The presence of the counter-anion can provide considerable stabilization by electrostatic interaction with the positively charged species. Evidence that ion-pairs intervene in almost all carbonium-ion reactions is extensive, and the reverse of the ionization step (ion-pair return) is an important factor governing the stereochemistry and formation of products of many reactions. In many cases product arises from ion-pairs without the intervention of a dissociated carbonium ion (see page 85).

$$\text{RX} \rightleftharpoons \overset{\text{Ion-pair}}{\text{R}^+\text{X}^-} \rightleftharpoons \overset{\text{Dissociated ions}}{\text{R}^+ + \text{X}^-} \rightarrow \text{ROH}$$
$$\downarrow$$
$$\text{ROH}$$

In solvolysis reactions where the nucleophile is solvent, first-order kinetics are observed. Variation of the concentration of the nucleophile to determine if this is involved in the rate-determining step is not possible. However, the investigation of the effect of the conjugate base of the solvent does give positive information, since in the S_N1 mechanism the nucleophilic species is not involved in the rate-determining step. For example, the hydrolysis of t-butyl bromide in water is independent of the concentration of hydroxide ion in fairly high concentrations, so that the mechanism does not involve direct displacement by hydroxide ion. A similar test for the involvement of solvent in the rate-determining step is not possible, and the use of mixed solvents so that the water concentration can be varied is not definitive. The rate of hydrolysis of t-butyl bromide is increased 40-fold when solvent is changed from 10 to 30% aqueous acetone. However, it is not possible to decide with sufficient accuracy the extent to which the increased polarity of the medium, or second-order nucleophilic attack by the larger amount of water, contributes to the rate increase. Kinetics, then, cannot be used in this way to determine if solvent is acting as a nucleophile.

Studies of the effect of added salts on solvolysis reactions have been used as strong evidence for the S_N1 mechanism. The rate of solvolysis of benzhydryl chloride (equation 9) in 80% aqueous acetone is decreased when 0.1N-lithium chloride is added owing to the recombination of chloride ions and reversal of the ionization step.[20] This 'common-ion effect', which can be used as a probe for carbonium ions, is only large, however, with the more stable carbonium ions. Substantial rate increases, attributable to the increased ionic strength which would favour the formation of carbonium ions, are observed when other salts are added to the solution. The addition of 0.1N-lithium bromide, sodium azide, or ammonium nitrate to p,p'-dimethylbenzhydryl chloride (RCl) in 85% aqueous acetone produces a 50% rate increase. Although all the salts produce a similar increase in rate, the interaction of each salt and the intermediate carbonium ion differs. Bromide ion would be expected to combine with the carbonium ion but the product RBr is more reactive than RCl and rapidly solvolyses, and no bromide is found as product. The combination of azide with the carbonium ion forms the stable RN_3 so that the final product consists of azide and ROH. Nitrate ions are too weakly nucleophilic to capture any of the carbonium ions but the ionic-strength effect is seen in the rate increase.

Stereochemistry of S_N1 Reactions

There is substantial theoretical and experimental evidence that the ideal structure for a carbonium ion has the three sp^2-orbitals in the same plane at angles of 120°.[21] Any change in geometry will be strongly resisted by a carbonium ion since both hybridization and repulsion energies are increased. Two

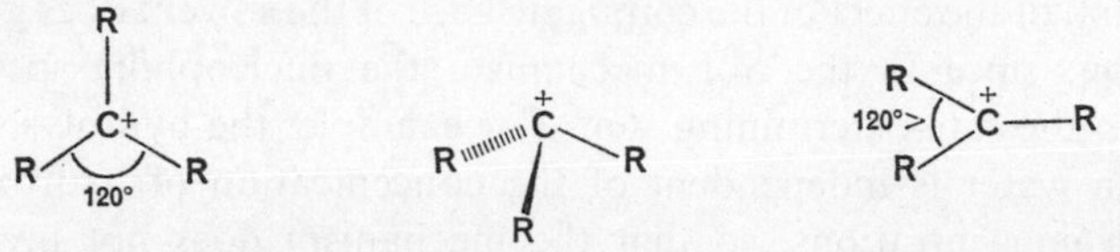

Strain-free   Out-of-plane strain   In-plane strain

Figure 2. Strain in carbonium ions.

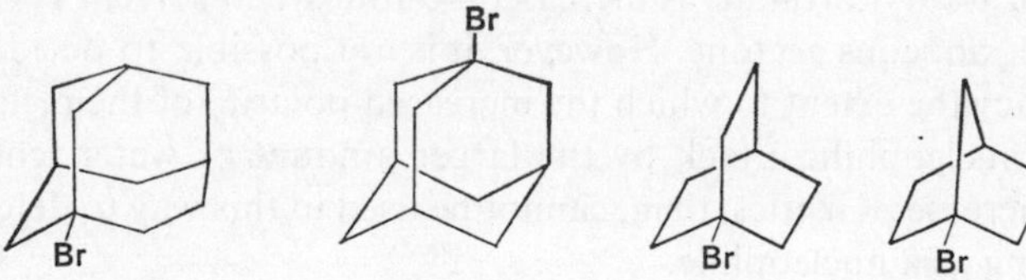

	(10)	**(11)**	**(12)**	**(13)**
Rel. rate	0.5	10^{-3}	10^{-6}	10^{-13}

Me$_3$CBr $= 1$

Figure 3. Relative reactivities of bridgehead bromides (80% ethanol; 25°).

types of distortion are possible: deformations in which the carbonium-ion carbon and the three attached atoms remain in the same plane, and distortions leading to non-planar structures (Figure 2). This latter type of distortion is inherently required for carbonium ion formation in some bridged ring systems in which, for structural reasons, planar ions cannot be formed. The reactivity of a wide range of ring systems with bridgehead substituents has been studied (Figure 3).[9] Both **12** and **13** are extremely unreactive, with the latter yielding alcohol only after treatment with aqueous silver nitrate at 150° for 2 days. **12** produces alcohol after reaction at room temperature for 4 hours. 1-Bromo-adamantane **(11)** can achieve a greater degree of planarity in the transition-state for ionization than **12** or **13** but is still substantially less reactive than t-butyl bromide.

A consequence of the formation of an sp^2-hybridized carbon from a tetrahedral carbon atom at the point of asymmetry of a system is the creation of a plane of symmetry in the carbonium ion, and subsequent racemic products would be expected. In practice, predominant inversion is found which increases with the decreasing thermodynamic stability of the carbonium ion; 70% net inversion is found in the solvolysis of 2-bromooctane in 60% aqueous ethanol. An explanation for the stereochemical course is that the carbonium ion reacts with the solvent soon after the energy barrier for ionization has been surmounted. Although ionization is complete, dissociation is not far advanced and the leaving group is still close to one face of the carbonium ion (as an ion-pair), so that solvent attack occurs predominantly from the opposite side leading to inversion of configuration. For tertiary ions, which are more stable with a longer lifetime, more dissociation is possible and subsequent attack from both sides produces greater racemization.

Stereochemistry can be seen to be no more than a useful guide to the mechanism of substitution reactions, and the observation of inversion does not distinguish between the occurrence of an S_N1 or S_N2 mechanism. A number of cases of nucleophilic substitution reactions appear to be accommodated by the concept of an S_N1 mechanism but with S_N2 character. The 'borderline' behaviour has been interpreted in several ways: (1) that a hybrid mechanism is operating, with characteristics intermediate between those of mechanisms S_N1 or S_N2, or (2) that there is a direct competition between distinct S_N1 and S_N2 processes, or (3) that there is intervention by ion-pairs (see page 91). These interpretations take account of the fact that, in solvolyses, a solvent molecule, possessing an unshared electron-pair, is involved in the transition-state on the opposite side to the leaving group. The interaction of the electron-pair and the developing p-orbital on carbon varies for each case. When weak, it can be called solvation, but the amount of interaction varies with the extent of reaction, the nucleophilicity of the solvent, and the substrate.

Ion Pairs

A consequence of the Debye–Hückel and Onsager theories of strong electrolytes is that ions of opposite charge can co-exist in solution as ion-pairs, kept together by electrostatic attraction. The detection of ion-pairs from conductivity data was followed by much kinetic and stereochemical evidence for the occurrence of ion-pairs as intermediates in solvolysis reactions. This was first clearly shown by the observation of Young, Winstein, and Goering[22] that acetolysis of 1,1-dimethylallyl chloride (**14**) is accompanied by rearrangement to 3,3-dimethylallyl chloride (16). The reaction rate decreases rapidly from the initial value to that expected for the primary chloride. Added chloride ions have no effect on the rate of acetate formation or isomerization, indicating that the reaction does not proceed through a dissociated carbonium ion and suggesting the intervention of an intermediate ion-pair (15). Since the rate of the reaction which leads to 16 is greater than that of solvolysis, the solution

becomes enriched in this product. Winstein has proposed the term 'internal return' for the recombination of ions from an ion-pair to form a covalent bond, in contrast to 'external return' which signifies the recombination of a carbonium ion and an anion from the solution.[23] Isomerization of 1,1-dimethylallyl chloride does not occur in ethanol which is more nucleophilic than acetic acid.

The question of the relationship between ion-pair return and solvent has also been examined by a different approach. The loss of optical activity (α) can also be used as well as acidimetric titration to follow a reaction velocity.

$$(+)\text{-RX} \xrightarrow[\text{SOH}]{k_\alpha} (\pm)\text{-ROS} + (\pm)\text{-RX}$$

$$k_\alpha = \frac{2.303}{t} \log \frac{\alpha_0}{\alpha_t}$$

$$k_t = \frac{2.303}{t} \log \left(\frac{a_0}{a_0 - a_t} \right)$$

a_0, determined by titration, refers to the concentration of RX at time zero and a_t at time t.

As a consequence of ion-pair return, k_α (polarimetric rate) is frequently larger than k_t (titrimetric rate). The relationship between k_α and k_t is dependent upon the solvent in a similar way to the rearrangement described in the previous paragraph. Thus, solvolysis of optically active 3-chloro-5-methyl-cyclohexene (17) in acetic acid demonstrates a k_α more than four times greater than k_t. Again it seems that ion-pair return only competes effectively with solvent when the solvent is a weak nucleophile since if the p-nitrobenzoate of the same system is solvolysed in 80% aqueous acetone, k_α/k_t is 1.7.[24] Products of both reactions are racemic, indicating that the reaction proceeds through a symmetric bridged ion-pair. The effect of solvent has been further demonstrated

(17)

in studies on the solvolysis of *threo*-3-phenyl-2-butyl *p*-bromobenzene-sulphonate (18) (see page 106 for further studies on this system); k_α/k_t is 1.18 in formic acid (25°), 2.05 in ethanol and 4.4 in acetic acid (75°). Measurements

(18)

in other solvents indicate that this ratio tends to increase as solvent dissociating power decreases, but other solvent properties such as hydrogen-bonding character and nucleophilicity are very important factors.

Studies of salt effects have proved extremely successful in investigation of ion-pairs. A 'special salt effect' discovered by Winstein and coworkers in the acetolysis of a number of alkyl arenesulphonates led to the proposal that at least two types of ion-pair were present in many solvolyses.[23] This effect, illustrated in Figure 4 for *threo*-3-*p*-anisyl-2-butyl brosylate (*p*-bromobenzene-sulphonate), is a strikingly rapid non-linear rise in the titrimetric rate for low lithium perchlorate concentrations, followed by a linear increase in rate constant with the concentration of added salt. The linear rate increase of k_t and a similar effect on the polarimetric rate constant, k_α, is described by the equations:

$$k_\alpha = k_\alpha{}^0(1 + b[\text{LiClO}_4])$$

$$k_t = k_t{}^0(1 + b[\text{LiClO}_4])$$

where b is the % change in k_t for 0.01M change in LiClO_4, and $k_\alpha{}^0$ and $k_t{}^0$ are the rate constants at zero salt concentrations; for secondary arenesulphonates the constant b varies in the range 10—40. It is clear that the special salt effect is concerned with the reduction of ion-pair return, and $k_{ext}{}^0$, the intercept produced by extrapolation of the linear portion, is the rate constant predicted in the absence of this return. The striking fact appears that $k_{ext}{}^0$ still falls

4

quite short of the ionization rate constant, k_α^0. It was proposed that for some solvolyses, together with initially formed 'intimate ion-pair' (**19**), a second 'solvent-separated ion-pair' intermediate (**20**) is present (Figure 5). The

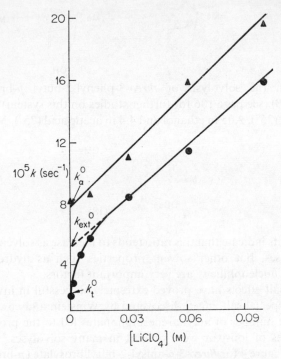

Figure 4. The effect of lithium perchlorate on the acetolysis of *threo*-3-*p*-anisyl-2-butyl brosylate at 25°.

$$RX \xrightarrow[\text{Internal return}]{\text{Ionization}} \underset{(\mathbf{19})}{R^+X^-} \underset{\text{Intimate}}{\rightleftharpoons} \underset{(\mathbf{20})}{R^+\|X^-} \xrightarrow{\text{Dissociation}} \underset{\text{Dissociated ions}}{R^+ + X^-}$$

Intimate · · · · · · Solvent-separated

Product   Product

Figure 5. Ion-pairs in solvolysis.

exact nature or form of these ion-pairs is not known but it appears that the solvent-separated ion-pair is captured by lithium perchlorate, thus eliminating part of the ion-pair return. Return from the intimate ion continues in the presence of salt and accounts for the greater polarimetric than titrimetric rate constant. (Some capture of the intimate ion-pair cannot be completely excluded.)

Suppression of return from the solvent-separated ion-pair by the addition of LiClO$_4$ is considered to be due to an exchange reaction between two ion-pairs:[25]

$$R^+\|X^- + Li^+ClO_4^- \rightleftharpoons R^+\|ClO_4^- + Li^+X^-$$

Owing to the low nucleophilicity of perchlorate ion and the rapid ionization of any covalent perchlorate ester (RClO$_4$) formed, the new perchlorate ion-pair rapidly breaks down to product, the ion-pair return to RX is reduced, and an enhanced k_t results. The 'special salt effect' is only associated with longer-lived carbonium ions and is also produced by other salts. The effect of salts on a series of arenesulphonates is summarized in Table 3.

Table 3. Salt effects in the acetolysis of arenesulphonates.[26]

	b[LiClO$_4$]	[LiClO$_4$]$_{\frac{1}{2}}$[a]	k_{ext}^0/k_t^0
exo-Norbornyl brosylate	38	—	1.0
threo-3-Phenyl-2-butyl tosylate	37	—	1.0
3-*p*-Anisyl-2-butyl brosylate	18	4×10^{-3}	3.1
2-*p*-Anisylethyl tosylate	11	3.4×10^{-4}	3.3
2-(2,4-Dimethoxyphenyl)ethyl brosylate	12	8×10^{-5}	2.2

[a] Concentration of LiClO$_4$ required to raise the titrimetric constant to half the value of k_{ext}^0.

Evidence has been discussed for the intervention of two kinds of ion-pair in the solvolysis, and the relative importance of these and also that of free ions in these reactions must be considered. The extent to which the two types of ion-pair and the free carbonium ion take part in solvolytic reactions depends upon the structure of the carbonium ion and the solvent. The longer the life-time of the carbonium ion intermediate the greater the likelihood of formation of the solvent-separated and dissociated ions. The observation of common-ion rate depression has already been mentioned as evidence for carbonium ions. A rate decrease would be observed if a salt such as Bu$_4$NOBs was added to a solvolysis medium containing ROBs if dissociated carbonium ions were present. No common-ion rate depression was found for the acetolysis of *threo*-3-*p*-anisyl-2-butyl *p*-bromobenzenesulphonate, which indicates the absence of dissociated ions. Both a special salt effect and a k_α/k_t of 4.1 indicated the intervention of the intimate and solvent-separated ion-pair so that product arises by the capture of these intermediates by acetic acid. The addition of acetic anhydride to the solvolysis medium steadily decreases the rate constant for formation of intimate ion-pairs (k_α^0) but the titrimetric constant passes through a maximum.[27] The effect of lithium perchlorate, indicated by k_{ext}^0/k_t^0, is inversely proportional to the amount of acetic anhydride in the medium, whilst common-ion rate depression, not observed in acetic acid, becomes increasingly apparent with increasing amounts of acetic anhydride (Table 4). These effects were discussed in terms of the ability of a solvent to promote ionization and to accommodate the dissociation of two oppositely charged ions. The results in Table 4 indicate that acetic acid is a better ionizing, but poorer dissociating, solvent than acetic anhydride. Thus, in acetic acid internal

Table 4. Kinetic data for the acetolysis of threo-3-p-anisyl-2-butyl p-bromobenzene-sulphonate.[27]

| | Solvent (Ac = MeCO) | | |
	AcOH (25°)	50% AcOH–Ac$_2$O (25°)	Ac$_2$O (50°)
$10^7 k_\alpha^0$ (sec^{-1})	798	578	442
$10^7 k_t^0$ (sec^{-1})	196	318	107
k_α^0/k_t^0	4.1	1.8	4.1
k_{ext}^0/k_t^0	2.58	1.56	1.2
% ROAc from solvent-separated ion-pair	100	21	3
% ROAc from dissociated ion	0	79	97

return from ion-pairs is very important and acetic anhydride reduces the rate of formation of ionic intermediates but increases the extent to which they are dissociated.

Significant advances in the knowledge of the stereochemistry of ion-pair return have been made by Goering and coworkers.[28] In solvolysing media, certain p-nitrobenzoates may undergo three processes involving alkyl–oxygen fission. These are solvolysis (21), randomization of the ester oxygen atoms followed by the use of ^{18}O (22), and racemization (23). The kinetics of these transformations are first-order and the last two are intramolecular. Studies with trans-1,3-dimethylallyl p-nitrobenzoate (24) indicated that racemization of the optically active ester in 60% acetone proceeded at the same rate as equilibration of ^{18}O. Resolution of the racemic alcohol obtained from the recovered ester, however, indicated that the recombination of the oxygen of the carboxylate ion to the allylic cation occurred preferentially between specific oxygen and carbon atoms. It was concluded that the oxygen atoms are not equivalent to each allylic carbon atom but each oxygen is more closely associated with one carbon atom (25).

$$RO_2C-\underset{}{\bigcirc}-NO_2 \xrightarrow{k_t} ROH + HO_2C-\underset{}{\bigcirc}-NO_2 \qquad (21)$$

$$RO-\underset{\overset{|}{{}^{18}O}}{C}-\underset{}{\bigcirc}-NO_2 \xrightarrow{k_{eq}} R^{18}O-\underset{\overset{|}{{}^{18}O}}{C}-\underset{}{\bigcirc}-NO_2 \qquad (22)$$

$$(+) RO_2C-\underset{}{\bigcirc}-NO_2 \xrightarrow{k_{rac}} (\pm) RO_2C-\underset{}{\bigcirc}-NO_2 \qquad (23)$$

(24)   (25)

The ion-pairs commonly formed in many carbonium ion reactions have significant effects on the products and stereochemistry of the reactions (for the effect of ion-pairs in elimination, see page 101). Their occurrence is not restricted to reaction where the carbonium ion has some special stabilizing factor (i.e. neighbouring-group participation). Streitwieser and Walsh[29] observed that, during the acetolysis of (+)-2-octyl toluene-*p*-sulphonate, the unsolvolysed sulphonate underwent racemization during the reaction owing to ion-pair return.

Sneen and Larsen have suggested that a single mechanism, involving as its most distinguishing feature the intermediacy of a configurationally stable ion-pair, could accommodate S_N1 and S_N2 as well as borderline behaviour.[30] Detailed studies of the kinetics and stereochemical features of the solvolysis of 2-octyl sulphonates in 75% aqueous dioxan, in the presence of sodium acetate, indicated the presence of ion-pair intermediates formed in the rate-determining step. In 25% aqueous dioxan, however, the reaction exhibited bimolecular kinetics and inversion of configuration, normally associated with an S_N2 mechanism, and it was demonstrated that the competitive reactions of water and added azide ion had a common ion-pair intermediate. It was argued

that the reaction occurred via a reversibly formed ion-pair which was attacked
in the rate-determining step by nucleophile:

$$RX \underset{k_{-1}}{\overset{k_1}{\rightleftharpoons}} R^+X^- \longrightarrow \begin{array}{l} \overset{k_s}{\longrightarrow} ROH \\ \\ \underset{k_n(Y^-)}{\longrightarrow} RY \end{array}$$

This scheme has also been shown to apply for a number of other systems;
the competitive reaction of water and sodium azide with p-methoxybenzyl
chloride in 70% aqueous acetone at 20° has been shown to proceed via the
same ion-pair.[31] It remains to be seen how far these generalizations apply but
experimental proof is required that ion-pairs take part in S_N2 reactions of
primary systems. Ion-pairs also occur in carbanion reactions (page 141).

Structural Factors Affecting Reactivity

Delocalization of charge by the inductive effect, conjugative effect, and, in
some cases, nucleophilic assistance from a group within the molecule (neighbouring-group participation) are the most important factors which favour

Table 5. Relative solvolysis rates of alkyl halides.[5]

Compound	Rel. rate (H$_2$O, 50°)	Compound	Rel. rate (40% EtOH–60% Et$_2$O)
MeBr	1.05		
MeCH$_2$Br	1.00	PhCH$_2$Cl	1.0
Me$_2$CHBr	11.6	Ph$_2$CHCl	2000
Me$_3$CBr	1,200,000	Ph$_3$CCl	30,000,000

the generation of carbonium ions. Any substituent capable of electron-supply
increases the rate of S_N1 reaction, and Table 5 summarizes the reactivity of a
series of alkyl- and phenyl-substituted halides. The effect of a methyl group
is, in fact, understated in Table 5 in that the solvolysis of methyl bromide
goes by an S_N2 displacement and the S_N1 rate of the primary compound will
be minute, if at all visible. Benzyl chloride solvolyses about as readily as
isopropyl chloride so that the conjugative effect of a phenyl group is equivalent
to the effect of two methyl groups. (Again the solvolysis reaction of these
two substrates involves a substantial S_N2 component.)

Coplanarity of groups attached to the central carbon atom is important for the accommodation of charge. Overlap of π-orbitals in the substituents with the p-orbital of the carbonium ion is most efficient if the p-orbitals are parallel. An example of this effect is found in the comparison of the relative rates of solvolysis in 80% ethanol of the compounds PhCHRCl for different R groups: methyl 540; ethyl 125; isopropyl 27; t-butyl 1.[32] These rate differences have been ascribed, in part, to steric repulsions between the alkyl group and the *ortho*-hydrogens of the ring, which increase with size of alkyl group, and prevent coplanarity between the aromatic ring and the developing vacant orbital. A 4-cyclopropyl substituent has a larger rate-enhancing effect in the solvolysis of the t-cumyl chloride (a) in aqueous dioxan than a 4-isopropyl substituent (b) owing to a conjugative interaction between the cyclopropyl ring and the developing carbonium-ion centre in the transition-state

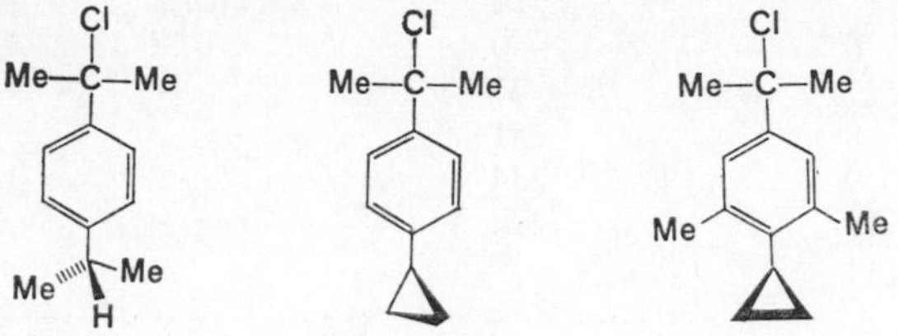

| $10^5 k$ (sec^{-1}) | 221 | 1947 | 47.3 |
| | (a) | (b) | (c) |

Figure 6. Solvolysis of t-cumyl chlorides in aqueous dioxan at 25°.

(Figure 6).[33] The substitution of two methyl groups into the system reduces the rate of solvolysis substantially owing to the conformation in which (c) is forced to exist.

If, in a carbonium ion precursor R_3C—X, the R groups are bulky with substantial non-bonded interaction, strain (termed 'B-strain' by H. C. Brown) will be relieved on ionization[34] (Chapter 1, page 10). Since the R groups are farther apart in the transition-state than in the ground-state, enhanced reactivity will result. This factor must contribute to the reactivity of t-butyl chloride in solvolysis reactions and will be even greater as the size of R increases. The converse situation exists in small or constrained rings, where the ground-states have angles of 100° or less around the reaction site. The corresponding carbonium ions prefer 120° geometry; strain (termed 'I-strain') increases on ionization and decreased reactivity results. Cyclopropane derivatives are extremely unreactive in solvolysis reactions (Table 6). Tertiary cyclobutane derivatives are in line with the expected order of reactivity due to angle strain but the secondary cyclobutane derivatives are unexpectedly

reactive.* Five-membered rings show a greater reactivity than acyclic derivatives since, although a small constraint is imposed from angle considerations,
the rehybridization of an sp^3-carbon to the sp^2-carbonium ion is accompanied
by a decrease of four axial carbon–hydrogen torsional interactions. A similar
effect is absent for cyclohexyl systems and solvolytic reactions proceed at

Table 6. Relative reactivities of cycloalkyl systems.[5]

$(\overbrace{CH_2)_{n-1}}$ CRX	R = H, X = OTs $MeCO_2H$, 60°	R = H, X = Cl 50% EtOH, 95°	R = Me, X = Cl 80% EtOH, 25°
Me$_2$CRX	1.0	1.00	1.00
$n = 3$	2×10^{-5}	<0.005	—
4	12	15	0.74
5	14	5.2	44
6	0.88	0.36	0.35
7	27	—	38
8	251	—	100
9	234	—	15.4
10	474	—	6.2
11	59	—	4.2

about the same rate as for acyclic analogues. The solvolyses of 1-chloro-1-
methylcycloalkanes in aqueous ethanol with ring sizes of 7 to 11 are enhanced
compared to those for t-butyl chloride; these differences in reactivity have been
attributed to the relief of carbon–hydrogen repulsions in the formation of the
transition-state for the medium-sized ring compounds. Rate enhancements,
compared to isopropyl tosylate, are also observed in the acetolysis of the
cycloalkyl tosylates with an eight-, nine-, and ten-membered ring, owing
predominantly to relief of non-bonded interactions (Chapter 1, page 10).

Cyclohexyl derivatives have received much attention, and these studies
have added substantially to the chemical understanding of the reactions of

* Extensive discussion of σ-bond participation (see page 111) in cyclobutyl and cyclopropylcarbinyl systems can be found in refs. 3 and 7.

conformationally mobile systems. Winstein and Holness[35] suggested the equation:

$$k = k_e N_e + k_a N_a$$

where N_e and N_a are the mole fractions of the substituted cyclohexane in the equatorial and axial conformations respectively. The equation indicates that the specific reaction rate for a substituted cyclohexane depends on the mole fractions in the equatorial and axial forms and on the specific rate at which these individual conformations react. An appropriate measure of the relative reactivities of axial and equatorial substituents was obtained by studies on the 4-t-butylcyclohexyl system in which the t-butyl group is constrained on steric grounds to an equatorial position. The relative solvolysis reactivities of cis- and trans-4-t-butylcyclohexyl tosylates (26 and 27) was $k_{\text{axial } (cis)}/k_{\text{equatorial } (trans)}$

(26)   (27)

(28a)   (28b)

3.9, 3.2, and 3.6 in ethanol, acetic acid, and formic acid respectively. Non-bonded interaction of the axial leaving group and the 3-axial hydrogens can be seen to have a substantial effect on the reactivity of the cis-isomer. Extensive investigations of the reactivity of trans-4-t-butylcyclohexyl tosylate now indicate that this solvolyses through a twist-boat or flexible conformation, rather than a chair form.[36] The Winstein–Holness equation was rewritten in a general form to take into account reactions which take place through flexible isomers (28a and b):

$$k = k_e N_e + k_a N_a + k_f N_f$$

For the trans-isomer, $k_a N_a$ can be regarded as negligible but solvolysis can occur from the flexible and chair conformers. It is apparent that, whereas the 2-trans- and 6-trans-hydrogen atoms are not antiparallel to the C—O bond when in the chair conformer, in the flexible conformation the two hydrogens are not equivalent and one hydrogen atom is approximately anti-parallel to the C—O bond. Evidence in agreement with this analysis was published by Shiner and Jewett in that the substitution of deuterium in the 2- and 6-equatorial position did not produce a cumulative effect on the rate of solvolysis.[37] The formation of a substantial amount of olefin and products arising as a consequence of hydride shift (page 112) provides strong evidence for the role of the flexible conformer in the solvolysis reactions of **27** since both of these processes would be favoured by *anti*parallel alignment of the H—C and C—O bonds.

Torsional effects have been suggested to be one factor contributing to the faster solvolysis of tertiary *exo*-2-norbornyl derivatives (**29**) than of their *endo*-isomers (**30**)[38] (page 113). In the initial state for the *exo*-isomer (**29**), the C(2)—X and C(1)—H bonds are partially eclipsed and the resulting torsional

(29)   (31)

(30)   (32)

(R = alkyl or aryl group; X = leaving group)

strain will therefore be reduced in the transition-state (**31**), when the C(2)—X bond is partly broken. Furthermore, another consequence of carbonium ion formation is relief of the partial eclipsing of the C(2)—R and C(1)—C(6) bond and a decrease in the non-bonded repulsion between the R-group and the *endo*-6-hydrogen. For the transition-state of the *exo*-isomer, then, both torsional and non-bonded interactions decrease. The opposite situation occurs in the transition-state for the *endo*-isomer (**30**), and eclipsing of the C(2)—R and C(1)—H bonds is produced when the 2-*exo*-R-group moves towards the *endo*-side. The torsional and non-bonded energy differences in the two transition-states are likely to be sufficiently great to account for the observed *exo*/*endo* ratios (200—500) in the solvolysis of tertiary norbornyl derivatives. It follows, also, by microscopic reversibility that attack on a 2-norbornyl cation should be strongly favoured from the *exo*-direction. This effect is not confined to carbonium ion chemistry and, among other examples, torsional factors supply an attractive explanation for the stereospecific base-catalysed deuterium exchange from the *exo*-side in 2-norbornanones.[39]

Whilst relief of steric strain accompanying the ionization of sterically crowded tertiary systems provides a driving force for carbonium ion formation, the opposite effect has now been recognized in some rigid bicyclic structures. In some cases (**33**), owing to the geometry of the system, the leaving group encounters appreciable steric interactions as the transition-state for ionization is formed.[40] This effect has been termed 'steric inhibition to ionization' and produces a substantial decrease in rate. The solvolysis rate for **34** is 4300 times greater than that for **33** in 60% aqueous acetone.

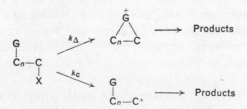

(33)   (34)

$(X = p\text{-}O_2NC_6H_4COO^-; Y = Me)$

Participation by Neighbouring Groups

One of the major advances in the study of carbonium ion reactions was the recognition of neighbouring-group participation.[7] The only requirement is that a substituent, G, possessing lone-pairs of electrons is located in the molecule in such a position as to become bonded or partially bonded to the reaction centre. If the carbonium ion is stabilized in this way, an increased rate of reaction is the result. If an increased reaction rate is observed due to the stabilizing influence of a neighbouring group on the transition-state this substituent is said to provide 'anchimeric' assistance. The anchimerically assisted reaction, with rate constant k_Δ, is frequently accompanied by the component of the reaction proceeding without participation, k_c (Figure 7). The rate of acetolysis of *trans*-2-acetoxycyclohexyl toluene-*p*-sulphonate (**35**) is 2000 times greater than that of the *cis*-isomer.[41] Confirmation that the

Figure 7. Neighbouring-group participation.

symmetrical ion is the intermediate in the reaction is the formation of racemic product when optically active material is used. Further indication of a reaction involving a neighbouring group is that a product is produced which differs from that which would be expected in the absence of participation. In some cases the participation occurs after the rate-determining ionization, so that, although the structure of the product is affected, no anchimeric assistance is observed. This may be a product with retained configuration, a ring-closed product, or one in which the participating group has migrated. Owing to the incursion of bromonium ions in the reaction of concentrated hydrobromic acid with (±)-*erythro*- and (±)-*threo*-3-bromobutan-2-ol (36 and 37) products[42]

are formed with retention of configuration; *meso*- and (±)-2,3-dibromobutane are produced, respectively. Generally, nucleophilic participation increases in the order Cl < Br < I, which is also the order of the ease with which these elements increase their valency. Participation by methoxy groups is common; the acetolysis products of 4-methoxy-1-pentyl (38) and 4-methoxy-2-pentyl *p*-bromobenzenesulphonate (39) are identical, indicating a common intermediate.[43] Participation has been observed with a wide range of other substituents such as hydroxy, amines, carboxy, thioethers, esters, amides, and aldehydes.[7] The subject of nucleophilic participation by σ- and π-bonds is discussed later.

The probability of neighbouring-group participation is dependent upon the size of ring formed in the transition-state. Increases in the number of degrees of rotational freedom with the length of carbon chain cause a greater decrease in entropy on going from the initial-state to the transition-state for the cyclization reaction. Additionally, for ring sizes 3–6, the strain-energy decreases with size of the ring formed. The balance of strain-energy and entropy factors favours neighbouring-group effects for the formation of three-, five-, and six-membered rings. Although, in most cases, neighbouring-group participation is most favoured for cyclic transition-states involving five atoms, this generalization is by no means true in every case (Table 7).

Table 7. Effect of ring size on the rates of neighbouring-group participation.[7]

Hydrolysis of $NH_2(CH_2)_nBr$					
Ring size $(n + 1)$	3	4	5	6	7
$10^4 k$ (sec^{-1})	6.0	0.083	5000	83	0.17
Acetolysis of $MeO(CH_2)_nOBs$					
Ring size $(n + 1)$	3	4	5	6	7
Rel. k	0.28	0.63	657	123	1.16
Hydrolysis of $PhS(CH_2)_nCl$; 20% aqueous dioxan					
Ring size $(n + 1)$	3	4	5	6	
Rel. k	9.2	0.048	1.8	0.025	

Participation by carbon–carbon bonds in carbonium ion reactions is an extremely important facet of neighbouring-group participation, and this is discussed extensively in the section on rearrangements and bridged ions (page 105).

Reactions of Carbonium Ions

Since carbonium ions are highly energetic intermediates, a number of reaction paths are possible (Figure 8). Competition between substitution and elimination frequently occurs for the reactions of carbonium ions. Proton loss particularly occurs for alkyl cations when substantial steric interactions are encountered in the transition-state for substitution; 16% and 40% olefin is produced in the solvolysis of Me_3CCl and Et_3CCl in aqueous ethanol, respectively.[44] A factor in the production of more olefinic product from the branched cations is the greater hyperconjugative stabilization of the double

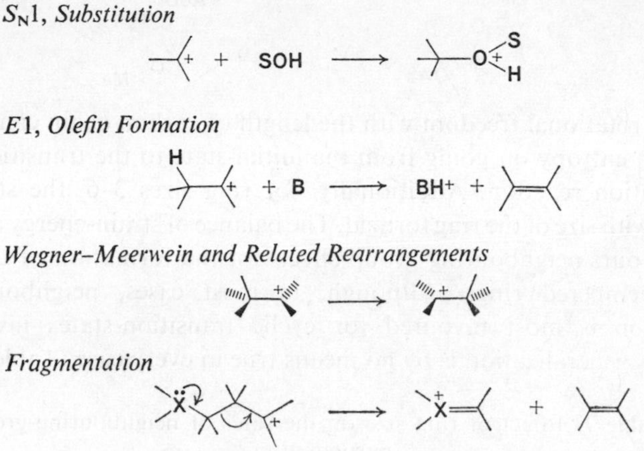

S$_N$1, Substitution

E1, Olefin Formation

Wagner–Meerwein and Related Rearrangements

Fragmentation

Figure 8. Reactions of carbonium ions.

bond by the alkyl substituents (**40**). This concept provides a satisfactory explanation for the production of the most highly substituted olefin (Saytzeff orientation) from carbonium ions. The stability of the products is thus reflected in the transition-states leading to their formation. Steric factors must also play a part in determining the stabilities of the products and in olefin production

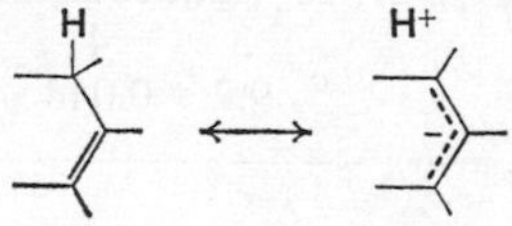

(**40**)

from $Me_2CHCH_2CH(OBs)Me$ in acetic acid the *trans*-2-ene/*cis*-2-ene ratio is 2.0.[45] Thus eclipsing interactions in the *cis*-olefin are reflected in the product-forming transition-state.

Olefin formation is also favoured when the carbon–hydrogen bond being broken is coplanar with the vacant *p*-orbital at the carbonium centre. This arrangement can be achieved in most acyclic carbonium ions, but with reactive species coordination with solvent may occur before the most favoured elimination conformation has been obtained. A high degree of olefin production from the axial cyclohexyl derivatives is due, to a large extent, to the favourable coplanar arrangement of the neighbouring hydrogen and the vacant orbital of the carbonium ion. In the solvolysis of the 2-decalyl derivative **41** in acetic acid, the product contains 59% and 26% of the olefins **43** and **44**; less than 13% of a mixture of acetates is formed.[36] For the corresponding solvolysis of **42**, where a less favourable arrangement of the carbon–hydrogen

| (41) | (42) | (43) | (44) |

bond and reaction centre exists, 40% and 24% of the same olefins were produced together with the 35% of acetate product. Generally, more olefin product is produced from the carbonium ion reactions of 'axial' than of 'equatorial' derivatives and similar elimination–substitution ratios are found for the acetolysis of *cis*- and *trans*-4-t-butylcyclohexyl tosylates.

Both solvent and leaving-group have a marked effect on olefin production in carbonium ion reactions (Table 8). In good ionizing solvents the nature of the leaving group has little or no effect on the amount of olefin produced. For less polar media, however, ion-pairing is important and the counter-ion remains in contact with the carbonium ion and has an appreciable effect on the elimination–substitution ratio. Formation of olefin is also favoured in good ionizing but poorly nucleophilic solvents such as formic acid; 95% and 65% of a mixture of cholest-2- and -3-enes are produced from 5α-cholestan-3β-yl tosylate in a 1:1 mixture of formic and acetic acids, and pure acetic acid, respectively.[47]

Table 8. Influence of the solvent and leaving group on % of olefin from the solvolysis of t-butyl derivatives.[46]

Solvent Temp.	H_2O^a 25°	H_2O^a 75°	EtOH^a 75°	MeCO$_2$H^b 75°
Leaving group				
Cl	5 ± 1	7.6 ± 1	44.2 ± 1	73 ± 2
Br	5 ± 1	6.6 ± 1	36.0 ± 1	69.5
I	4 ± 1	6.0 ± 1	32.3 ± 1	—
Me$_2$S$^+$ClO$_4^-$	—	6.5 ± 1	17.8 ± 1.4	11.7 ± 1
$^+$OH$_2$	3	4.7	—	—

^{a} 2,6-Lutidine present. ^{b} Sodium acetate present.

Fragmentation of Carbonium Ions

Fragmentation of a carbonium ion to an olefin and a new carbonium ion is possible in certain circumstances.[48] The new ionic centre on the γ carbon must be well stabilized, for example by a heteroatom (Figure 9). Two mechanistic possibilities exist: (a) fragmentation occurs from a preformed ion; this implies no acceleration of the overall reaction rate if the first (ion-forming step) is rate-

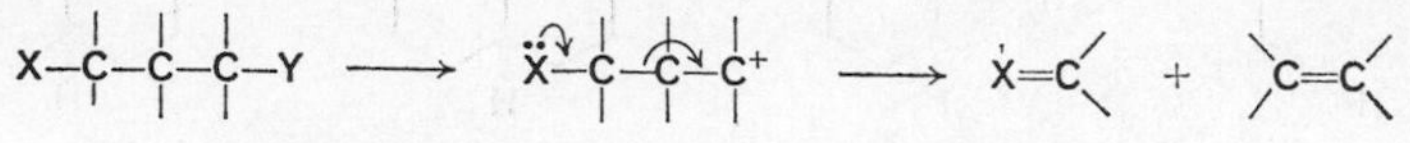

$$X-\overset{|}{\underset{|}{C}}-\overset{|}{\underset{|}{C}}-\overset{|}{\underset{|}{C}}-Y \longrightarrow \overset{..}{X}-\overset{|}{\underset{|}{C}}-\overset{|}{\underset{|}{C}}-\overset{|}{\underset{|}{C}}{}^{+} \longrightarrow \overset{.}{X}=C\Big\langle + \Big\rangle C=C\Big\langle$$

(Y = halogen, OTs, etc.; X = NR$_2$, OH, etc.)

Figure 9. Fragmentation of carbonium ions.

limiting; (b) fragmentation is concerted with ionization and an accelerated reaction rate results. In case (a) other products typical of carbonium ion decomposition (substitution, elimination, cyclization) may also be formed.

One of the earliest examples is the acid-catalysed fragmentation of tetramethylpentane-2,4-diol (45) into acetone and dimethylbut-2-ene. Some typical results for the solvolysis of the compounds RCH$_2$CH$_2$C(Cl)Me$_2$ in 80%

(45)

aqueous ethanol are given in Table 9. The rate of the reaction is almost independent of the nature of the γ-substituent but this does largely control the amount of fragmentation. This absence of any large effect on the rate of reaction by the substituents is an essential requirement for the proposed two-step mechanism involving an initial rate-determining ionization, followed by fragmentation. In some cases fragmentation becomes possible only after

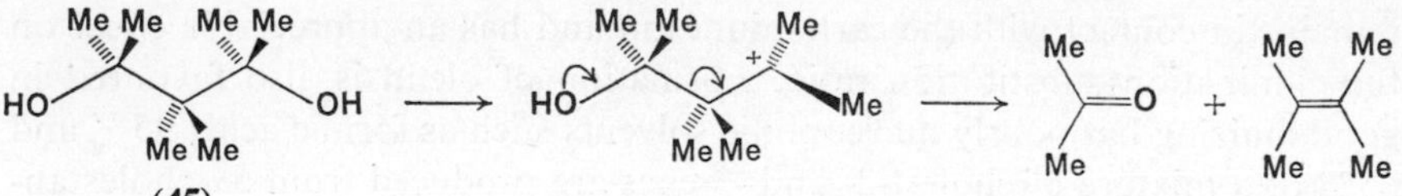

Table 9. Relative rates and products of solvolysis of RCH$_2$CH$_2$C(Cl)Me$_2$ in 80% aqueous ethanol at 56°.[49]

R	Me$_2$CH	NH$_2$	Me$_2$N
$10^4 k_1$	4.73	4.66	3.55
Rel. rate	1.0	0.99	0.75
Fragmentation (%)	0	20	50

rearrangement of the initially formed carbonium ion. Thus the solvolysis of 4-(toluene-*p*-sulphonyloxymethyl)quinuclidine (**46**) involves the rearrangement of a primary cation to the bridgehead tertiary carbonium ion followed by fragmentation.[50] Although a similar rearrangement is found in the solvolysis of the non-nitrogenous analogue of **46**, no fragmentation occurs. The rates of solvolysis of the two systems are comparable, the only difference being due to the small retarding inductive effect of the nitrogen atom.

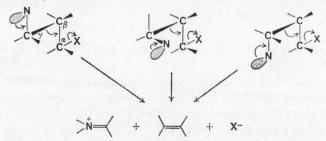

A second class of fragmentation reactions is also known with different mechanistic characteristics; many of the stereochemical studies have been again with γ-amino-compounds. This is a one-step process and the γ-substituent is involved in the rate-determining step.[48] A consequence is that if the rate of reaction is compared to that of a model compound (similar but not containing the γ-substituent) a rate enhancement, termed frangomeric acceleration is observed. For these synchronous reactions the preferred stereochemistry is considered to be that in which the C—X bond and the orbital of the unshared electron pair are both *anti*-periplanar to the fragmenting C_β—C_γ bond (Figure 10). In the three conformations shown, overlap of the *p*-orbitals at C_α, C_β, and C_γ, and the lone-pair is a maximum in the transition-state.

Figure 10. Arrangements for synchronous fragmentation reactions.

Fragmentation with acceleration is observed in the solvolysis of **47** in aqueous ethanol, which has a fully eclipsed conformation of the nitrogen atom and leaving group. For comparison, 1-bromobicyclo[2.2.2]octane solvolyses 5.3×10^4 slower at 40° and the expected substitution product is observed.[28] A skew conformation also satisfies the stereoelectronic requirements and 1-alkyl-4-chloropiperidine (**48**) undergoes solvolytic fragmentation in 80%

aqueous ethanol 34—178 times faster than the solvolysis of cyclohexyl chloride.[51]

The stereochemical requirement for synchronous fragmentations is further illustrated by the solvolysis reactions of 3β- and 3α-tropanyl chlorides (**49** and **50**). With **49** the geometrical requirements for the concerted mechanism are satisfied and fragmentation is the only observable reaction. Furthermore, the rate is 2.5×10^4 times faster than that of cyclohexyl chloride at 62°. In **50** the axial C—Cl bond is not parallel to the C_β—C_γ bond and no fragmentation is observed. The fact that the main product, 3α-tropanol, has a retained configuration, and that the rate is still 645 times that for cyclohexyl chloride, indicates that ionization is assisted by neighbouring-group participation of the amino group (51).[48]

Rearrangement and Formation of Bridged Carbonium Ions

The interactions between π-orbitals and positive centres have provided the most spectacular cases of anchimeric accelerations; rearrangement of structure takes place in many cases.[7] Early work by Shoppee showed that nucleophilic substitutions on cholesteryl chloride occur with complete retention of configuration, while the saturated cholestanyl chloride reacts normally with inversion.[52] The retention of configuration was ascribed to participation of the adjacent double bond. It was observed that the methanolysis of cholesteryl tosylate (52) leads to the formation of the methyl ether (53) with retention of configuration, while the same reaction carried out in the presence of potassium acetate yields the i-ether (54).[53] The formation of the two products and the stereochemical course of the reaction was explained by the intervention of the ion 55 as an intermediate. In the case of cholesterol, the geometrical requirements are favourable and quantum-mechanical calculations indicate a

stabilization of this non-classical ion owing to overlap of the double-bond
π-orbitals and the p-orbitals of the carbonium ion (55); anchimeric acceleration consistent with this analysis is found, and 52 acetolyses 123 times faster
than cholestanyl tosylate. Product studies are of equal importance in providing
evidence for the mechanism of reactions; 53 and 54 are consistent with capture
of the intermediate by nucleophilic attack of solvent at the 3β- and 6β-positions
on the opposite side to the delocalized π-electrons. Retention of configuration
in unrearranged product is therefore identified as a general phenomenon of
neighbouring-group participation.

The requirement for effective participation of the double bond is clearly a
favourable electronic arrangement in relation to the p-orbital of the carbonium
ion. The additional steric constraint found in cyclic systems compared to
acyclic compounds has produced examples where participation is particularly
favourable; the reverse is also found. One of the most striking examples of
double-bond participation is found in the acetolysis of *anti*-norborn-2-en-7-yl
tosylate (56) which proceeds 10^{11} faster than that of the analogous saturated
compound, norborn-7-yl tosylate. The π-electrons of the double bond are
particularly well placed to interact with the developing carbonium ion at the
7-position to yield the ion (57); indicative of the intervention of this intermediate is the reaction product, *anti*-norborn-2-en-7-yl acetate (58) obtained

TsO⟍   AcO⟍

(56)   (57)   (58)

with retention of configuration.[54] Numerous examples exist of neighbouring-
group participation of the π-electrons of a double bond, of unsaturated carbon–
carbon bonds (Figure 11), and also cyclopropane rings.[59]

Although the nature of double-bond participation has been firmly established, the evidence for participation by neighbouring aryl groups has involved
considerable speculation and study.[60, 61, 7] The proposed intermediate formed
by participation of a remote aryl group in a solvolysis reaction was called
a phenonium ion (61). This term was proposed by Cram after studies
on the acetolysis of optically active *erythro*-3-phenyl-2-butyl tosylate (60)
which yielded the optically active *erythro*-acetate (62) with 96% retention of
configuration; this product results from solvent attack at either C_α or C_β of
the intermediate.[62] Similarly, the optically active *threo*-isomer (63) yields
almost racemic *threo*-acetate (64) since the phenonium ion produced in this
case has an axis of symmetry. Substantial ion-pair return was detectable in
the solvolysis from the evidence of a $k_\alpha/k_t = 5$ (see page 86) which indicates

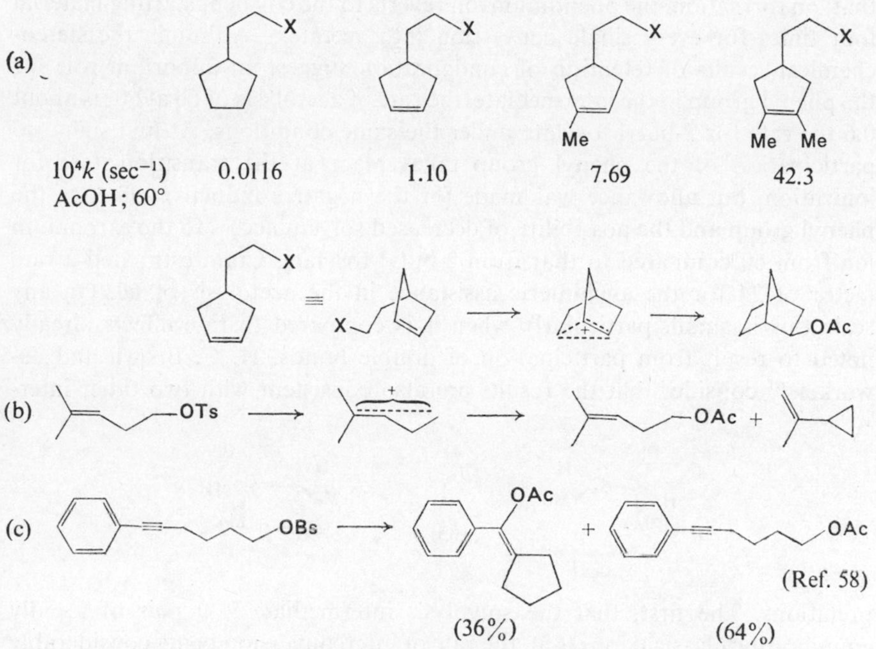

(a)

$10^4 k \ (\text{sec}^{-1})$ 0.0116   1.10   7.69   42.3
AcOH; 60°

(b)

(c)

(Ref. 58)

(36%)   (64%)

Figure 11. Participation by double and triple bonds. The influence of successive methyl substitutions in (a) shows that a symmetrical ion is involved.[56,57]
(b) Rate of acetolysis at 45° is 1200 times greater than that of ethyl tosylate.[55]

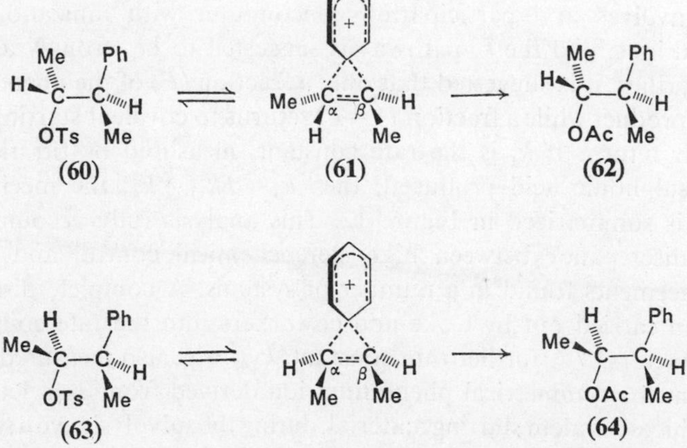

that, on formation, the phenonium ion reverts to the covalent starting material four times for every single conversion into acetate.[63] Although the stereochemical results of retention of configuration suggest an important role for the phenyl group in the intermediate, the rate of acetolysis of **60** at 50° is about 0.6 the rate for 2-butyl tosylate under the same conditions. At first sight, no participation of the phenyl group takes place at the transition-state for ionization, but allowance was made for the negative inductive effect of the phenyl group and the possibility of decreased solvent access to the carbonium ion from **60** compared to that from 2-butyl tosylate. Cram estimated a rate factor of 24 for the anchimeric assistance in the acetolysis of **60**. On any scale this is small, particularly when it is compared to rate effects already noted to result from participation of double bonds. H. C. Brown and coworkers[61] consider that the results are also consistent with two other inter-

(65)

pretations. The first, that the solvolysis intermediate is a pair of rapidly equilibrating classical ions (**65**), the rate of interconversion being considerably faster than the rate of rotation about the central carbon–carbon bond, thus preventing solvent attack from the phenyl side of the ions; secondly that the mechanism is ionization to an essentially unbridged ion-pair, followed by bridging prior to attack of solvent.

Exhaustive work has led to an explanation, consistent with the experimental facts for the solvolysis of 2-arylalkyl arenesulphonates, that there are two competing paths for displacement of tosylate designated k_Δ and k_s.[64-68] The k_Δ path involves aryl participation concomitant with ionization to give phenonium ions, and the k_s pathway is suggested to be strongly assisted by solvent. Further, it is suggested that only a fraction (F) of the phenonium ion goes on to product while a fraction ($1 - F$) returns to covalent starting material by ion-pair return. If k_t is the rate constant, measured by titration of the toluene-p-sulphonic acid produced, then $k_t = Fk_\Delta + k_s$; the mechanism of solvolysis is summarized in Figure 12. This analysis fully accounts for the apparent discrepancy between high stereochemical control and yet small rate enhancements found in a number of systems. A complete dissection of k_t has been carried out by Coke and coworkers into the rate constants k_Δ, k_s, and F-factors.[64] A further rate constant, k_{14}, was also measured. Ion-pair return from the symmetrical phenonium ion derived from 2-arylethyl derivatives produces covalent starting material, during the solvolysis, with scrambling of the two aliphatic carbon atoms, and this can be followed by the use of ^{14}C. This rate of scrambling $k_{14} = (1 - F)k_{\Delta}$. Schleyer *et al.*[65, 66] have also estimated Fk_{Δ} and k_s for acetolysis of 1-aryl-2-propyl tosylates by a number

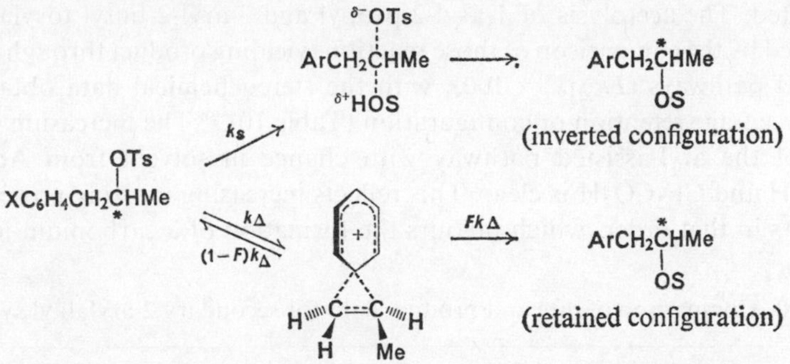

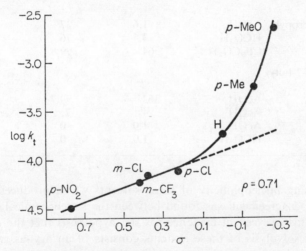

Figure 12. Mechanism of solvolysis involving phenonium ions.

Figure 13. Plot of $\log k_t$ against σ for solvolysis of 1-aryl-2-propyl tosylates in acetic acid at 100°.[65] (A more refined treatment of the data has been reported.[67])

of methods. An example of one of these is that $\log k_t$ was plotted against the σ-value of the substituent in the phenyl ring[65] (Figure 13). Deviation from the straight line was found for the substituents *p*-methyl, *p*-methoxy, and also for the unsubstituted phenyl ring, and this was taken as a measure of the anchimerically assisted process Fk_{Δ}. $\log k_t$ for the derivatives with substituents

p-nitro, p-trifluoromethyl, m-chloro, and p-chloro correlated linearly with σ, and the rate constants were considered to be a direct measure of k_s in these cases with no contribution from an aryl-assisted process. Extrapolation of the line allowed the rate constants k_s for other substituted derivatives to be estimated. The acetolysis of 1-aryl-2-propyl and 3-aryl-2-butyl tosylates was analysed by the comparison of those reactions yielding product through phenyl-assisted pathways ($Fk_\Delta/k_t \times 100$), with the stereochemical data obtained of the percentage retention of configuration (Table 10).[68] The increasing importance of the aryl-assisted pathway with change in solvent from AcOH to HCO_2H and CF_3CO_2H is clear. This reflects increasing ionizing power of the solvents in that order, which favours the formation of a carbonium ion, and

Table 10. Comparison of rate and product data for secondary 2-arylalkyl systems.[68]

Sulphonate	Solvent	Rate enhancement (k_t/k_s)	% Assisted pathway ($Fk_\Delta/k_t \times 100$)	% Yield ester with retained confign.
1-Phenyl-2-propyl	AcOH	1.6	37	25
	HCO_2H	4.5	76	68
	CF_3CO_2H	564	99.8	~100
threo-3-Aryl-2-butyl Ar =				
p-MeOC$_6$H$_4$	AcOH	363	99.7	99.7
Ph	AcOH	4	75	78
p-O$_2$NC$_6$H$_4$	AcOH	1.0	~0	1
	HCO_2H	1.0	~0	4

the decreasing nucleophilicity of these solvents which produces smaller k_s values. Good agreement was found between the stereochemical and kinetic data, and this is excellent evidence for the hypothesis that the overall rate process for solvolysis of these systems consists of an aryl-assisted and an aryl-unassisted component. Furthermore, this agreement also confirms that these two processes are completely independent of each other with no inter-conversion ('cross over'). If this were not so, product studies would provide stereochemical data with no direct correlation (unless fortuitous) with rate data. This conclusion is rationalized on the basis that, for primary and secondary derivatives, the k_s component is strongly assisted by solvent and would require a high activation energy for interconversion to a phenonium ion. The striking consequence follows that, since aryl-participation has to compete with this solvent-assisted process, *any* rate enhancement due to anchimeric assistance is indicative of strong participation of a neighbouring group, even though the enhancement appears to be small. Illustrating this point, it can be seen (Table 10) that for *threo*-3-phenyl-2-butyl tosylate in acetic acid a rate enhancement of only 4 is observed but 75% of the reaction proceeds through the phenonium ion. The introduction of a substituent into the phenyl ring is effective in making changes in the reaction course. Thus, no phenonium ions are formed in the solvolysis of *threo*-3-*p*-nitrophenyl-2-butyl tosylate whilst *threo*-3-*p*-methoxyphenyl-2-butyl toluene-*p*-sulphonate solvolyses almost completely through a phenonium ion.

Phenonium ions have, in fact, been positively identified by spectral techniques; **66**, **67**, and **68** and other ions have been generated in highly acidic

(66)   (67)   (68)

solvents and positively characterized by n.m.r.[69] Although useful in themselves, evidence from these studies cannot be directly applied to those in solvolytic medium since in these non-nucleophilic solvents no competition from a solvent-assisted process is possible. Participation by neighbouring aryl groups occurs in the solvolysis of the conjugate bases of 2-(*p*-hydroxyphenyl)ethyl bromides (**69**) and 4-(*p*-hydroxyphenyl)butyl brosylate (**71**); dienones (**70** and **72**) have been isolated and, although **70** is very unstable, both have been identified by spectroscopic techniques. It was estimated that **69** undergoes solvolysis in ethanol at 25° 10^6 faster than the rate expected for unassisted solvolysis of the bromide.[70]

(69)   (70)   (71)   (72)

σ-Bond Participation

The question of σ-bond participation has caused even more controversy than that associated with aryl groups.[7, 8] Since the nucleophilicity of a σ-bond would be predicted to be substantially less than that of double bonds and phenyl groups, any anchimeric assistance would be expected to be relatively small. Migrations of alkyl groups, e.g. pinacol and related rearrangements and hydride shifts, are common in carbonium ion chemistry but the kinetic evidence for rate-enhancement is sparse. It is important to consider the intermediate that is required for a bridged carbonium ion involving a σ-bond (74). This type of intermediate which has delocalized bonding σ-electrons is termed a non-classical ion. A broken line is employed to represent the overlap of two atomic orbitals, each of which contributes to a multicentre molecular orbital containing two electrons. A useful comparison can be made here of boron compounds with carbonium ions. BH_3 and CH_3^+ are isoelectronic since both have an incomplete outer shell of a sextet of electrons. The tendency of boron compounds to accommodate bridging bonds is well established and participation of a σ-bond in carbonium ion chemistry can be considered with this in mind.

$$Me_3CCH_2Cl \xrightarrow{EtOH} Me_2C\!\!\cdots\!\!\overset{CH_3}{\overset{+}{C}}H_2 \longrightarrow \begin{array}{c} Me_2C(OEt)CH_2Me \\ + \\ Me_2C\!\!=\!\!CHMe \end{array}$$

$$(73) \qquad\qquad (74)$$

Substantial rearrangement occurs in the reaction of neopentyl chloride (73) in ethanol, and evidence has been found for a striking enhancement of the rate of ionization of neopentyl tosylate in trifluoroacetic acid.[71a] The rate of solvolysis in a series of solvents was compared to that of ethyl tosylate (Table 11). The large rate difference in the most ionizing solvent was attributed to methyl participation with concomitant formation of the tertiary substituted product and olefin, and the discussion in the previous section on phenonium ions of a competition between two processes k_Δ and k_s can also be applied here. In general, however, there is little evidence for rate enhancements, and even where small accelerations have been observed a convincing argument can also be made that this is caused by steric accelerations. A further note of caution must also be made since the discussion of rates is only meaningful if the products of the reaction are known. In the above example, rearrangement of the neopentyl skeleton is incomplete in ethanolysis and small amounts of 1,1-dimethylcyclopropane and unrearranged products are also formed.[71b] (Detailed product studies would also be desirable in the other solvents because frequently the last 5% of a reaction mixture is more interesting than the major 95%, since the energetic preference of various pathways becomes clear.)

Migration of hydrogen (hydride shift) is widespread in carbonium ion reactions but definite evidence for participation in the rate-determining step has been found in only a few cases. This type of participation is more likely

Table 11. Solvolysis rates[a] of neopentyl tosylate, 75°.[71a]

Reactant	MeCH$_2$OH	MeCO$_2$H	HCO$_2$H	CF$_3$CO$_2$H
MeCH$_2$OTs	356	9.2	226	6.4
Me$_3$CCH$_2$OTs	0.2	1.0	227	1.018

[a] Rates are relative to neopentyl tosylate in acetic acid.

when charge is transferred from a secondary to a tertiary carbon atom. Neomenthyl chloride and arenesulphonates (75) solvolyse 40—80 times faster than the corresponding menthyl derivatives, and this has been ascribed to hydrogen participation (76).[72] A further kinetic method of investigation is

available for hydrogen participation. Large β-deuterium isotope effects have been observed in the solvolysis of cis-4-t-butylcyclohexyl tosylate (26); the introduction of an axial β-deuterium gives $k_H/k_D = 1.436$ in 50% aqueous ethanol whilst equatorial deuterium gives $k_H/k_D = 1.096$.[37] The larger isotope effect has been attributed to participation of the β-hydrogen in the ionization. β-Deuterium isotope effects and the amount of product produced by hydride shift have been observed to be markedly dependent on the dihedral angle between the C—H(D) and C—X bonds, being greatest when the angle is 180°. When hydride shift occurs the substitution product is frequently formed with retention at the new site (77).[73] This has been explained as the consequence of hydrogen-bonding from the migrating hydrogen atom to incoming solvent (78) or of the continued presence of the counter-ion on the opposite side of the molecule (79).

After earlier studies[74, 75] the question of σ-bond participation was placed on a firm experimental basis by Winstein and Trifan[76] who showed that exo-2-norbornyl brosylate (80) solvolysed in acetic acid 350 times faster than the endo-isomer (82). Furthermore, exo-brosylate produced exo-acetate with complete stereospecificity, and when optically active brosylate was employed

(77)

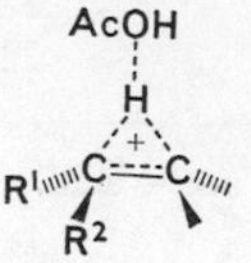

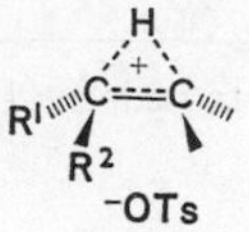

(78)   (79)

the product was found to be completely racemic. The solvolysis of optically active *exo*-2-norbornyl brosylate is accompanied by racemization of the unreacted brosylate, due to ion-pair return, which is strong evidence for a symmetrical intermediate. In all solvents studied, the rate of racemization exceeded the rate of solvolysis, as determined by the rate of formation of *p*-bromobenzenesulphonic acid; the ratio varied from 3.46 in acetic acid to 1.40 in 75% aqueous acetone. The solvolysis of *endo*-2-norbornyl brosylate (82) also yielded products with solely *exo*-configuration but, in contrast with the *exo*-epimer, optically active 82 yielded product in which 7—8% of the optical activity was retained. In this case, no significant difference between the rate of racemization and the rate of formation of *p*-bromobenzenesulphonic acid was observed.

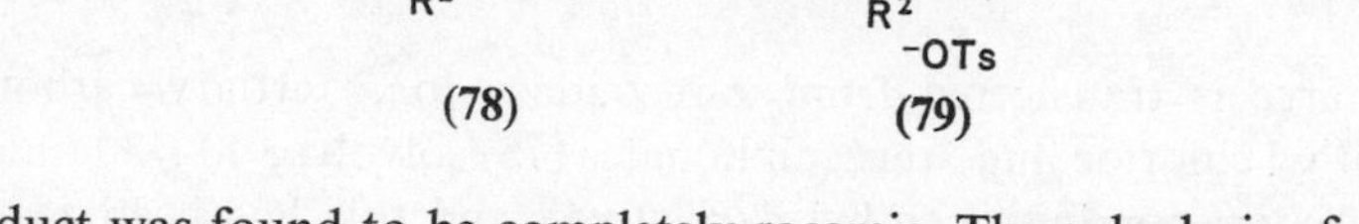

(80)   (81)

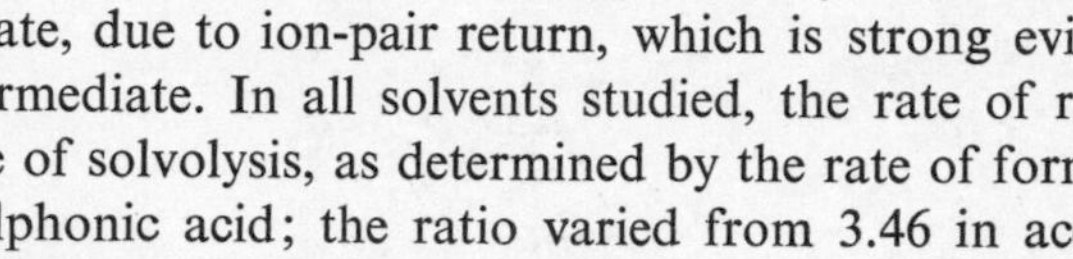

(82)

It was argued that the enhanced rate of solvolysis of the *exo*-epimer relative to that of *endo*-2-norbornyl brosylate could only be explained in terms of an intramolecular displacement reaction involving attack by the C(6)—C(1) σ-bond during the ionization of the *exo*-epimer. No such displacement is possible for the *endo*-epimer. The bridged symmetrical ion (81) was also suggested to be the direct product of ionization of the *exo*-brosylate, existing as an intimate ion-pair, and capable of collapsing to give racemic brosylate or, less frequently, with solvent to form product. A non-classical ion such as **83** was regarded as a resonance hybrid between several structures (**83a—c**), and evidence has been presented that the same ion is formed from the solvolysis of **84** by a 'π-route' rather than a 'σ-route'. Stereospecific product formation and ion-pair return was considered to be a consequence of the bridged intermediate since nucleophilic attack would be expected from the rear on one of the atomic orbitals involved in the multicentre bond. Attack in this manner at either C(1) or C(2) yields only *exo*-2-norbornyl acetate. Solvolysis of *endo*-2-norbornyl brosylate was postulated to proceed through a classical norbornyl cation coordinated with solvent on the side opposite the departing group.

[cf. Figure 11]

An alternative view has been taken by H. C. Brown that it is possible to explain the stereochemical and the kinetic results on the basis of classical carbonium ions.[77, 78] A large amount of evidence has been presented that, in the solvolysis of *exo*-norbornyl derivatives, the unbridged ion (**83a**) is initially formed which interconverts with the rearranged ion (**83b**) at a rate which is more rapid than reaction of either cation with the solvent. Particular attention has been paid to the effect of substituents in the 1- and 2-positions of the norbornyl system on the rates and products. If delocalization of the 1,6-bond is complete in the transition-state the effect of a 1- and 2-substituent on the rate should be almost identical. It was found, however, that whereas 2-phenyl-*exo*-norbornyl chloride (**85**) undergoes ethanolysis 3.9×10^7 times faster than the *exo*-norbornyl chloride, with the 1-phenyl isomer (**86**) the corresponding rate enhancement was only 3.9.[79] Provided that there is no steric inhibition of resonance between the 1-phenyl substituent and the developing cationic centre, this is strong evidence against the formation of a non-classical ion in this case. Similarly, 1-methyl-*exo*-2-norbornyl tosylate (**87**) reacts only 50 times faster than the *exo*-norbornyl tosylate.[78] Brown has also questioned that a non-classical intermediate is necessary to explain the high *exo/endo* rate ratio for 2-norbornyl derivatives. It is widely accepted that the demand on a neighbouring group from a carbonium ion decreases with the change from a secondary to a tertiary system. On this basis the observations that the *exo/endo* rate ratios for the solvolysis of the tertiary *p*-nitrobenzoates (Figure 14) are similar to that for the 2-norbornyl derivative were used as evidence that the intermediates for all these reactions are classical.[78] For several secondary derivatives, however, the *exo/endo* ratios are small and the high ratios for the tertiary derivatives can be rationalized in terms of torsional and other steric effects (see page 96).[38]

The high *exo/endo* rate for the secondary 2-norbornyl derivative was explained by Brown as due, not to a large rate for the *exo*-isomer, but to a small rate for the *endo*-isomer. This was attributed to steric inhibition to ionization resulting from interactions between the departing leaving group and the C(5) and C(6) *endo*-hydrogens. Although this factor is certainly

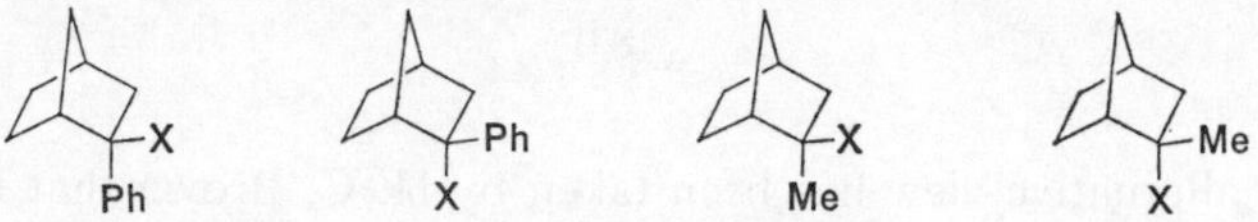

(85a)　　　　　　(88)　　　　　　(89)　　　　　　(90)

$$(X = OCOC_6H_4NO_2\text{-}p)$$

Rate ratio　　　85a/88 = 260　　　　　　　　　89/90 = 580

Figure 14. *exo/endo* Rate ratios for tertiary norbornyl *p*-nitrobenzoates.

important for the system **16** and other systems with large 5,6-substituents (page 97), any interactions between the *endo*-hydrogens and leaving group for *endo*-norbornyl tosylate must be very small. This illustrates a common problem in organic chemistry in the discussion of rates of reactions, in that suitable model systems are not easily found. An approach which has met with considerable success was followed by Schleyer and Foote who chose the solvolysis of cyclohexyl arenesulphonates as a model, and calculated the rates of solvolysis of a large number of secondary arenesulphonates relative to the model system by employing a number of empirical correlations.[80] The effect of bond-angle strain was estimated from the equation:

$$\log_{10} \text{Rel. rate} = 0.125 \, (1715 \text{ cm}^{-1} - \nu_{co} \text{ cm}^{-1})$$

where ν_{co} is the carbonyl stretching frequency of the ketone. Further allowances in the estimates were made for torsional, non-bonded interactions, and inductive effects. By this approach the observed rate of solvolysis of *exo*-2-norbornyl tosylate was found to be 10^3 greater than the calculated value.

Although this approach has had considerable success, Schleyer has pointed out that it suffers from the disadvantage that it fails to take into account the variable amount of solvent participation in the solvolysis of secondary systems.

A further original investigation into the nature of these carbonium ion intermediates was conducted by Goering and Schewene.[81] They measured the rate of perchloric acid catalysed loss of optical activity of *exo*-norbornyl acetate (91) and its isomerization to the *endo*-isomer (92). The former is a measure of the rate of ionization of the *exo*-acetate and the latter the rate of capture of the ion by *endo*-attack. The mechanism of this reaction is thought to involve reversible protonation of the substrate, followed by heterolysis of the conjugate acid of the acetate. It was estimated that *exo*-attack predominates to the extent of 99.99% at 25°. From these results and the equilibrium constant

for *exo–endo* conversion, a potential-energy diagram (Figure 15) was constructed. It can be seen that the energy of activation for capture of the norbornyl cation from the *exo*-direction is 4.4 ± 0.7 kcal mole^{-1} less than from the *endo*-direction, and that this, less the difference in initial-state energy, is also the energy of activation for the ionization of *exo*- and *endo*-norbornyl acetate. This work illustrates that the controlling factor determining the stereochemistry of substitution is the difference in the energies of the transition-states for reaction by *exo*- and *endo*-attack. A corresponding difference also applies for the ionization of the *exo*- and *endo*-acetates. The lower free-energy of formation of the carbonium ions from *exo*-derivatives and of its *exo*-capture are apparent and are best explained in terms of some σ-participation. Strictly, however, it is only valid to draw conclusions about the structures of the transition-states from these kinds of results. Extrapolation to the structure of the intermediate involves the assumption that this structure is closely

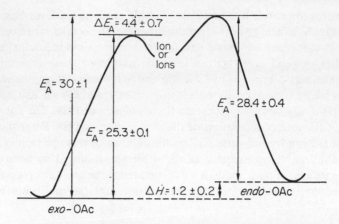

Figure 15. Equilibration of *exo*- and *endo*-norbornyl acetate.

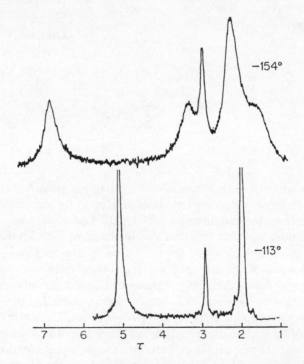

Figure 16. N.m.r. spectrum (100 MHz) of the norbornyl cation in SbF_5–SO_2ClF–SO_2F_2 solution.

related to that of the transition-states. There seems little doubt that some anchimeric assistance by the σ-bond is present in the solvolysis of secondary norbornyl derivatives. In some respects the controversy has revolved around the question of whether the non-classical structure is a transition-state or an intermediate.

A number of physical techniques have become available for the direct observation of intermediate cations derived from norbornyl derivatives. The norbornyl cation has been generated from 2-norbornyl fluoride in SbF_5, SbF_5–liquid SO_2, and SbF_5–SO_2ClF–SO_2F_2, and the 100 MHz n.m.r. spectrum recorded at various temperatures (Figure 16); spectra have been obtained at temperatures as low as $-154°$ in the latter medium. In the temperature range $-5°$ to $+37°$ the n.m.r. spectrum consisted of a single broad band at -3.75 p.p.m., indicating that all the protons are equivalent. At $-60°$, the spectrum separated into three bands: -5.35 p.p.m., 4 protons at C(1), C(2), and C(6); -3.15 p.p.m., 1 proton at C(4); -2.20 p.p.m., 6 protons at C(3), C(5), and C(7). On cooling to $-120°$ no further change occurred.[82] These results provided no definitive evidence for a non-classical structure since the observations could be explained in terms of a Wagner–Meerwein rearrangement (**93** to **94**) together with a $6\rightarrow2$ hydride shift (**93** to **95**) which occurs rapidly over the

C(1) $\equiv$ C(2); C(3) $\equiv$ C(7) (**94**)

C(2) $\equiv$ C(6); C(3) $\equiv$ C(5) (**95**)

C(1) $\equiv$ C(4); C(2) $\equiv$ C(3); C(5) $\equiv$ C(6) (**96**)

temperature range studied, and a $3\rightarrow2$ hydride shift (**93** to **96**) which is fast at room temperature but slow below $-23°$. Significant changes were observed, however, in the spectrum at temperatures between $-125°$ and $-154°$.[83] The peak due to the four equilibrating protons at 5.35 p.p.m. became broader as the temperature was lowered and separated into two signals each of relative area 2 at 3.05 p.p.m. and 6.59 p.p.m. The high-field signal due to the six methylene protons also became broader and developed a shoulder but that due to the single bridgehead proton remained unchanged. These observations

5

are consistent with the conclusion that the exchange between the protons at C(1), C(2), and C(6) had been 'frozen out' at the lower temperatures and that the ion is non-classical. Olah represents the ion as **97**, which is essentially the same as **83** but also shows its structural relationship to the protonated cyclopropane nortricyclene (see page 187). Further confirmation of this view has been provided from studies of the Raman and ^{13}C magnetic resonance spectra of the norbornyl cation.[84]

(97)

CARBANIONS[85-96]

We may begin discussion of the chemistry of negatively charged carbon by consideration of the equilibrium between a hydrocarbon and its conjugate base in a basic medium:

$$CH_4 + B^- \rightleftharpoons CH_3^- + BH \qquad (98)$$

A number of factors contrive to make methyl anion an energetically unfavourable species, and this equilibrium will be to the left even with strong bases. In fact, methyl and alkyl anions are too reactive to have any well characterized solution chemistry, and even when the counter-ion is a very electropositive metal like lithium the carbon–metal bond is largely covalent. Most of our knowledge of carbanion chemistry arises from the study of more stable species, and it is obviously of prime importance to be able to quantify carbanion stability. A fundamental approach has been to measure such acid–base equilibria in which both components are carbon acids; when the anions are highly coloured their concentrations may be measured spectrometrically or even colorimetrically. In the 1930's McEwen and Conant and Wheland drew up a hydrocarbon pK_a scale for ethereal solutions. For example, a solution of the orange sodium salt of diphenylmethane, $Ph_2CH^-Na^+$, on treatment with triphenylmethane, was gradually converted into diphenylmethane by proton-transfer and the solution assumed the red colour of triphenylmethide, Ph_3C^-. It could be ascertained that triphenylmethane was the more acidic by at least 2 pK_a units. By using at one extreme a compound whose pK_a could be independently measured, say in aqueous solution, this kind of technique was used to produce data for a range of pK_a values from 15 to 30. Much more refined work has been carried out recently by Streitwieser in the polar protic solvent cyclohexylamine which has very low acidity,[97] but this has served to reinforce rather than contradict earlier studies.

Dipolar aprotic solvents, and particularly dimethyl sulphoxide, have been widely used in carbanion studies, and a number of attempts have been made to obtain hydrocarbon pK_a values in this solvent. One approach takes advantage of the fact that the basicity of sodium methoxide in methanol is mitigated by hydrogen-bonding from solvent molecules to the anion, but since dimethyl sulphoxide has no tendency to form hydrogen-bonds, sodium methoxide will be much more strongly basic in this medium. Accordingly the basicity of a solution of sodium methoxide at standard concentration in mixed methanol–dimethyl sulphoxide media will increase as the proportion of dimethyl sulphoxide increases, and a basicity scale H_- may be set up along the lines described in Chapter 1, page 58, by the indicator technique. Other methods have been applied, and in particular potentiometry using a glass electrode[98] offers the opportunity to obtain pK_a data for a variety of hydrocarbons. There is an inherent limitation that species less acidic than dimethyl sulphoxide itself ($pK_a = 31.3$) cannot be studied since the equilibrium:

$$RH + {}^-CH_2SOMe \rightleftharpoons R^- + CH_3SOMe \tag{99}$$

will lie over to the left.

Direct measurement of hydrocarbon pK_a is the soundest and most reliable method for the interrelation of carbanion stabilities but is inapplicable to a large number of species of interest. With very weakly acidic compounds, either the equilibrium with solvent conjugate base (99) is unfavourable, or the carbanion may only be obtained under conditions where substantial covalent bonding to its ion-partner prevails. In these situations, our knowledge of relative carbanion stability has been obtained indirectly. We might for example study the equilibrium between two organometallic compounds where bonding is polar, but substantially covalent:

$$RMgR + R'HgR' \rightleftharpoons R'MgR' + RHgR \tag{100}$$

Dessy and coworkers[99] made the assumption that the more stable carbanion would associate with magnesium, which is the more electropositive metal, and measured the equilibria 100 in a number of competitive reactions by an n.m.r. technique. The results obtained may be 'normalized' so that carbanion stability of RH is assessed relative to toluene as a reference (Figure 18, page 124).

A more general probe into carbanion stability is afforded by the *rates* of proton removal by base. If an equilibrium such as 99 is too far over to the left for direct measurement, the rate of proton abstraction from the hydrocarbon may still be rapid. Clearly, if a substrate such as 2-phenylbutane is dissolved in a solution of base (potassium t-butoxide) in deuteriated dimethyl sulphoxide, the possibility exists for isotope exchange according to 101; the rate of proton abstraction, k_1, should be equal to the rate of deuterium incorporation at the 2-position of 2-phenylbutane.

$$(101)$$

It would be very useful if a quantitative correlation existed between rates of deprotonation and equilibrium acidities, and certain important requirements must first be met. If a free carbanion is intermediate in the exchange process and has sufficient lifetime to achieve equilibrium with its environment, then the rate of deuterium exchange of optically active 2-phenylbutane should be equal to its rate of racemization (the carbanion centre will be close to planar). This condition is met when dimethyl sulphoxide or cyclohexylamine is used as solvent, but not in less polar or more strongly proton-donating media (see Chapter 3, page 183). Furthermore, if k_1 is rate-determining, then a kinetic isotope effect should be detectable in the appropriate experiment. Thus in cyclohexylamine containing caesium cyclohexylamide and a small amount of N-tritiated solvent, α-trideuteriotoluene incorporates tritium nearly ten times more slowly than toluene, implying that the bond to hydrogen is being broken at the rate-determining transition-state.[100] In contrast, exchange reactions of this type in dimethyl sulphoxide often show a rather low isotope effect due to an internal return ($k_{-1} \approx k_2$) similar to that encountered in solvolysis reactions (102).[101]

$$(102)$$

(If $k_{-1} \gg k_2$, no isotope effect
will be observed)

To relate isotope exchange rates in cyclohexylamine with acidities, it is necessary to invoke a linear free-energy relationship of the type:

$$\log k_1 = \alpha \log K_a + C \tag{103}$$

where K_a is the equilibrium acidity, k_1 is the rate of ionization, and α and C are constants. α, termed the 'Brønsted coefficient' by analogy with the classic equation derived by Brønsted (Chapter 1, page 67) interrelating acidity and catalytic efficiency, may be considered to be a measure of the extent of proton-transfer at the deprotonation transition-state. If the structure of that transition-state is close to that of the carbanion, α should be close in value to unity. Equation 103 may be rewritten:

$$\log k_{-1} = \left(\frac{\alpha - 1}{\alpha}\right) \log k_1 + \frac{C}{\alpha} \tag{104}$$

since $K_a = k_1/k_{-1}$, where k_{-1} is the rate of carbanion reprotonation.

There is good reason for expecting that α will increase towards a limiting value of 1 as the hydrocarbon studied becomes increasingly unreactive. It

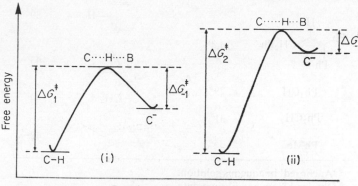

Figure 17. Free-energy profiles for (i) deprotonation to form a fairly stable carbanion, and (ii) deprotonation to form a highly reactive carbanion.

may be seen from 104 that $(1 - \alpha)$ is a measure of the sensitivity of k_1 to changes in k_{-1} (Figure 17), and therefore to be expected that as the carbanion intermediate becomes increasingly reactive the rate of reprotonation will increase to the point of diffusion-control, and no further. At that stage changes in pK_a will only reflect changes in k_1. The implication is that α will vary according to the difference in pK_a between the acid and base in a proton-transfer reaction, and this is borne out by Eigen's studies[89] on rapid proton-transfers. A relationship such as 103 can be useful in comparing the acidity of a range of hydrocarbons of similar structure and acidity,[102] but obviously generalizations should be made with caution.

Using both equilibrium and kinetic data, a compendium of hydrocarbon acidities may be obtained (Table 12). Even the most acidic compounds shown form anions which are more strongly basic than aqueous sodium hydroxide!

Table 12. Approximate values of hydrocarbon pK_a.

Compound	pK_a	Compound	pK_a
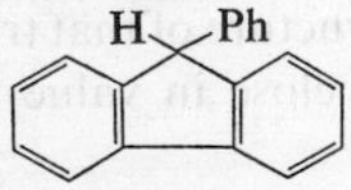	16.4[a]	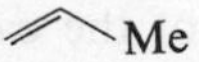	35[c]
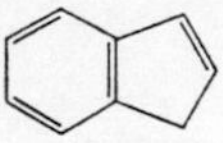	17.8[b]	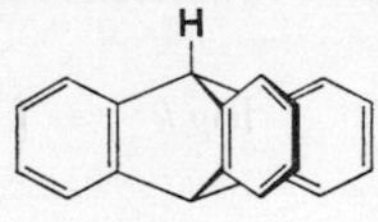	39[c]
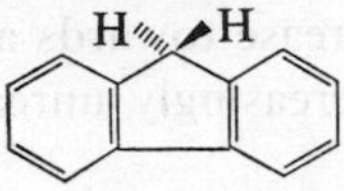	20.6[b]	$PhCHMe_2$	39[c]
		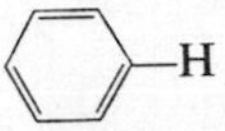	39[c]
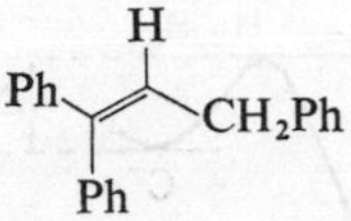	24.4[b]		43[c]
Ph_3CH	29[b]	CH_4	44[c]
Ph_2CH_2	31[b]		
PhMe	35[b]	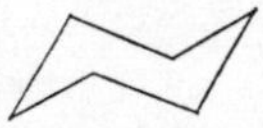	48[c]

[a] 'Anchored' in aqueous solution.
[b] Obtained from equilibrium data.
[c] Extrapolated from isotope exchange rates.

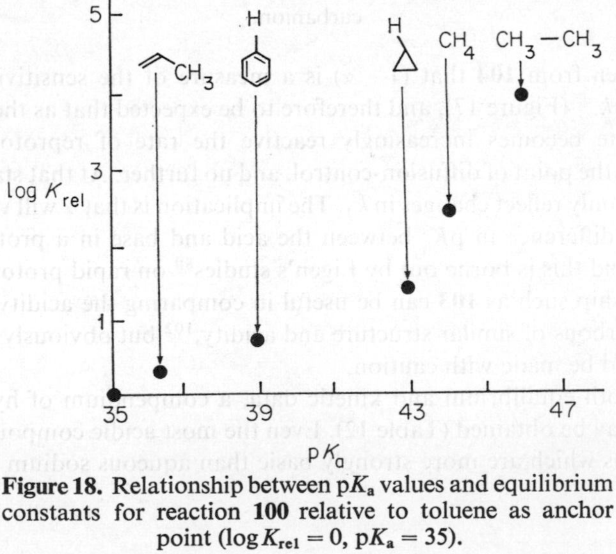

Figure 18. Relationship between pK_a values and equilibrium constants for reaction **100** relative to toluene as anchor point ($\log K_{rel} = 0$, p$K_a = 35$).

Stabilization of Anionic Carbon

Hybridization Effects

Shapes and electron-density distributions for $2s$- and $2p$-carbon orbitals and various hybrid mixtures may be calculated quite accurately. It is known that as the proportional s-character of a carbon sp^n-hybrid orbital increases, the average distance between electron and nucleus decreases, and the Coulombic term relating to stabilization by electron-cloud/nucleus electrostatic inter-action increases. Couched in different language, the electronegativity or desire of carbon to attract electrons increases with increasing s-character. In a C—H bond orbital the nucleus is effectively shielded from much of the electron-cloud by the proton, but in a carbanion the coulombic stabilization is much more apparent. Hence the acidity of a C—H bond increases with increasing s-character, a phenomenon best known in the facile conversion of acetylenes into acetylide anions, but one which has wider and more general implications.

There is a high degree of bond-angle strain inherent in the structure of cyclopropane, and this is evident in the fact that the (as yet hypothetical!) reaction **105** is exothermic by 54 kcal mole^{-1}. The nature of bonding involved

$$2 \triangleright \longrightarrow \diagup\!\!\diagdown\!\!\diagup \tag{105}$$

$$\Delta H_f = 2 \times 12.7 \qquad \Delta H_f = -29.4 \qquad \text{kcal mole}^{-1}$$

in strained rings has been of interest to theoretical chemists. Since the orbitals are not constrained to follow linear interatomic paths, it has been deduced[92] that the best bonding between carbon atoms ensues when the orbitals point away from the interatomic line, thereby compensating the opposing considera-tions of strain and overlap. This turns out to be equivalent to the effect of mixing p-character into the C—C σ-bond, so that this has a *higher percentage* p-*character* than a normal single bond. Since the total hybridization at saturated carbon must average to sp^3, this must mean that the C—H bonds in cyclopropane have enhanced s-character. Indeed, both theoretical and spectro-scopic approaches suggest hybridizations of C—C $= sp^5$ and C—H $= sp^2$. Similar considerations will apply to the bonds in any strained ring, and the more pronounced the strain, the higher the s-character in the C—H bond (Figure 19).

Hybridization is a theoretical concept which has received experimental support from determination of the lengths of carbon–carbon single bonds by X-ray and electron diffraction. On this basis Dewar has suggested[103] that the length of any such bond is given by the empirical equation:

$$r_{C-C} = 1.692 - 0.0051 \text{ (mean } \% \text{ } s\text{-character) Å} \tag{106}$$

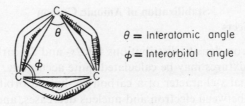

$\theta =$ Interatomic angle

$\phi =$ Interorbital angle

Figure 19. Relief of strain in cyclopropane by 'banana bond' formation.

which fits well for the central bond in cyanoacetylene, where both atoms are sp (1.387 Å), in butadiene where both atoms are sp^2 (1.476 Å), and in butane where both atoms are sp^3 (1.533 Å). It is particularly interesting that the centrosymmetric bond in 1,1'-binorbornyl (**107**)[104] is shorter than an average C—C bond by 0.018 Å. The norbornane ring structure can only be made with considerable angle-strain, and this will be reflected in increased s-character of all the exocyclic bonds.

(107)

For our purposes a more important correlation exists. We can see that an increase in s-character changes the interaction between carbon and hydrogen in a C—H bond, and s-character also plays a major role in determining the n.m.r. spin–spin coupling constant between ^1H and ^{13}C.* Using acetylene, ethylene, and ethane as anchor points, there is an experimental relationship between percentage s-character and $J_{^{13}C-H}$:

$$J_{^{13}C-H} = 500 - 5(100 - \% \, s_{C-H}) \tag{108}$$

In consequence the s-character of any carbon–hydrogen bond may be assessed if the ^{13}C satellite splitting of its proton magnetic resonance signal can be determined. Streitwieser and coworkers have measured the rates of hydrogen isotope exchange of cycloalkanes in cyclohexylamine containing caesium cyclohexylamide[105] and found an excellent correlation with $J_{^{13}C-H}$.

* ^{12}C has a spin quantum number of 0, and therefore causes no n.m.r. effect. ^{13}C is present at 1% at natural abundance, and its presence gives rise to 'satellite' signals of a C—H signal by spin–spin splitting.

Highly strained hydrocarbons are sufficiently acidic to be deprotonated by methyl-lithium in ether,[106] and reaction rates accord with $J_{13_{C-H}}$ values (Figure 20).

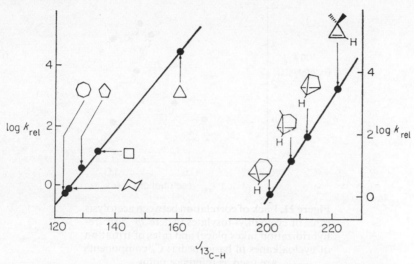

Figure 20. Correlation of $J_{13_{C-H}}$ with rates of detritiation of tritiated cyclo-alkanes with caesium cyclohexylamide in cyclohexylamine (left-hand plot) and with rates of deprotonation by methyl-lithium in ether (right-hand plot).

Conjugation Effects; $p\pi$-Conjugation

Our previous discussion indicated that the lone-pair orbital of a localized carbanion will be most stable when it has the highest s-character. We might therefore expect that an alkyl carbanion would have strong preference for sp^3-hybridization over sp^2 in which the lone-pair must occupy a p-orbital. Indeed the former configuration is predicted for methyl anion by theoretical calculations of the CNDO type, with a barrier to pyramidal inversion of around 15 kcal mole^{-1}. Theory is supported by experimental evidence, since it is well known that in reactions where a centre rehybridizes from sp^2 to sp^3 (borohydride reduction of ketones) medium-ring compounds react very slowly, and conversely in a reaction where the reacting centre rehybridizes from sp^3 to sp^2 (S_N1 solvolysis of tosylates) medium-ring compounds react more rapidly than acyclic analogues (see Figure 21). This general effect has its origin in the excessive steric crowding and non-bonded interactions existing in medium rings, which are somewhat relieved by the presence or production of sp^2 centres. Streitwieser has shown that medium-ring cycloalkanes do not under-go isotope-exchange with caesium cyclohexylamide any faster than with cyclohexane, and this indicates little rehybridization towards sp^2 has taken place at the transition-state for deprotonation, whose sp^3 geometry might be expected to be rather similar to that of the carbanion itself.[107]

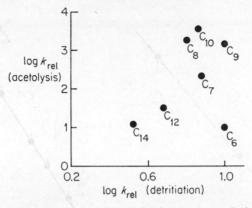

Figure 21. Lack of correlation between acetolysis rates of cycloalkyl tosylates (where $sp^3 \rightarrow sp^2$ rehybridization takes place) and rates of tritiation of cycloalkanes in basic media; C_6 components are used as reference point.

If the reader now considers the case of benzyl anion, two opposing effects are apparent. The tendency to maximize s-character will be opposed by the increased delocalization of negative charge possible if the carbanion is associated with a p-orbital. On an *a priori* basis there is no means of assessing which effect is dominant, although it might appear that $p\pi$-delocalization will be a strongly stabilizing factor. The experimental properties of benzyl anion are in accord with a highly delocalized species (**109**) and in particular the ^{13}C n.m.r. chemical shift and $J_{^{13}C-H}$ (138 Hz for the methylene group, compared to 91 Hz for MeLi)[108] suggest preference for coplanarity.

There is a considerable degree of stabilization attending delocalization in an sp^2 carbanion of this type. Inspection of Table 12 shows that the stepwise replacement of hydrogens in methane decreases the acidity by 9, 13, and

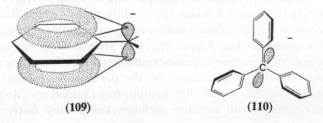

(**109**)   (**110**)

15 pK_a units, which implies that triphenylmethyl anion is 21 kcal mole^{-1} ($RT\log_e 10^{15}$) more stable than methyl anion at 25°C. There is a suggestion in the data that less energy is gained in each successive replacement of hydrogen by phenyl; for most favourable interaction between the phenyl group and carbanion the system should be coplanar, and this condition becomes increasingly difficult to maintain for steric reasons as the number of phenyl groups increases. In fact, triphenylmethyl anion (and cation) are thought to be propellor-shaped (110), a structure which effects the best compromise between conjugative and steric interactions.

Where a carbanion is adjacent to a $p\pi$-bond orbital whose remote atom is oxygen or nitrogen, a stable system may result. In species such as enolate anions, α-nitro-carbanions, and α-cyano-carbanions, it is to be expected that most of the negative charge will reside on oxygen or nitrogen owing to their

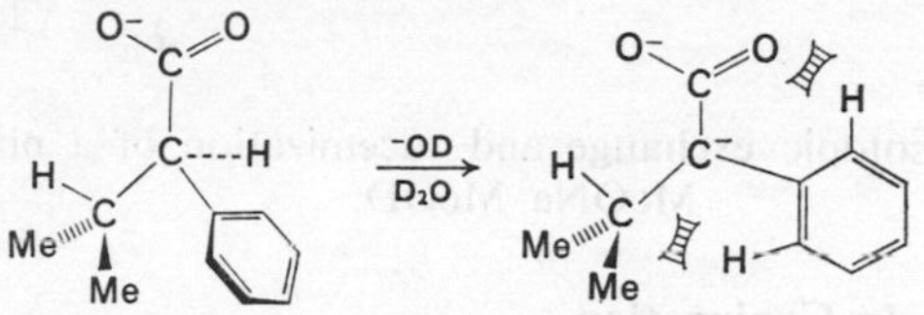

Relative exchange rate as a function of side-chain R:

$Pr^i = 1$; $Et = 18$; $Me = 155$; $H = 42,000$

Figure 22. Steric inhibition of resonance in planar carbanion formation.

greater electronegativity. The pK_a values obtained in this type of system are generally much lower than observed for hydrocarbons, for example:[109] CH_3NO_2 10.2; CH_3COCH_3 20; $CH_3SO_2CH_3$ 23; CH_3CN 25; CH_3CO_2Et 25; $CH_3CH{=}CH_2$ 35. The importance of conjugation in stabilizing anions of this type is nicely illustrated[110] by considering the rates of exchange of α-substituted sodium phenylacetates (Figure 22) in alkaline D_2O. In a planar sp^2 carbanion, there will be considerable steric interactions between the alkyl-group and ring hydrogens, and it is clear that these slow the reaction rate to an extent dependent on the steric bulk of the alkyl group.

A base-catalysed isotope exchange α to a nitro, keto, ester, or cyano group will normally result in loss of chirality at that site since the intermediate will be sp^2-hybridized at the carbanion centre. One interesting exception occurs when the acidic hydrogen is bonded to a cyclopropane[111] or cyclopropene[112] ring. In this situation a non-planar anion may well be preferred because of the extra stability inherent in a cyclopropyl anion (page 126) and also because of the extra angle-strain necessary to attain sp^2-hybridization in a three-membered ring. The nitrile **111** (Figure 23) undergoes isotopic exchange more than 4000 times faster than its racemization in MeONa–MeOD. In keeping with this result a related cyclopropyl-lithium was shown to be configurationally stable[113] under conditions where optically active alkyl-lithium compounds rapidly racemized.

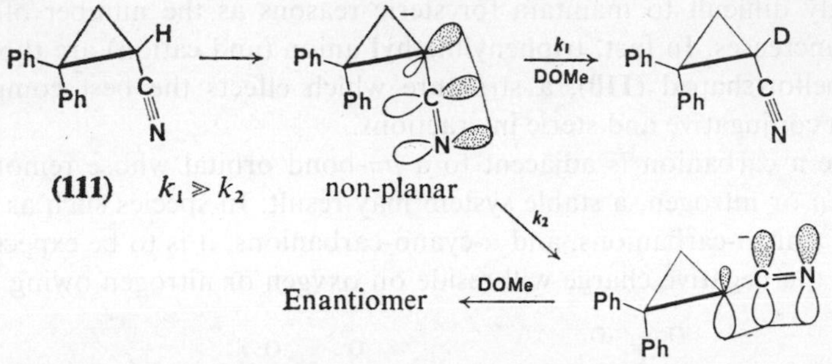

Figure 23. Isotopic exchange and racemization of a nitrile in MeONa–MeOD.

Conjugation Effects; $d\pi$-Conjugation

The ability of sulphur and phosphorus to achieve high valence states by using $3d$-orbitals is well recognized.[114] It is therefore possible that in the carbanion $MeSCH_2^-$ the negative charge might delocalize by participation of a vacant d_{xy}-orbital on sulphur, such that the S—C bond will shorten and the anionic carbon atom show a preference for sp^2 over sp^3 hybridization in order to achieve maximum overlap (Figure 24).

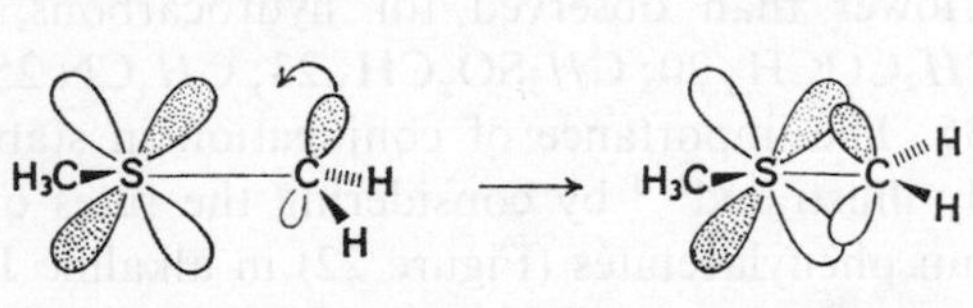

Figure 24. Carbanion stabilization by $d\pi$–$p\pi$ overlap.

The existence and importance of this kind of bonding in carbanions adjacent to second-row elements has been the subject of much vigorous debate, opponents of the idea suggesting that $3d$-orbitals are too diffuse to overlap well with $2p$-orbitals. In addition to classical chemical approaches to this kind of question, fundamental quantum mechanical calculations have been done in the hope that definitive evidence might be obtained. Two specific areas will serve to illustrate the general problem.

Ylids

Very strong bases are required to effect deprotonation of tetramethylammonium salts $Me_4N^+X^-$, and the colourless product Me_3N^+—CH_2^- has limited existence, being eventually consumed by fragmentation or rearrangement. In contrast, phosphonium salts such as $Me_4P^+X^-$ are deprotonated by moderately strong base, and the yellow ylid Me_3P=CH_2 (b.p. 210°) is indefinitely stable in the absence of air! The existence of stable phosphorus and sulphur ylids forms the basis of the Wittig and related reactions (Chapter 5). A crystal structure analysis[115] on Ph_3P^+—CH_2^- shows a P—CH_2 bond-length of 1.66 Å, i.e. 0.18 Å shorter than a normal P—C single bond and quite comparable to the C—C contraction between ethane and ethylene. It was not possible to say from this study whether the methylene group was sp^2-hybridized.

No pK_a data are available for simple P, S, and N ylids; however, in terms of our pK_a scales we might guess that Ph_3N^+—CH_3 has a pK_a around 35—40, whereas Ph_3P^+—CH_3 has a pK_a in the range 22—26. Exchange experiments carried out on ylids fully support a difference of this magnitude.

α-Sulphoxide and α-Sulphone Anions

$3d$-Orbitals are well screened from the nucleus by inner-shell electrons, thus accounting for their diffuseness. Any factor which reduces the screening will cause contraction, and may help to bring the wave function into the appropriate region to accept electrons effectively from an α-carbanion lone-pair. Putting a positive charge on the second-row atom (cf. ylids) is one way of reducing this screening; another is to have a second-row atom carrying electron-withdrawing groups, as in a sulphone.

Experimental evidence gained from study of base-catalysed isotopic exchange in optically active sulphones has provided a basis for a great deal of discussion. In solvents such as dimethyl sulphoxide, base-catalysed isotopic exchange at an optically active site usually leads to racemization but exchange α to a sulphone occurs with 98 % retention of optical activity (Figure 25).

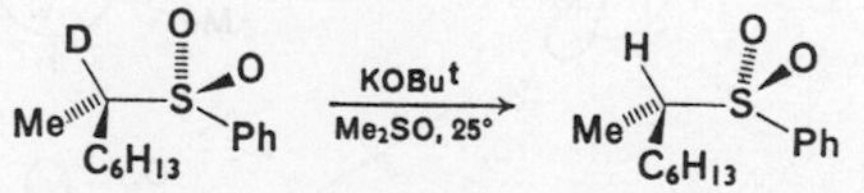

Figure 25. Retention of configuration in α-sulphone exchange.

This may be explained in one of two ways. If the carbanion centre is pyramidal, and rotation around the C—S bond relatively free, optical activity may arise from a barrier to pyramidal *inversion*. Alternatively, the carbanion might have a fixed conformation with a barrier to *rotation*. Evidence against the former possibility is derived from an experiment of Corey which defines more clearly the stereochemical course of protonation[116] (Figure 26). It is

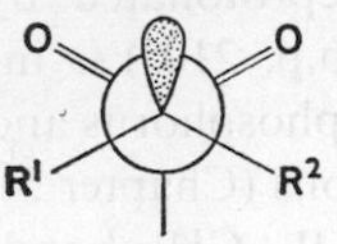

(112)

N.B.   *Any* pyramidal α-sulphone anion with a barrier to inversion is chiral (see **112a**)

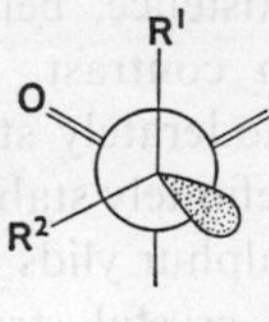

(112a)

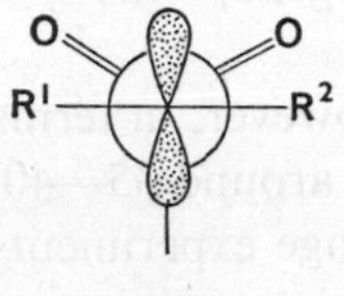

(113)

A chiral planar α-sulphone anion may racemize via **113a**

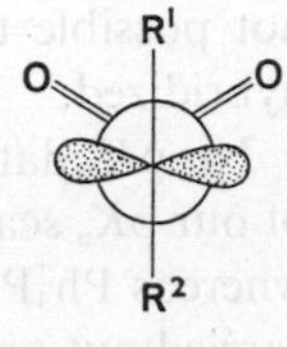

(113a)

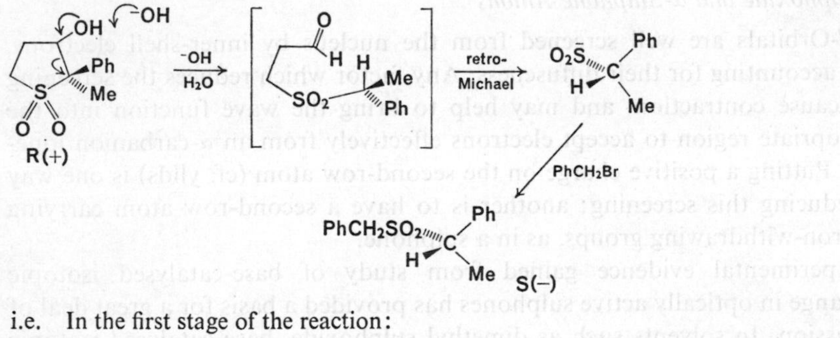

i.e.   In the first stage of the reaction:

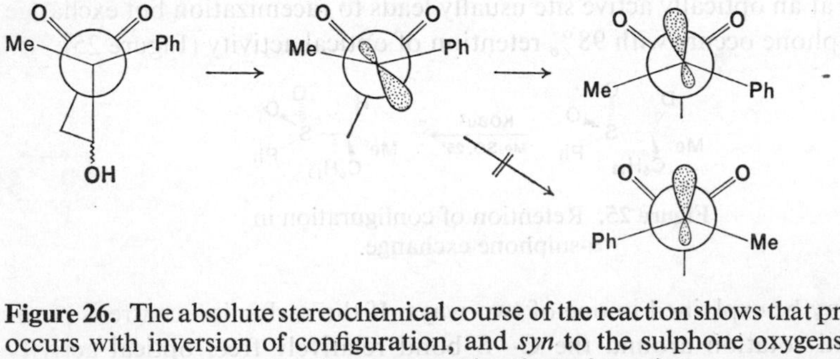

Figure 26. The absolute stereochemical course of the reaction shows that protonation occurs with inversion of configuration, and *syn* to the sulphone oxygens. There is therefore no great barrier to inversion.

important to note, however, that an experiment of this kind does not distinguish between sp^2- and sp^3-hybridization at the carbanion centre, since the result could be explained by an intermediate of either structure **112** or **113**.

A contribution to our understanding of anions of this type has been made through non-empirical self-consistent-field calculations. The energy of the (hypothetical) anion $HSO_2CH_2^-$ was calculated as a function of the angle of rotation and hybridization at carbon. The result of these calculations[117] (Figure 27) was to indicate a strong preference for a conformation with the carbanion lone-pair *syn* to sulphone oxygens and a best HCH angle of 115°,

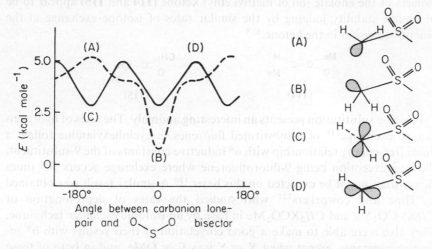

Figure 27. Calculated energy of an α-sulphone anion as a function of conformation and bond-angle; — = planar, --- = pyramidal.

suggestive of hybridization intermediate between sp^3 and sp^2. More surprisingly, the results did not require *d*-orbital participation and the stabilization of α-sulphone carbanions could be explained by field effects. Whether this will be the last word on the question remains to be seen, of course! Exchange α to a sulphoxide is also stereoselective in appropriate systems.

Inductive Effects

Almost all substituents which stabilize carbanions by conjugation also offer significant stabilization by field and inductive effects. It seems, however, that alkyl, fluoroalkyl, and fluoro substituents interact with carbanions by field and inductive mechanisms alone. Thus alkyl substitution at an sp^3-carbanion causes very marked destabilization, and although there is not a great deal of useful quantitative data, it is instructive to compare the stabilities of methyl and t-butyl-lithium. The former is stable for months in ether solution, whereas the latter can be obtained only in hydrocarbon solvents, inflames in air, and even reacts with nitrogen. Conversely trifluoromethyl substituents are strongly carbanion-stabilizing and $(CF_3)_3CH$ has an estimated pK_a value of 11. The situation may be rather more complex with sp^2-hybridized carbanions—here a methyl substituent may operate a destabilizing inductive effect and also offer stabilization, since alkyl substituents are known to stabilize sp^2-carbon. These effects may roughly counterbalance, and the two possible structural isomers of the enolate ion of methyl ethyl ketone (**114** and **115**) appear to be of similar stability judging by the similar rates of isotope exchange at the two possible sites in the ketone.[118]

Me—C=C—H
⁻O   Me

(114)   (115)

Fluorine substitution presents an interesting anomaly. The rates of hydrogen isotope exchange[119] of 9-substituted fluorenes in cyclohexylamine follow a linear free-energy relationship with σ^* inductive constants of the 9-substituent, the lone exception being 9-fluorofluorene where exchange occurs 10^4 times slower than might be expected on this basis.[120] A similar result was obtained by Hine and coworkers[121] who studied the rates of deprotonation of $CHXYCO_2Me$ and CH_2XCO_2Me in base by an isotopic exchange technique. They also were able to make a good correlation of their results with σ^* inductive constants, except when X or Y was F or OMe, and in both of these cases exchange occurred very much more slowly than expected (Figure 28).

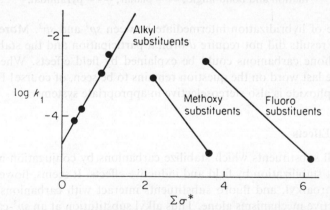

Figure 28. Rates of isotope exchange of $CHXYCO_2Me$ plotted against σ^* for different X,Y substituents.

There are probably two factors which contribute to the apparent destabilization of fluoro-substituted sp^2-carbanions. The increasing electronegativity of carbon on going from sp^3- to sp^2-hybridization might be responsible for weakening of the C—F bond, since carbon then competes better with the highly electron-withdrawing fluorine for the bonding electrons, to the detriment of the bond-strength. Additionally, there may be severe repulsive interactions between $p\pi$-electrons at the carbanion centre and lone-pair electrons on fluorine. The effects of fluorine attached to a carbon possessing filled *p*-orbitals can be quite striking. For example, t-butylacetylene is indefinitely stable at room temperature, but attempts to prepare t-butylfluoroacetylene lead to products arising from its spontaneous trimerization (**116**).[122]

$$(116)$$

Reactions Involving Carbanions

Isomerizations and Rearrangements

We have seen how the production of an intermediate carbanion may result in hydrogen isotope exchange or racemization under appropriate conditions. If that carbanion is delocalized, with charge density on several sites, then reprotonation may result in structural isomerization. For this latter reaction, application of Hammond's postulate suggests that the transition-state will be much closer in structure to the reactive carbanion than to the protonation product, and that therefore the site of protonation will be governed by charge-density distribution in the anion rather than by the relative thermodynamic stabilities of possible products. In the base-catalysed interconversion of cyclohexa-1,3-diene and cyclohexa-1,4-diene (**117**), a detailed kinetic analysis[123] shows that the intermediate anion protonates much more readily at the central position to give the less stable isomer. This is in accord with the charge-density distribution shown, which was calculated by a Hückel self-consistent-field method.[124]

$$(117)$$

The isomerization of olefins in base can be a useful synthetic procedure, and one remarkable example is the Varrentrapp reaction[125] (118) which has been known since 1840. Despite the brutal conditions used in this transformation, it can be remarkably specific, and in the case cited the product of at least nine successive isomerizations and a subsequent base-catalysed cleavage is produced in 80% yield.

$$\text{(118)}$$

The stereochemistry of base-catalysed olefin isomerizations is of interest, since allyl ethers are isomerized to *cis*-vinyl ethers in strongly basic media[126a] and terminal olefins react, for example, with potassium t-butoxide in dimethyl sulphoxide to produce internal olefins of predominantly *cis*-configuration (119).[126b] The *cis*-isomer is of course the thermodynamically less stable product, but one must remember that the rate-determining stage involves deprotonation

$$\text{(119)}$$

preferred pathway

by base to form a reactive anion, and it is the relative stabilities of the isomers of this anion which determine the overall stereochemical course of reaction. It is thought that the *cis*-allylic anion is more stable than the *trans*-isomer, which may be a general phenomenon. This contrasts with a preference for *trans*-allylic cations (Chapter 1, page 27), and may originate in a minimization of dipole–dipole repulsion between the allyl system and the electron-releasing alkyl group. The base-catalysed isomerization of acetylenes may produce a complex mixture of products.[127a] An interesting application is the synthesis

of an annulene by treatment of a polyacetylene precursor with potassium
t-butoxide in t-butyl alcohol (120).[127b]

(120)

(2% overall)

In carbonium-ion chemistry, the most striking single feature is the pro-
pensity for deep-seated skeletal rearrangements by alkyl migration to the
electron-deficient charged centre. There is virtually no analogy for this behaviour in carbanion chemistry, although rearrangements may occur in
certain specific systems. Probably the most important of these is the rearrangement discovered by Wittig and bearing his name, in which an organolithium
compound bearing an α-ether substituent OR is converted into a lithium
alkoxide by migration of R (121).[128] A number of possible mechanisms might

(121)

be envisaged for this transformation but in the example shown an intramolecular migration is indicated by the retention of configuration in the alkyl
group; the transition-state can be imagined as a tight π-complex between a
ketone radical anion and an alkyl radical. Evidence linking migration aptitudes
and radical stabilities supports this.[129] In the rearrangement of 2,2,2-
triphenylethyl-lithium to 1,1,2-triphenylethyl-lithium in tetrahydrofuran at
0° (122) an intramolecular mechanism related to the Wittig rearrangement
above is suggested by the failure of an attempted trapping experiment where
the reaction was run in the presence of ^{14}C-labelled phenyl-lithium, and the
rearrangement product shown to be quite lacking in radioactivity. The
specific rearrangement of 2,2,3-triphenylpropyl-lithium to 1,1,3-triphenyl-
propyl-lithium (123) appears to be a very similar reaction but nevertheless
involves quite a different pathway, since here a dissociative mechanism was
demonstrated by partial equilibration of the benzyl group with external
radioactive benzyl-lithium.[130]

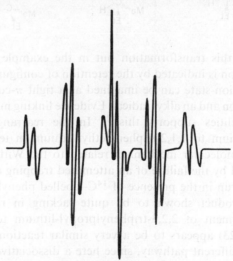

Reductions by Electron Transfer

If a solution of anisole in tetrahydrofuran is placed in contact with a potassium metal film at −80° the solution gradually becomes coloured and acquires an electron spin resonance signal. The hyperfine splitting pattern of this signal (Figure 29) shows that the anion-radical of anisole (124), formed by transfer of an electron from potassium into an antibonding orbital of the arene, has been formed.[131]

A solution of potassium in liquid ammonia containing a little ethanol is rapidly decolorized on addition of anisole, and 1,4-dihydroanisole is produced

Figure 29. E.s.r. spectrum of **124** in tetra-hydrofuran at −80°.

in high yield. The reaction is quite general for aromatic compounds (Birch reduction)[132] and is considered to require the initial formation of **124** which is then protonated by ethanol. Further electron-transfer to the radical so produced gives a cyclohexadienide anion, which we have already seen to undergo preferred protonation at the central position, thereby accounting for the observed product.

(124)   **(125)**

7%   93%

Dissolving-metal reductions of this kind, particularly in liquid ammonia, afford a general method for the reduction of conjugated systems where low-lying antibonding orbitals are available to accept an electron. In some circumstances electron-addition may result in expulsion of a stable anion, as in the conversion of benzyl ethers into methylbenzenes. Here the intermediate benzyl anion could be found in one of two ways (**126**) depending on whether one or two electrons are added in the rate-determining stage. Ether cleavage is one of the classic methods for the preparation of carbanions in solution; and benzylpotassium may be readily prepared in tetrahydrofuran solution by the interaction between benzyl methyl ether and sodium–potassium alloy.

(126)

Reactions of Ambident Anions[133a]

In a species like an α-nitro-carbanion or enolate anion, it is to be expected that most of the charge will reside on oxygen, and in many reactions with electrophiles this is the exclusive site of attack. There remains, however, the possibility of reaction at carbon even though this atom carries a minor part of the negative charge-density. The reader will recall that stabilized carbanions such as the sodium salts of ethyl acetoacetate and acetylacetone react with alkyl and acyl halides to form carbon–carbon rather than carbon–oxygen bonds although proton-transfer from acids to these sodium salts occurs at oxygen. The rationalizations which may be advanced to account for this dichotomy of behaviour have their origins in differential solvation effects. In a protic medium such as methanol, solvation of the enolate ion of acetylacetone will largely occur by hydrogen-bonding from the solvent to

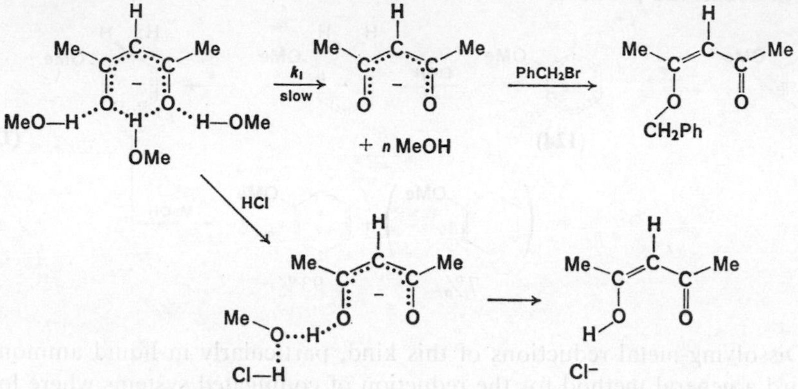

Figure 30. Protonation compared with alkylation at enolate oxygen.

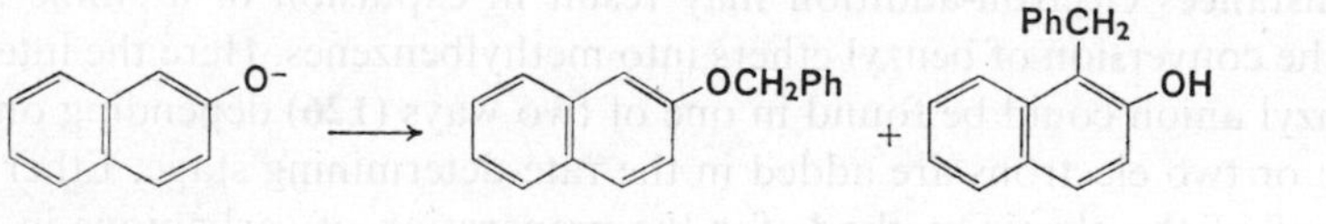

Solvent	%	%
HCONMe$_2$	97	0
MeOCH$_2$CH$_2$OMe	70	22
MeOH	57	34
H$_2$O	10	84
CF$_3$CH$_2$OH	7	85

Figure 31. Effect of solvent on the course of phenoxide alkylation.

anionic oxygen, and considerable solvent reorganization must proceed before oxygen alkylation can ensue (Figure 30). The charge at carbon will remain relatively unsolvated, and may react with alkyl halide without reorganization. In contrast, protonation at oxygen may occur via a relay mechanism in which the acid protonates a solvating molecule whilst this transfers a proton to the oxygen atom of the substrate.

It is interesting to compare the effects of various solvents on competitive carbon and oxygen alkylation. In dipolar aprotic solvents, the consideration of hydrogen-bonding solvation effects is no longer relevant, and here oxygen alkylation may prevail. Kornblum and coworkers[133b] have studied various aspects of the alkylation of phenoxide ions (formally enolates) and shown that product proportions vary with solvent, according to its hydrogen-bonding ability rather than any other property such as dielectric constant (Figure 31). In the case of β-diketones and β-keto-esters, the proportion of oxygen alkylation of the conjugate anion is enhanced in dipolar aprotic solvents, but carbon alkylation still generally predominates.

Stable Carbanions in Solution

In order to interpret the nature of salt effects in S_N1 solvolyses, Winstein and Robinson found it necessary to postulate the existence of two kinds of ion-pair, one where the electrostatic interactions between ion-partners were unaffected by solvent, and the other in which each partner was surrounded by its own sheath of solvent (page 88). It might be expected that similar situations would exist in carbanion chemistry, but good evidence for this has been obtained only recently.[134] Smid and coworkers measured the conductance of solutions of sodium fluorenylide in tetrahydrofuran, showing that only ion-pairs, and not free ions, were present. However, the ultraviolet spectra of such solutions showed remarkable changes between 25° and −50°, which

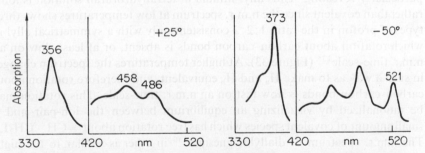

Figure 32. Absorption spectrum of fluorenylsodium in tetrahydrofuran as a function of temperature. The band at 356 nm is specific to the contact ion-pair; those at 373 and 521 nm are specific to solvent-separated ion-pairs.

were independent of concentration and unchanged by the addition of common-ion salts. The authors concluded that the only reasonable interpretation of these observations was that equilibrium between contact and solvent-separated ion-pairs was being established. Solvent-separated ion-pairs, which absorb at longer wavelengths than contact ion-pairs, are favoured by low temperature, small electropositive cations ($Li > Na > Cs$), and by addition of cation-binding solvents. Even in pure dimethyl sulphoxide, where contact ion-pairs are absent, there is no appreciable dissociation of solvent-separated ion-pairs into free ions since the solutions are not strongly conducting. One complicating factor in the interpretation of these results was that the absorption maximum for the contact ion-pair depended on the nature of the metal cation, increasing in the series $Li^+ < Na^+ < Cs^+$.

In polar solvents, therefore, Group IA derivatives of relatively stable carbanions might be expected to exist as ion-paired species. With less stable carbanions, the structure may be relatively complex. Lithium alkyls in solution often show apparent molecular weights which increase with concentration, and a combination of physicochemical measurements in solution and X-ray structure determination in the solid state suggests that the structure of these aggregated species varies with the nature of the alkyl group. Less is known about the importance of aggregation in the *reactions* of alkyl-lithium reagents, but it is probable that a major part of the activation energy in the reaction between, say, methyl-lithium and acetone in ether may be dissociation of the tetrameric cage structure of methyl-lithium.

The factors which determine whether a given polar organometallic compound exists as an aggregate, a covalent monomer, or as ion-pairs in solution may be fairly finely balanced, and n.m.r. has been most useful in this area of structure determination. Study of substituted allyl-metal derivatives has been particularly revealing. Thus allyl-lithium in tetrahydrofuran solution is ionic rather than covalent since the n.m.r. spectrum at low temperatures shows three types of proton in the ratio $1:2:2$ consistent only with a symmetrical allyl in which rotation about carbon–carbon bonds is absent, or at least slow on an n.m.r. time-scale[135] (Figure 33). At higher temperatures the spectrum changes in such a way as to make H_a and H_b equivalent, and therefore rotation about carbon–carbon bonds is now fast on an n.m.r. time-scale. This rotation may be rationalized by visualizing an equilibrium between the ion-pair and a small amount of covalent species which has free rotation about $=CH-CH_2Li$. The n.m.r. spectrum of diallylmagnesium[136] in ether is similar to the high-temperature spectrum of allyl-lithium but is temperature-independent. Since diallylmagnesium is thought to be covalent rather than ionic, it might be wondered why the terminal methylene groups have equivalenced chemical shifts. This could be due to a rapid intramolecular shift of magnesium, or to equilibrium between the covalent species and a small amount of ion-pair (the latter has C_{2v} local symmetry).

In contrast to the delocalized ionic structure of allyl-lithium, alkyl-lithium compounds appear to be largely covalent in non-polar solvents. At low

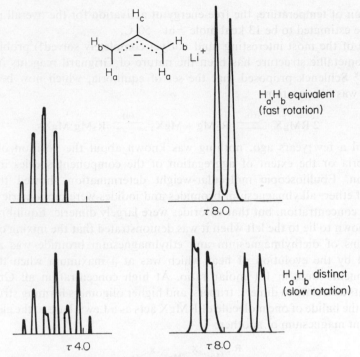

H$_a$H$_b$ equivalent
(fast rotation)

τ 8.0

H$_a$H$_b$ distinct
(slow rotation)

τ 4.0   τ 8.0

Figure 33. Schematic representation of the n.m.r. spectrum of
allyl-lithium.

temperatures the methylene hydrogens of CH$_2$Li in 2-methylbutyl-lithium are
non-equivalent; being next to a centre of chirality they are diastereotopic.[137]
H$_a$ and H$_b$ can, however, be made equivalent by a dissociation–inversion
sequence (Figure 34). By following the rate of this reaction by n.m.r. as a

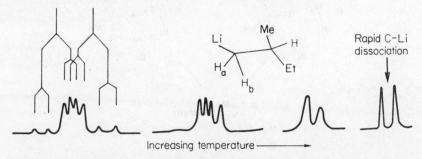

Rapid C–Li
dissociation

Increasing temperature ⟶

Figure 34. Schematic representation of the H$_a$,H$_b$ portion of the n.m.r.
spectrum of 2-methylbutyl-lithium as a function of temperature.

function of temperature, the free-energy of activation for the overall process may be estimated to be 13 kcal mole^{-1} at $-50°C$.

One of the most interesting (and not yet completely solved!) problems in organometallic structure has been the nature of Grignard reagents in solution.[138] Schlenck proposed that the set of equilibria, which now bears his name, was in operation:

$$2\,RMgX \underset{}{\overset{(i)}{\rightleftharpoons}} R_2Mg + MgX_2 \underset{}{\overset{(ii)}{\rightleftharpoons}} R_2Mg{:}MgX_2 \qquad (127)$$

Until a few years ago, nothing was known about the position of these equilibria or the extent of aggregation of the component species in ether solution. Ebullioscopic molecular-weight determination showed that, in diethyl ether, alkylmagnesium bromides and iodides were monomeric at low (0.1M) concentration, but that chlorides were largely dimeric. Equilibrium (i) was shown to lie to the left when it was demonstrated that the mixing of ether solutions of diethylmagnesium and ethylmagnesium bromide was accompanied by the evolution of heat which was at a maximum when the two components were in 1:1 molar ratio. At high concentration all Grignard reagents associate to dimers, trimers, and higher oligomers forming structures where the halide of one molecule of RMgX acts as a Lewis base to the electron-deficient magnesium of another:

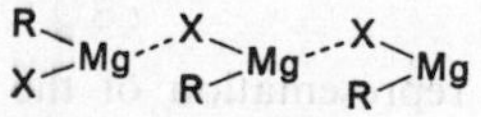

$$\qquad (128)$$

Detailed knowledge of various equilibria in Grignard reagents tells us nothing about the actual *reactant* in their reactions, and this subtle question

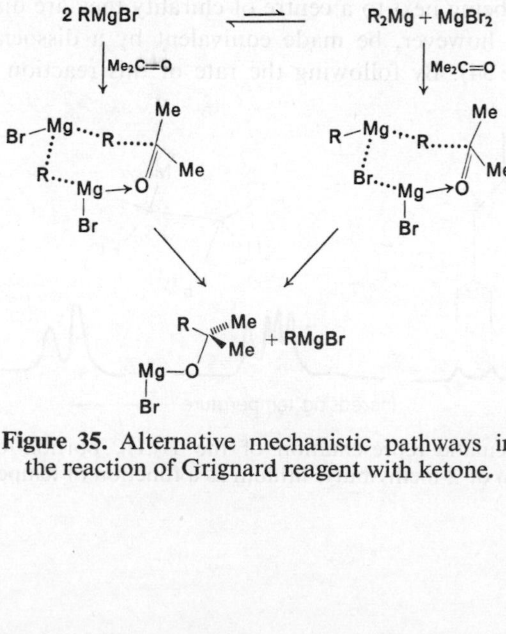

Figure 35. Alternative mechanistic pathways in the reaction of Grignard reagent with ketone.

is not entirely resolved. What appears to be well established is that two molecules of Grignard reagent appear in the rate-determining stage of the reaction with ketone and that one of these is required for complexation. A major problem is that there are two kinetically equivalent pathways (Figure 35). Dialkylmagnesium compounds are known to be much more reactive towards ketones than are alkylmagnesium bromides so that the former might be the reactive intermediate even though its concentration is small.

CARBENES[139-145]

Historical aspects of the chemistry of bivalent carbon present an amusing example of a concept which was totally erroneous but nevertheless made a contribution to the development of the science. In the latter part of the nineteenth century, many authors such as Nef and Geuther viewed the structure of chloroform as CCl_2,HCl, similar to an amine hydrochloride, and attempted to interpret its chemistry on this basis; not surprisingly, the idea did not become generally adopted. In 1950 and later Hine and coworkers[146] studied the hydrolysis of polyhalogenomethanes in strongly basic media very carefully. Many interesting results were uncovered, the most relevant being the following.

(a) Chloroform is hydrolysed by base much more rapidly than is either methylene chloride or chloroform; similarly, every bromine-containing haloform is more than 600 times as reactive towards hydroxide ion as CH_2BrCl or CH_2Br_2.

(b) The hydrolysis of chloroform is slowed down by Cl^-, and even more by Br^- or I^-; therefore halide-ion is being produced in a reversible rate-determining stage.

(c) Aqueous sodium thiophenoxide does not react with chloroform at an appreciable rate in aqueous media; if sodium hydroxide is added, rapid reaction occurs and the product is $HC(SPh)_3$.

(d) A kinetic isotope effect $k_H/k_D = 1.74$ is observed in the hydrolysis of CHBrClF.

These and other observations are uniquely accommodated by an α-elimination mechanism (129) in which a dihalogenomethylene is formed as a reactive intermediate.

$$H-C\overset{X}{\underset{Y}{\overset{\cdots}{Z}}} \underset{k_{-1}}{\overset{k_1}{\rightleftharpoons}} \quad C\overset{X}{\underset{Y}{\overset{\cdots}{Z}}} \underset{k_{-2}}{\overset{k_2}{\rightleftharpoons}} \quad \overset{O}{\underset{X\quad Y}{C}} \overset{-OH}{\longrightarrow} \quad \begin{matrix} CO \\ + \\ HCO_2H \end{matrix} \tag{129}$$

In their classic series of papers, Hine's group was able to measure the overall rate of hydrolysis, and also the rate of carbanion formation (k_1) by following the rate of deuteriation (since $k_{-1} > k_2$). Carbanion formation was often thousands of times faster than the overall hydrolysis rate, and the order of halogen effectiveness in enhancing the former was $I \sim Br > Cl > F$. In their relative effect on the overall hydrolysis rate, however, an order of $F > Cl > Br > I$ operated. The ability of fluorine to stabilize bivalent carbon (page 147) is such that for $CHClF_2$, $CHBrF_2$, and $CHIF_2$ k_2 is faster than k_{-1}, and the reaction effectively a concerted α-elimination of proton and fluoride ion.

Structure of Bivalent Carbon Intermediates

Methylene, CH_2, is the simplest organic molecule and has been subjected to numerous quantum-mechanical studies;[147] there is some agreement between theory and experiment on the essential structural details. In any structure of CH_2, there are two non-bonding electrons on carbon which may or may not be spin-paired. Two extreme situations are represented by **130a** and **130b**.

sp^2 singlet

(130a)

sp triplet

(130b)

In the linear triplet ($^3\Sigma_g$ is the spectroscopic state description), C—H bonds are strong, being sp-hybridized, but the unpaired electrons are in relatively diffuse $2p$-orbitals with weak coulombic stabilization. In the bent singlet state (spectroscopically termed $^1A_1$), the lone-pair and C—H bonds are sp^2-hybridized, improving coulombic interactions for the former at the expense of weakening C—H binding. Both singlet and triplet methylene have been detected in gas-phase photolysis of diazomethane, and from analysis of the rotation and vibrational fine-structure of the electronic adsorption spectrum molecular dimensions were obtained: $^3\Sigma_g$, $r_{C-H} = 1.07$ Å, $HCH = 140°$; $^1A_1$, $r_{C-H} = 1.12$ Å, $HCH = 102°$. The general consensus is that the triplet is considerably more stable than the singlet state, probably by as much as 20 kcal mole^{-1}. It is worth noting that the experimental bond-angles suggest that there is some deviation from the 'ideal' hybridization in both cases.

It has proved feasible to generate many substituted methylenes by flash-photolysis of a reactive precursor in a solid matrix at very low temperatures. In such experiments it is possible to distinguish singlet and triplet carbenes by electron spin resonance, since only the latter state has unpaired spin. In a typical and highly informative experiment 1-naphthyldiazomethane was photolysed in a benzophenone matrix at 77°K.[148] The e.s.r. spectra of *two* triplet species were detected and used in evidence of a 'bent' triplet state, which would give rise to geometrical isomers.

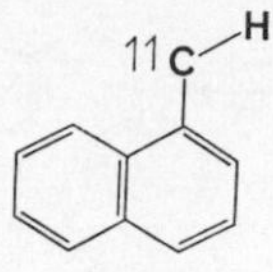

(131a) 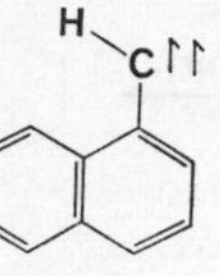   (131b)

It is fruitful to consider triplet carbenes as diradicals, and singlet carbenes as internally compensated carbonium-ion–carbanions with an electron-deficient centre carrying an electron-pair. Consequently any substituent which might donate electrons to an unoccupied orbital will militate in favour of a singlet ground-state. This may be achieved, for example, by halogen substitution where back-donation will be important (Figure 36). Stabilization

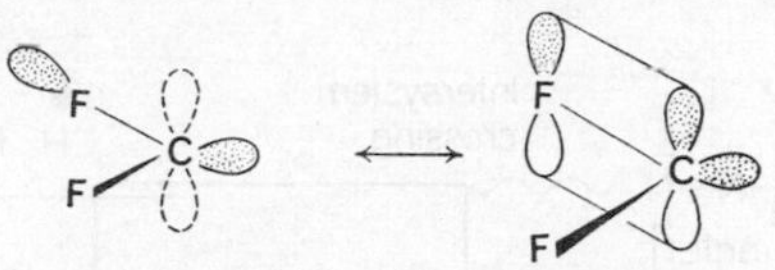

Figure 36. Resonance stabilization of difluorocarbene by back-donation from fluorine lone-pair orbitals.

of this nature will best be achieved by fluorine whose lone-pair electrons in 2 sp^3-orbitals are much better suited to overlap with carbon than the more diffuse lone-pair electrons of heavier halogens. All halogenomethylenes studied appear to have singlet ground-states, and difluorocarbene seems much the most stable having a half-life of several milliseconds in solution (with respect to dimerization to tetrafluoroethylene).[149] Support for this back-donation concept comes from analysis of the microwave spectrum of CF_2, which gives FCF = 105° and $r_{C-F} = 1.30$ Å, compared to the 'normal' C—F bond-length of 1.37 Å.

We shall see that the overall electron-deficiency of carbenes makes them highly reactive electrophiles and that this is the major feature of their chemistry. Certain species have been postulated as intermediates which are formally bivalent carbon compounds but whose chemistry is primary nucleophilic. A typical example is the reaction of triethylamine with a benzothiazolium salt in which a dimeric species is produced by a sequence whose key step is nucleophilic attack of the 'carbene' on starting material. Although formally bivalent,

this intermediate and related species[150] are best described in terms of a canonical form in which electron-deficiency is completely removed from carbon (132).

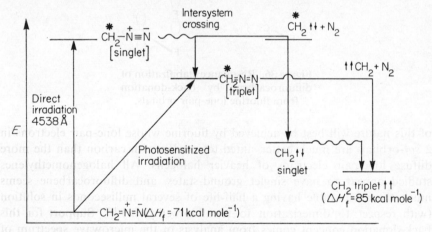

(132)

Reactions of Carbenes

Methylene

Methylene may be generated as a transient intermediate either in the gas-phase or in solution by the photolysis of ketene ($CH_2{=}C{=}O$), or diazomethane ($CH_2{=}N{=}N$) or its cyclic isomer diazirine. Consideration of an energy diagram (Figure 37) for the photolysis of diazomethane makes it clear that it may be formed in either the singlet or the triplet state.

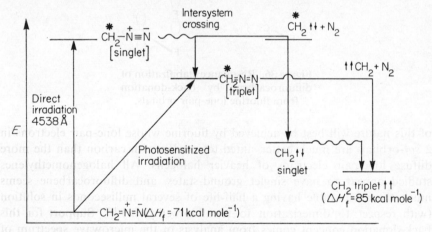

Figure 37. Production of singlet and triplet methylene in the photolysis of diazomethane.

Whether singlet or triplet methylene is generated under any given set of experimental conditions is subject to chemical test, requiring an answer to the question: "Can we make a distinction between singlet and triplet CH_2 on the basis of their chemical reactions?". Fortunately it is possible to generate triplet methylene selectively in the photochemical reaction, using mercury vapour as a sensitizer to provide triplet–triplet energy transfer. In general, methylene generated in the triplet state is rather unreactive towards saturated C—H bonds but will add across olefinic double bonds to form cyclopropanes; it is known to react very rapidly with oxygen (itself a triplet molecule). Methylene produced in direct photolysis, in the gas-phase or in solution, is thought to exist largely in the singlet state initially, and is an extremely reactive and unselective reagent (described by Doering as "the most indiscriminate species in organic chemistry") which undergoes insertion into C—H bonds in competition with its addition to double bonds.

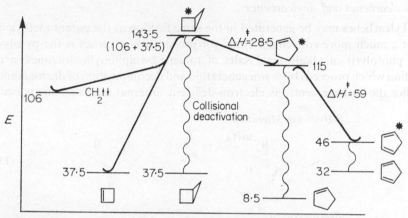

Reaction of cyclobutene with triplet methylene.

Figure 38. Contrasting reactivities of singlet and triplet methylene towards cyclobutene. The diagram shows the energetics of some pathways available the cyclobutene–singlet-methylene system. Numbers refer to enthalpy of formation (or reaction) in kcal mole^{-1}.

The reaction of methylene (produced by gas-phase photolysis of ketene) with cyclobutene serves to illustrate the different reactions of the two spin-states (Figure 38).[151] With triplet methylene formed by mercury sensitization, the major product is vinylcyclopropane. It will be apparent that the reaction cannot be concerted, since a spin inversion is required to achieve the final bond. Evidently the rearrangement step is more rapid than this spin inversion. In direct photolysis, some vinylcyclopropane is produced, indicating the possibility that a minor amount of methylene formed directly is in the triplet state. The major products appear to arise from addition and insertion reactions of singlet CH_2, although they are not those expected on simple considerations.

The explanation lies in the high energy of singlet methylene, whose heat of formation has been calculated to be around 106 kcal mole^{-1}. From its heat of combustion, cyclobutene is known to have ΔH_f 37.5 kcal mole^{-1}. Since the activation energy for addition will be small, and bicyclopentane has $\Delta H_f = 37.5$ kcal mole^{-1}, energy conservation decrees that the initial product has 106 kcal mole^{-1} of excess thermal energy (enough to cause rearrangement to cyclopentene, which is still sufficiently excited to undergo further reaction giving cyclopentadiene and hydrogen). Excess energy may be dissipated by collision, and at high pressures of inert gas the yield of products from these vibrationally excited rearrangements is reduced. In addition 1- and 3-methylcyclobutene are formed by insertion reactions but these are also in vibrationally excited states, and rearrange to penta-1,3-diene.

Alkylcarbenes and Acylcarbenes

Alkylcarbenes may be generated in the same fashion as the parent methylene, but a much more convenient route for preparative purposes is the pyrolysis or photolysis of alkali-metal salts of toluene-p-sulphonylhydrazones, a reaction which proceeds by *in situ* generation and decomposition of diazoalkanes. Since the carbene centre is electron-deficient, internal rearrangements occur

β-Hydrogen Migration

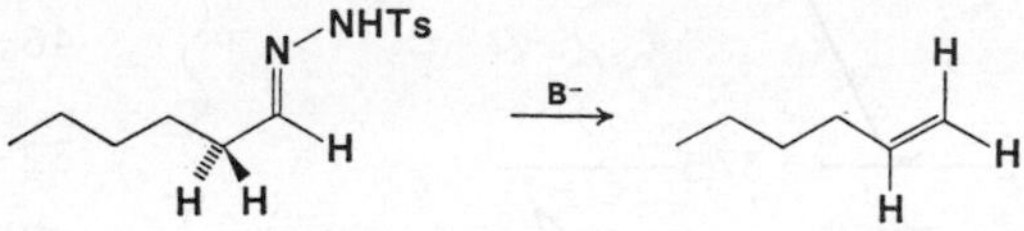

(133)

Rearrangement

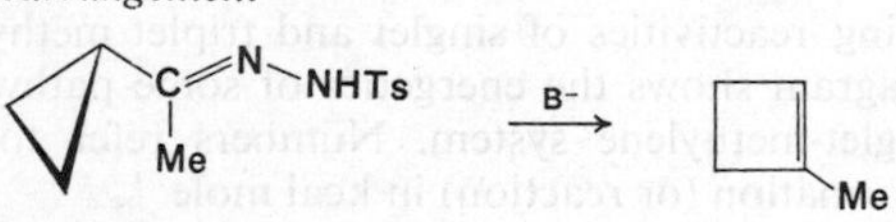

(134)

γ-Hydrogen Insertion

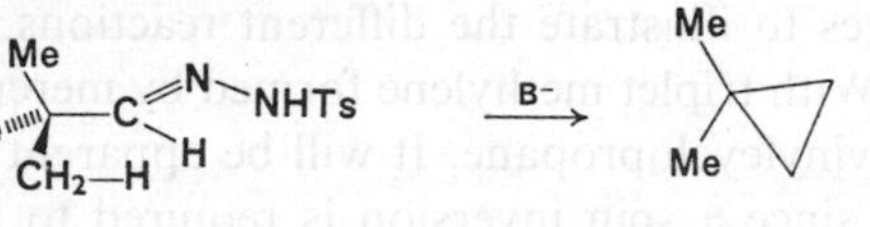

(135)

Fragmentation

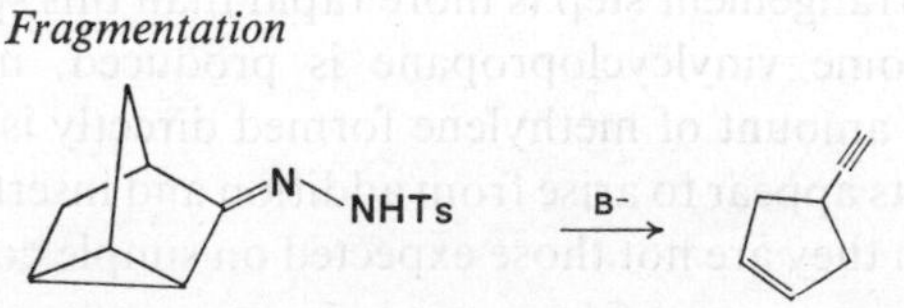

(136)

readily, and far more attention has been devoted to these than to intermolecular reactions with olefins and other substrates. Examples of the type of reaction which may be observed are shown in **133—136**.[152]

A typical condition for tosylhydrazone decomposition is pyrolysis of the dry sodium salt, and it is of interest to compare the intramolecular selectivity under these conditions with that observed in the production of the same alkylcarbenes by photolysis of alkyldiazirenes in the gas-phase (Figure 39).[153] It will be seen that the pyrolysis route produces a species which is quite selective towards α-hydrogen insertion (secondary > primary). In the photolysis route there is very much less selectivity, making it evident that the carbene is reacting in a state where it possesses a large excess of vibrational energy, and here the product of β-hydrogen insertion is also detectable.

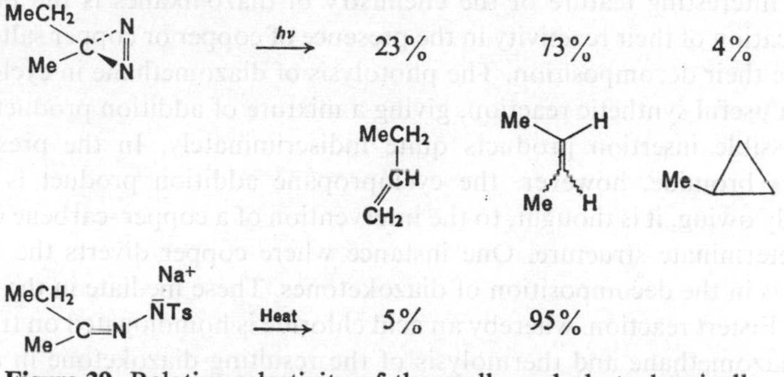

Figure 39. Relative selectivity of thermally and photochemically produced ethylmethylcarbene in intramolecular insertion reactions.

The decomposition of tosylhydrazones in solution (Bamford–Stevens reaction) may be more complex. A diazoalkane intermediate may be considered to be the conjugate base of a diazonium ion:

$$R_2C\overset{+}{=}\overset{-}{N}{=}N \underset{}{\overset{H^+}{\rightleftharpoons}} R_2CH{-}\overset{+}{N}{\equiv}N$$

This equilibrium appears to be set up with great facility. Since diazonium ions are known to collapse rapidly to carbonium ions and nitrogen (even *methyl*diazonium ion has proved to be non-isolable), the apparent 'carbene' reaction may in fact be a carbonium-ion reaction, particularly when a deficiency of base is used or the reaction is carried out in protic solvents. This is illustrated by the decomposition of camphor tosylhydrazone (**137**), where tricyclene, produced by γ-hydrogen insertion, is the product obtained via a carbene precursor, and camphene, whose formation requires a Wagner–Meerwein rearrangement, is the result of diazonium-ion decomposition.[154]

6

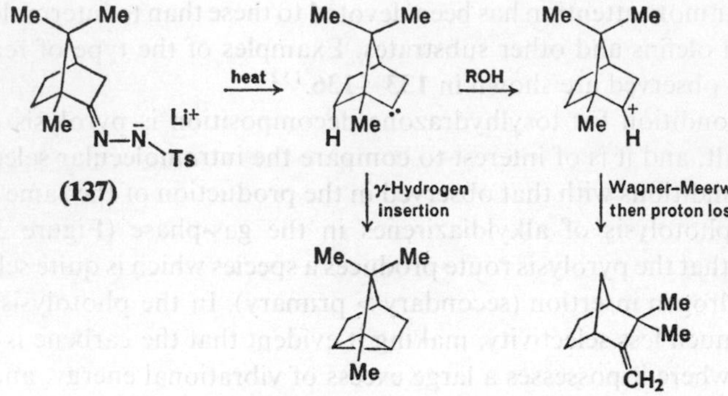

One interesting feature of the chemistry of diazoalkanes is the profound modification of their reactivity in the presence of copper or copper salts, which catalyse their decomposition. The photolysis of diazomethane in cyclohexene is not a useful synthetic reaction, giving a mixture of addition product and all the possible insertion products quite indiscriminately. In the presence of cuprous bromide, however, the cyclopropane addition product is formed uniquely owing, it is thought, to the intervention of a copper–carbene complex of indeterminate structure. One instance where copper diverts the reaction course is in the decomposition of diazoketones. These mediate in the familiar Arndt–Eistert reaction, whereby an acid chloride is homologated on treatment with diazomethane and thermolysis of the resulting diazoketone in aqueous solution containing silver ions. Intramolecular capture of the keto-carbene, which is an entirely different course of reaction, may be uniquely promoted by the presence of copper ions but does not occur under other conditions (138).[155]

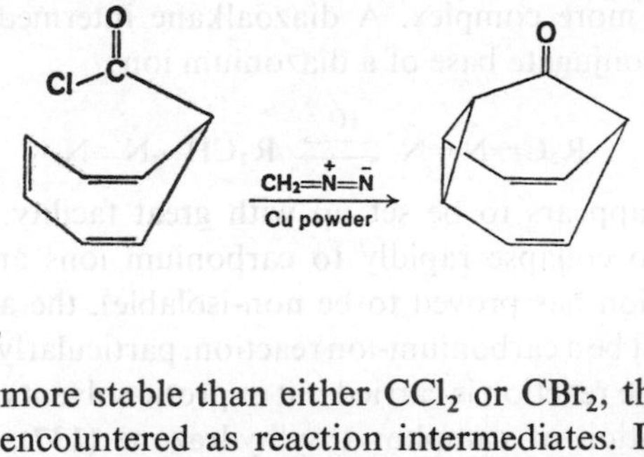

(138)

Halogenocarbenes

Although CF_2 is more stable than either CCl_2 or CBr_2, the latter are much more commonly encountered as reaction intermediates. In a classic paper, Doering and Hoffmann showed that dichlorocarbene generated in the addition of chloroform to basic media could be intercepted by reaction with cyclohexene to give the dichlorocyclopropane (139),[156] unaccompanied by C—H insertion products. This experiment stimulated much additional work, particularly by Skell, who established the electrophilic nature of the reactive intermediate in competition experiments between pairs of nucleophilic olefins and also the stereospecific course of reaction. Skell reasoned that if a triplet carbene were involved, then bond rotation might compete with spin inversion, producing a *trans*-cyclopropane from *cis*-but-2-ene.[157] In fact none was observed in the CBr_2 reaction, although triplet methylene is known to add olefins non-stereospecifically (Figure 40).

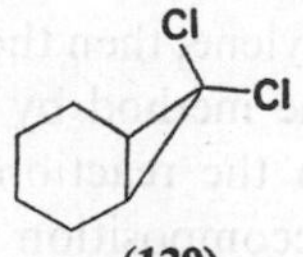

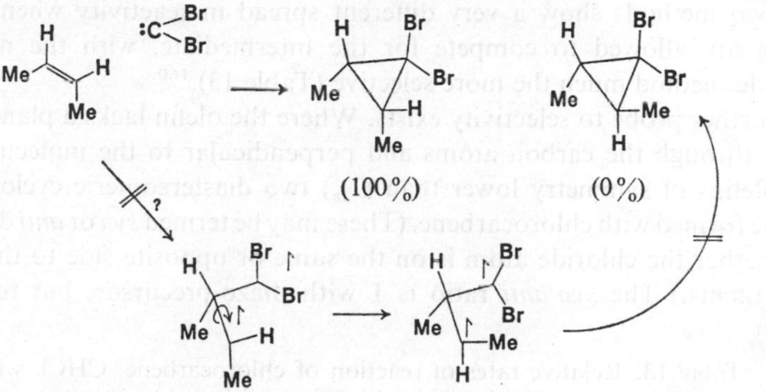

Figure 40. Stereospecific addition of CBr_2 to *cis*-but-2-ene.

There has been some discussion on the question of whether reactions which formally proceed through a carbene actually involve a free bivalent carbon intermediate. The reader may recall the high reactivity and low selectivity of methylene and consider in this light the reagent introduced by Simmons and Smith.[158] Their iodomethylenezinc iodide reacts slowly with olefins to produce cyclopropanes and has no tendency to undergo the alternative reaction in which C—H insertion products are formed. Clearly cyclopropane formation does not involve free CH_2, and the accepted mechanism (Figure 41) invokes a nucleophilic displacement on the reagent by the olefin. There is independent evidence that α-halogenomethyl derivatives are prone to displacement of halogen by nucleophiles, since chloromethylethylmercury reacts with iodide ion by an S_N2 mechanism 10^6 times faster than does n-butyl chloride.[159]

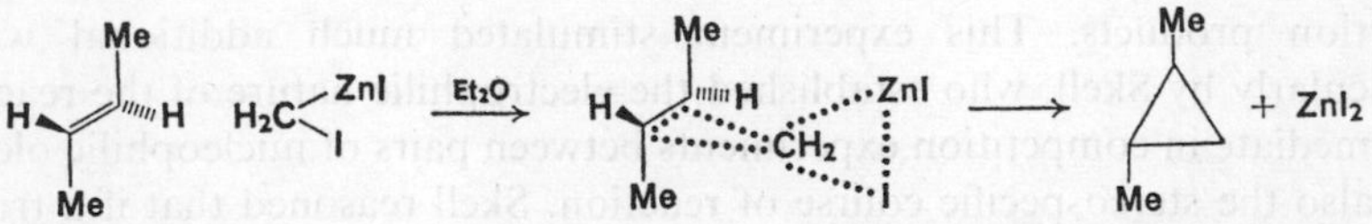

Figure 41. A carbenoid mechanism for the addition of iodomethylenezinc iodide to *trans*-but-2-ene.

Iodomethylenezinc iodide may be considered as a 'carbenoid', a species capable of reacting like a carbene. If carbenoid intermediates intervene in the reactions of any given methylene, then the relative reactivity of that methylene should be dependent on the method by which it is generated. Thus chlorocarbene may be formed in the reaction of n-butyl-lithium with methylene chloride at $-30°$, or by decomposition of the highly reactive chlorodiazomethane at $-30°$ under identical conditions of solvent and concentration. The two methods show a very different spread in reactivity when pairs of olefins are allowed to compete for the intermediate, with the methylene chloride method much the more selective (Table 13).[160]

A further probe to selectivity exists. Where the olefin lacks a plane of symmetry through the carbon atoms and perpendicular to the molecular plane (i.e. olefins of symmetry lower than C_{2h}) two diastereomeric cyclopropanes may be formed with chlorocarbene. (These may be termed *syn* or *anti* depending on whether the chloride atom is on the same or opposite side to the bulkier substituents.) The *syn/anti* ratio is 1 with diazo-precursor, but reaction is

Table 13. Relative rates of reaction of chlorocarbene, CHCl, with olefins at $-30°$ determined by competition experiments.

Olefin	Base route	Diazoalkane route
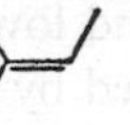	2.8	1.2
	1.7	1.2
	1.0	1.0
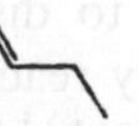	0.45	1.09
	0.23	0.77

Table 14. *syn/anti* Rate ratios for chlorocarbene additions at −30°.

Olefin	Base route	Diazoalkane route
⌲	1.6	1.0
⬡	3.2	1.0
⌇	5.5	1.0
⌁	3.4	1.0

stereoselective in favour of the *syn*-form when dichloromethyl-lithium is involved (Table 14).[160] Clearly there are different transition-state structures in the two routes, since in dilute solution the lack of selectivity and stereo-selectivity of the diazo-route is unlikely to be due to vibrational excitation. The accepted explanation is that, in this and related systems, bonding to lithium halide at the transition-state of the halogenomethyl-lithium–olefin reaction is still important. The overall reaction is therefore a bimolecular nucleophilic substitution with olefin as nucleophile. On the other hand, the diazo-route involves a free carbene.

In contrast to these results, the relative reactivity of dichlorocarbene towards different olefins is independent of the method by which the carbene is generated.[161] Experiments with a variety of very different precursors such as $PhHgCCl_2Br$, $KOBu^t/CHCl_3$, and $LiCCl_3$ (at −70°) show similar selectivity between olefins. In particular, Skell and Cholod[162] have pyrolysed CCl_4 and $CHCl_3$ at 1500° and allowed gaseous CCl_2 so produced to condense into, and react with, a mixture of olefins in the liquid phase at low temperatures, and again observed reactivity rates compatible with other methods of carbene generation. It therefore appears that carbenoids need not be postulated to explain the chemistry of $:CCl_2$ but that they seem to be important in less stable systems such as $:CHPh$, $:CHCl$, and $:CH_2$.

FREE RADICALS[163−171]

The chemistry of organic free-radicals dates back over seventy years. In 1900, Gomberg attempted to make hexaphenylethane by reaction of triphenyl-methyl chloride with bivalent metals in an inert atmosphere. Instead he obtained a highly coloured solution which reacted rapidly with oxygen, nitric oxide, or halogens, and which he later showed to contain the triphenylmethyl radical (140) in equilibrium with a dimer. For many years this dimer was thought to be hexaphenylethane, and only recently has it been shown to have a different structure.* Since the unpaired electron in 140 is delocalized over all three rings, there will be appreciable spin-density in *para*-positions. This explains why 141 rather than 142 is the product of dimerization, since any transition-state leading to 142 will be subject to severe steric hindrance.

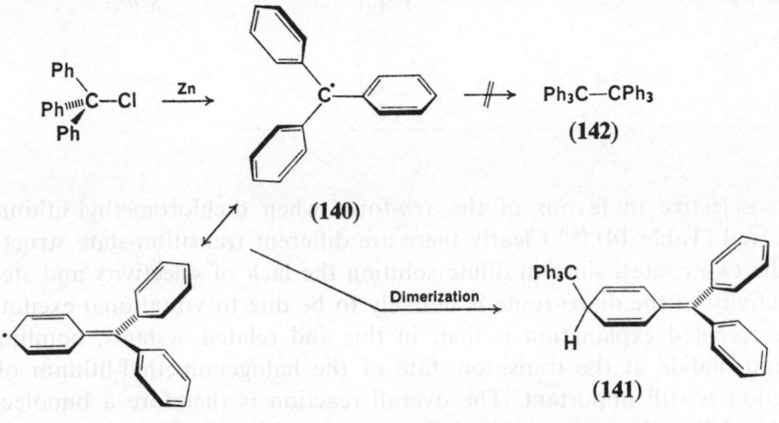

Figure 42. Dimerization of triphenylmethyl radical.

Triphenylmethyl radical is greatly stabilized by extensive delocalization, as are its anionic and cationic counterparts. Most free-radicals which may be encountered are highly reactive transient intermediates which can only be examined spectroscopically by special techniques at low temperatures. Such radical intermediates occur in a very wide range of organic reactions, of which typical examples are shown in Figure 43 ((143)–(146)), and there are many common features of mechanism.

The first step in a radical reaction is the initiation, in which a species with unpaired spin is formed. This may involve oxidation, thermal cleavage of a weak bond, or a photochemical bond-breaking:

Initiation $\qquad\qquad$ R—R $\rightarrow$ ·R + R· $\qquad\qquad$ (147)

The fate of these newly formed radicals depends largely on the reaction conditions; they may react with a substrate to produce a new radical or with

---

* Although hexaphenylethane (142) was the generally accepted structure until 1968, Jacobsen[172a] clearly preferred structure 141, and it remained for later workers[172b] to prove the latter form.

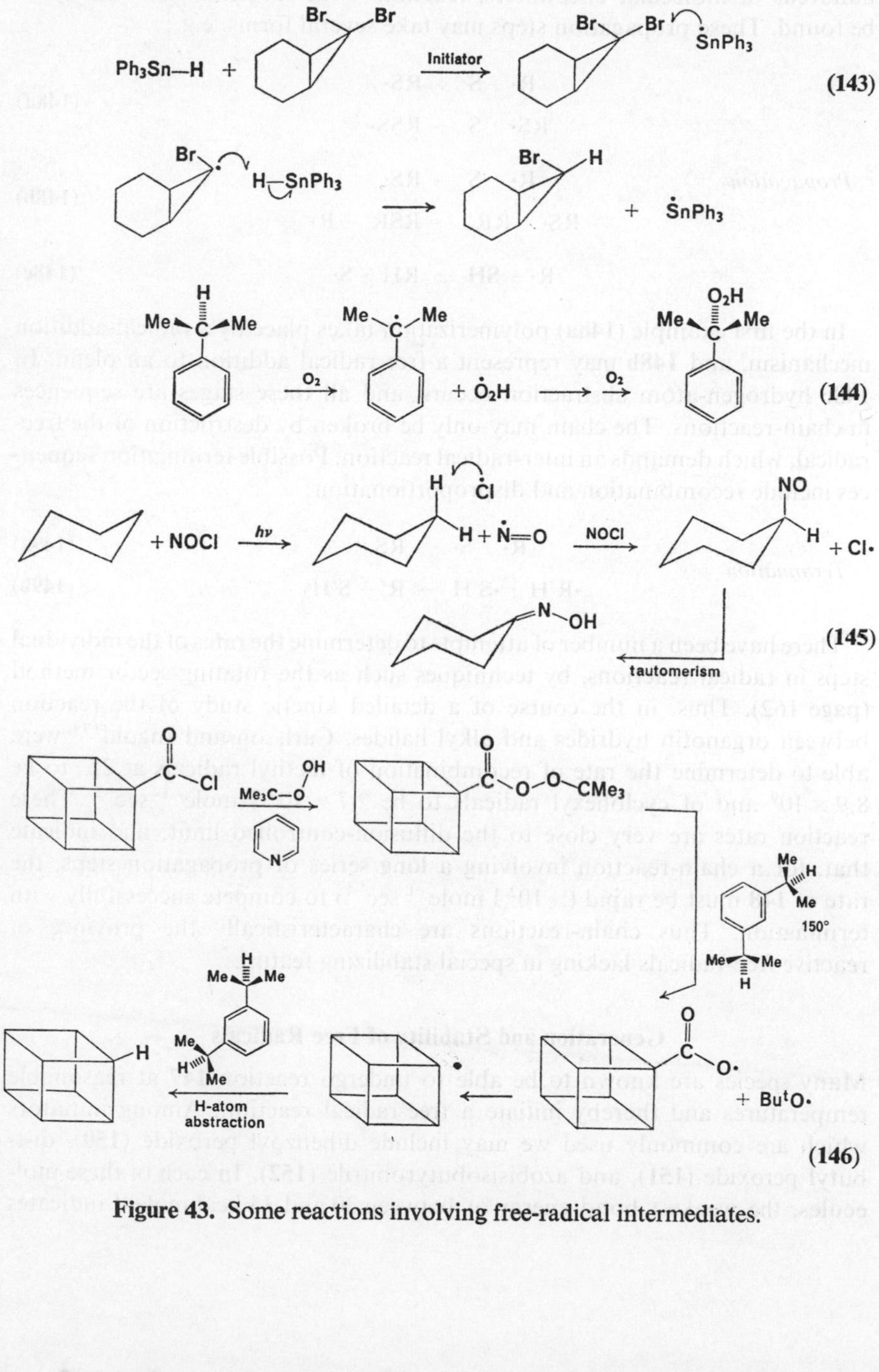

Figure 43. Some reactions involving free-radical intermediates.

another radical to produce a spin-paired product. Since a reactive intermediate at high dilution is unlikely to meet another such intermediate for many hundreds of molecular encounters, reactions with substrates are usually to be found. These propagation steps may take several forms, e.g.:

$$R\cdot + S \rightarrow RS\cdot$$
$$RS\cdot + S \rightarrow RSS\cdot$$
(148a)

Propagation

$$R\cdot + S \rightarrow RS\cdot$$
$$RS\cdot + RR' \rightarrow RSR' + R\cdot$$
(148b)

$$R\cdot + SH \rightarrow RH + S\cdot$$
(148c)

In the first example (148a) polymerization takes place by a radical-addition mechanism, and 148b may represent a free radical addition to an olefin. In 148c hydrogen-atom abstraction occurs, and all these stages are sequences in chain-reactions. The chain may only be broken by destruction of the free-radical, which demands an inter-radical reaction. Possible termination sequences include recombination and disproportionation:

Termination

$$R\cdot + S\cdot \rightarrow RS$$
(149a)
$$\cdot R'H + \cdot S'H \rightarrow R' + S'H_2$$
(149b)

There have been a number of attempts to determine the rates of the individual steps in radical reactions, by techniques such as the rotating-sector method (page 162). Thus, in the course of a detailed kinetic study of the reaction between organotin hydrides and alkyl halides, Carlsson and Ingold[173] were able to determine the rate of recombination of methyl radicals at 25° to be 8.9×10^9 and of cyclohexyl radicals to be 2.7×10^9 1 mole^{-1} sec^{-1}. These reaction rates are very close to the diffusion-controlled limit, and indicate that, for a chain-reaction involving a long series of propagation steps, the rate of 148 must be rapid ($\geqslant 10^2$ 1 mole^{-1} sec^{-1}) to compete successfully with termination. Thus chain-reactions are characteristically the province of reactive free-radicals lacking in special stabilizing features.

Generation and Stability of Free Radicals

Many species are known to be able to undergo reaction 147 at reasonable temperatures and thereby initiate a free-radical reaction. Among initiators which are commonly used we may include dibenzoyl peroxide (150), di-t-butyl peroxide (151), and azobisisobutyronitrile (152). In each of these molecules, the weakest bond energy of between 30 and 35 kcal mole^{-1} indicates facile homolysis around 50°C to 100°C. Photochemical initiation may also be used, and is perhaps most commonly encountered in free-radical reactions of halogen atoms produced by irradiation of the parent halogen.

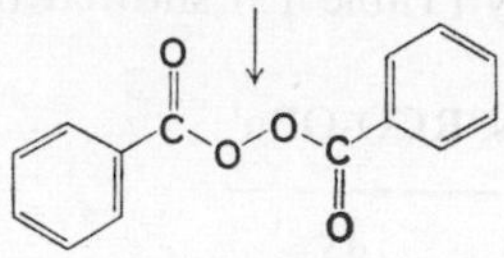

(150) 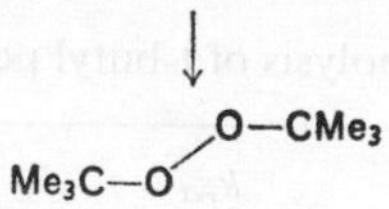   (151) 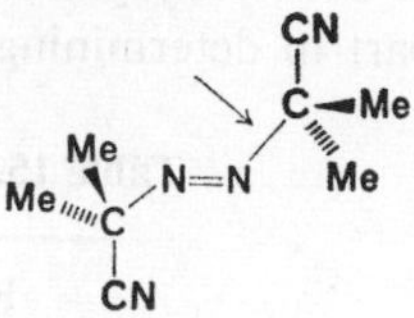   (152)

There is another important and interesting aspect of the mechanism of bond-homolysis. In all the examples cited, two radicals are produced in the decomposition of one molecule of substrate, and in solution these may recombine rather than diffuse away into the reaction medium, so that the efficiency of radical generation is less than 100%. This 'cage effect' may be evaluated by using a stable free-radical such as galvinoxyl[174] (153) (Figure 44) to scavenge all those radicals which escape and do not undergo recombination.

In the decomposition of the perester (154) there is a secondary isotope effect of 1.14 on replacing the benzyl hydrogens by deuterium.[175] This suggests that breakage of the CH_2—CO bond is concerted with the fracture of the weaker O—O bond, and that there is some measure of rehybridization in $-CH_2$ at the transition-state. If this is the case, then the CH_2 will have radical

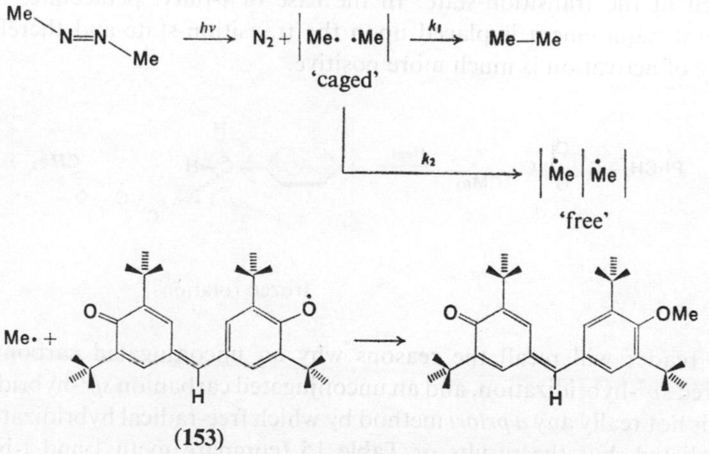

(153)

% Cage recombination = $k_1/(k_1 + k_2)$

Figure 44. The cage effect and radical trapping.

character at that transition-state, and the relative rates of decomposition of a series of t-butyl peresters RCO_2OBu^t should therefore reflect the relative stabilities of radicals R·. Bartlett and Hiatt studied this reaction[176] and showed that both conjugation and the degree of alkylation at the radical site played a part in determining the rate. Their detailed study (Table 15) showed that

Table 15. Thermolysis of t-butyl peresters, RCO_2OBu^t

R	k_{rel}	ΔH^*	ΔS^*
Me	1	38	17
Ph	17	33.5	7.8
CH$_2$Ph	290	28.7	3.9
CMe$_3$	1700	30.6	13
CHPh$_2$	19,300	24.3	−1.0
CMe$_2$Ph	41,500	26.1	5.8
CH$_2$CH=CHPh	125,000	23.5	−5.9

not only were there considerable rate enhancements when a stabilized radical was being produced, but that the Arrhenius parameters varied in a predictable way. This is readily seen by comparing the values for the t-butyl peresters of acetic and phenylacetic acid. In the phenylacetate case a conjugated radical is being produced and there is a demand at the transition-state for the two bond-rotations indicated (154) to be frozen so that maximum overlap is achieved at the transition-state. In the case of t-butyl peracetate, no such structural requirement is placed upon the transition-state and therefore the entropy of activation is much more positive.

frozen rotations

The reader will recall the reasons why an unconjugated carbonium-ion preferred sp^2-hybridization, and an unconjugated carbanion sp^3-hybridization. There is not really any *a priori* method by which free-radical hybridization may be predicted, but the results of Table 15 (compare methyl and t-butyl, or benzyl and α,α-dimethylbenzyl) suggest that a planar sp^2-conformation is preferred as the effects of alkyl substitution are similar for radicals and cations.

The situation is not so clear-cut as in carbonium-ions or carbanions. In particular there is some degree of correlation of rates of fragmentation of t-butyl peresters in which a cycloalkyl radical is produced with J_{13C-H} for the corresponding cycloalkane[177] (Figure 45, and see page 128). The rates of thermolysis of 1-cyano-azacycloalkanes on variation of ring-size correlate well with rates of solvolysis of cycloalkyl tosylates, and therefore suggest a planar sp^2-hybridized structure for 1-cyanocycloalkyl radicals.

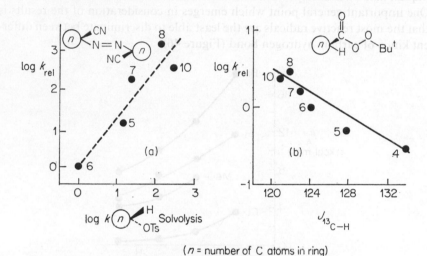

(n = number of C atoms in ring)

Figure 45. (a) Correlation of thermolysis rates of some cycloalkyl azo-compounds with solvolysis rates.

(b) Correlation of thermolysis rates of some t-butyl peresters with J_{13C-H}.

Reactions of Free Radicals

Hydrogen Atom Abstraction

Although the rate-constant and Arrhenius parameters for the attack of a transient free-radical on a substrate may not be measured directly, it is possible to determine these by indirect means. The method used may be illustrated by the decomposition of diacetyl peroxide in the presence of cyclohexane:

$$\text{MeCOO}_2\text{COMe} \xrightarrow{\text{heat}} 2\,\text{Me·} + \text{CO}_2 \tag{155a}$$

$$\text{Me·} + \text{C}_6\text{H}_{12} \xrightarrow{k_1} \text{CH}_4 + \text{C}_6\text{H}_{11}. \tag{155b}$$

$$\text{Me·} + \text{Me·} \xrightarrow{k_2} \text{C}_2\text{H}_6 \tag{155c}$$

In the gas-phase, where cage recombination does not occur, the relative values of k_1 and k_2 may be determined from the ratio of methane to ethane, since:

$$k_1^2/k_2 = [CH_4]^2/[C_6H_{12}]^2[C_2H_6] \qquad (156)$$

Now the absolute value of k_2 may be determined by the rotating-sector method,[178] and hence k_1 may be measured. By using this and similar techniques, data on a variety of C—H abstraction reactions have been obtained.[167] One important general point which emerges in consideration of the results is that the most reactive radicals are the least able to discriminate between different kinds of carbon–hydrogen bond (Figure 46).

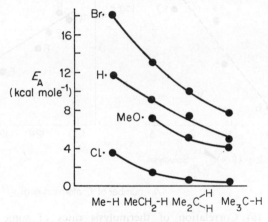

Figure 46. Activation energy for a variety of hydrogen-atom abstraction reactions.

It will be noted from Figure 46 that the activation energies for Cl· atom attack on any substrate are invariably lower than for Me· radical attack on the same substrate, and this in spite of the fact that the bond energies of Me—H and Cl—H are quite similar. It appears that the transition-state for attack by an atom of high electronegativity such as chlorine may have a significant stabilizing contribution from a polar canonical form:

$$Me-CH_2\overset{\delta+}{\cdots}H\overset{\delta-}{\cdots}Cl \qquad (157)$$

The effect of this may operate in a rather different way. Whereas the relative reactivities of different positions in a molecule towards H-atom abstraction by Me· will be largely determined by relative C—H bond energies, polar and inductive effects can be important in determining the site of reaction of polar radicals such as Cl·, CF_3·, and Bu^tO·. Thus when propionic acid is the substrate, the reactivity of the two types of C—H bond depends on the nature of the radical-abstracting reagent[179] (158).

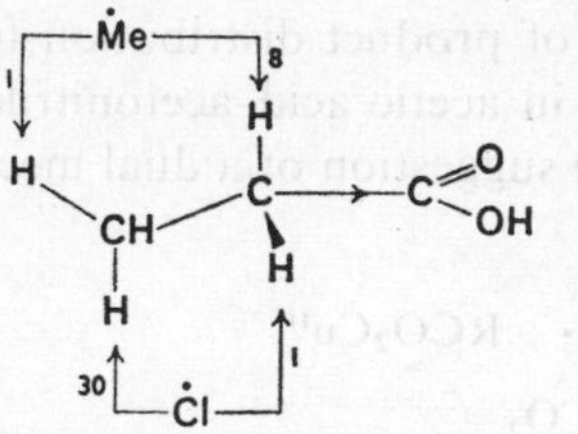

Methyl radicals abstract selectively at the position α to the carboxy group, and chlorine atoms at the β position

(158)

Further evidence on the dipolar nature of some radical transition-states stems from a study of substituent effects on the bromination of toluene[180] by N-bromosuccinimide (Figure 47). This reagent serves to produce a very low concentration of bromine atoms by homolytic decomposition and will convert toluene into benzyl bromide or cyclohexene into 3-bromocyclohexene. If a Hammett-type correlation is made for the relative rates of reaction of substituted toluenes, then σ^+-constants show a better fit than σ-values; a ρ-value of -1.39 likewise suggests considerable positive charge on carbon at the rate-determining transition-state.

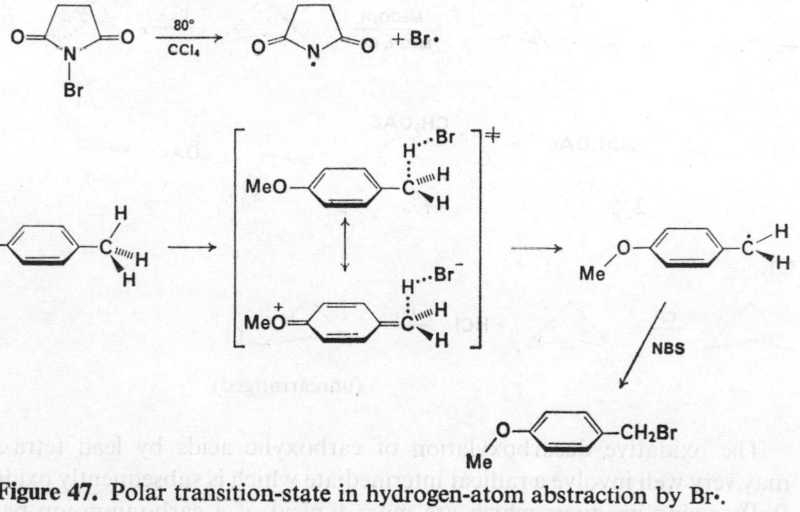

Figure 47. Polar transition-state in hydrogen-atom abstraction by Br·.

Electron Transfer

There are many reactions which apparently proceed through radical intermediates in which metal-ion catalysts are used. Among these, the cupric-ion catalysed decomposition of peresters and peracids is typical and has been studied in some detail by Kochi and coworkers.[181] In this system the products are typically those which might be expected from a carbonium-ion rather than a free-radical pathway, and a study of product distribution for the decomposition of various diacyl peroxides in acetic acid–acetonitrile catalysed by cuprous and cupric acetate led to the suggestion of a dual mechanism with a common intermediate:

$$RCOO_2COR \xrightarrow{Cu^I} RCO_2\cdot + RCO_2Cu^{II} \tag{159a}$$

$$RCO_2\cdot \longrightarrow R\cdot + CO_2 \tag{159b}$$

$$R\cdot + Cu^{II}(OAc)_2 \longrightarrow RCu^{II}(OAc)_2 \tag{159c}$$

$$ROAc \xleftarrow{(a)} R^+ \xleftarrow{(b)} olefin + AcOH + Cu^IOAc$$

Path (b) predominates in simple alkyl systems where there are no special features available to stabilize a carbonium ion R^+, but where such features exist the electron-transfer pathway (a) is followed. The behaviour of cyclobutyl systems is particularly striking, since cyclobutyl radicals maintain structural integrity but cyclobutyl cations rearrange in a characteristic fashion. The two reactions 160 and 161 illustrate the difference.

$$(160)$$

Cf.

(unrearranged)

The oxidative decarboxylation of carboxylic acids by lead tetra-acetate may very well involve a radical intermediate which is subsequently oxidized by Pb^{IV}, giving products which are more typical of a carbonium-ion pathway. Since 2-norbornyl radical formed as a transient intermediate in other reactions does not tend to racemize, the production of exo-2-norbornyl acetate which is largely racemized in the decarboxylation of exo-norbornane-2-carboxylic acid is indicative of the intervention of norbornyl cation[182] (162).

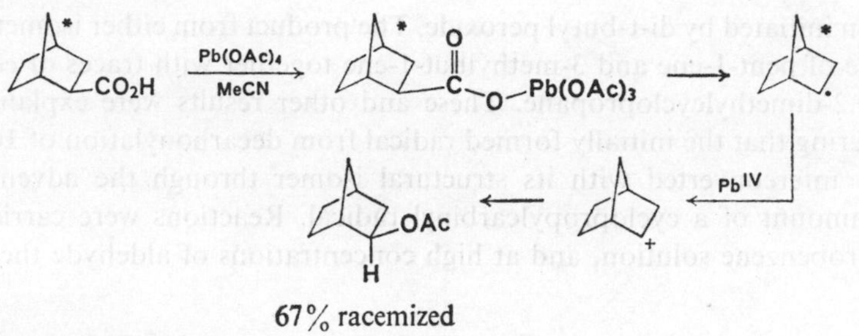

(162)

67% racemized

Intramolecular Reactions

There are a number of examples of hydrogen-atom abstraction by a radical from a remote site within the same molecule, and this is a reaction which can be put to good use in synthesis since it affords a potential method for the functionalization of a saturated, unactivated carbon atom. The most striking example of this transformation has become known as the Barton reaction;[183] in this procedure the photolysis of a nitrite ester affords an alkoxide radical capable of intramolecular hydrogen-atom abstraction (163). The main applications of the method have been in steroid chemistry and have allowed the synthesis of otherwise inaccessible functional derivatives.

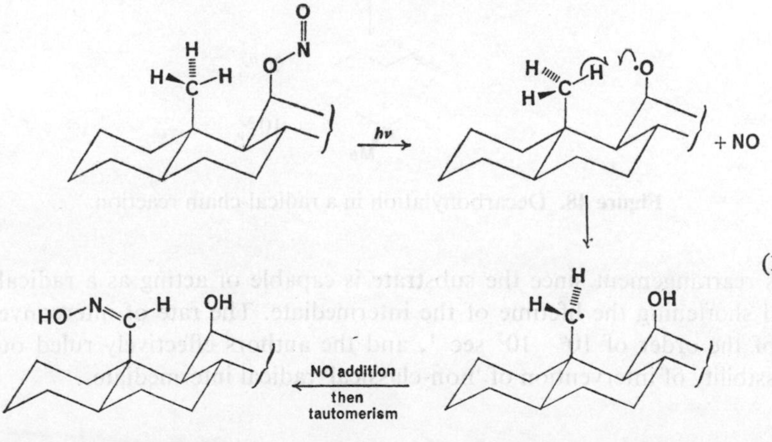

(163)

Examples are known of radical rearrangements involving 1,2-migration of phenyl groups, and of interconversion of acyclic and cyclic radicals. In the latter context it is informative to consider the behaviour of but-3-en-1-yl radicals labelled by methyl substitution. Thus Montgomery and Matt[184] studied the decarbonylation of 164 and 165 (Figure 48) in a radical-chain reaction initiated by di-t-butyl peroxide. The product from either isomer was a mixture of pent-1-ene and 3-methylbut-1-ene together with traces of *cis*- and *trans*-1,2-dimethylcyclopropane. These and other results were explained by considering that the initially formed radical from decarbonylation of **164** was rapidly interconverted with its structural isomer through the advent of a small amount of a cyclopropylcarbinyl radical. Reactions were carried out in chlorobenzene solution, and at high concentrations of aldehyde there was

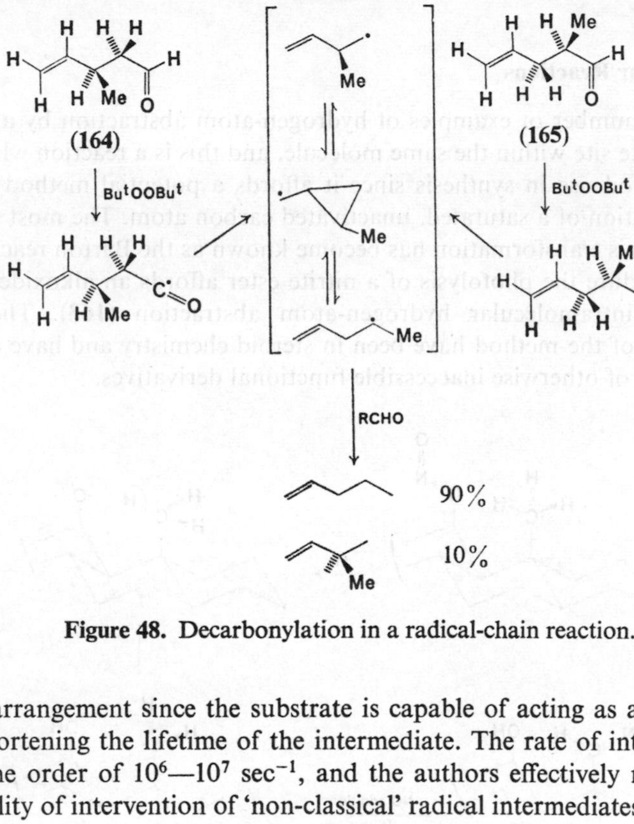

Figure 48. Decarbonylation in a radical-chain reaction.

less rearrangement since the substrate is capable of acting as a radical trap and shortening the lifetime of the intermediate. The rate of interconversion is of the order of 10^6—10^7 sec^{-1}, and the authors effectively ruled out the possibility of intervention of 'non-classical' radical intermediates.

Detection of Free Radicals

Species with unpaired electrons may be observed and characterized by electron spin resonance, which depends on the fact that in a magnetic field there will be two spin quantum states associated with the unpaired electron depending on whether the spin aligns with or against the magnetic field. With a 10,000 gauss magnetic field, the quantum transition between these states may be detected by adsorption of electromagnetic radiation at a specific point in the microwave region. Any element such as hydrogen or nitrogen which has a spin quantum number other than zero may be considered as a small nuclear magnet, and when in the region of the unpaired electron in a free-radical this will cause further splitting of the e.s.r. signal, since the transition energy will depend on the spin quantum state of the hydrogen or nitrogen nucleus. This 'hyperfine splitting' effect is a particularly useful guide to the structure of organic free-radicals, particularly for hydrogen splitting, since it is a measure of the extent of interaction between the H-nucleus and the unpaired spin, and consequently a measure of the amount of unpaired spin density in the region of that nucleus.

A variety of techniques have been utilized to provide radicals for e.s.r. examination. With highly unstable species such as alkyl radicals whose great reactivity precludes the use of normal sampling methods, flow techniques, radiolysis, or photolysis in argon matrices had been necessary until recently. Kochi and coworkers, however, developed a method of general utility in which a mixture of di-t-butyl peroxide and a hydrocarbon RH are photolysed directly

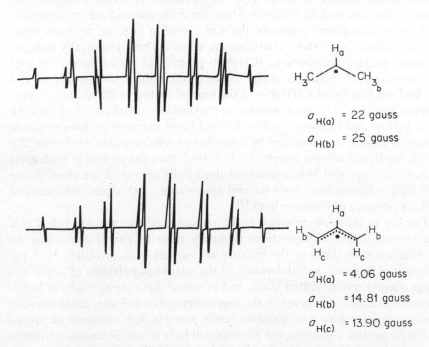

Figure 49. Schematic e.s.r. spectra of isopropyl and allyl radicals.

at low temperatures in the cavity of an e.s.r. spectrometer. Radicals are produced by the following reaction:

$$Bu^tOOBu^t \xrightarrow{h\nu} 2 Bu^tO\cdot \qquad (166)$$

$$Bu^tO\cdot + RH \longrightarrow R\cdot + Bu^tOH \qquad (167)$$

An analysis of the spectra obtained (Figure 49) from isopropyl and allyl[185] reveals the hyperfine splitting constants (a_H) indicated.

It will be apparent that the β-hydrogens of isopropyl radical interact rather more strongly with the unpaired electron than the α-hydrogen. A probable reason for this is that in a planar radical the non-bonding p-orbital will be orthogonal to the α-C—H bond, thereby minimizing their electronic interaction. No such stereoelectronic restriction operates in the interaction with β-hydrogen atoms.

Detection of Free Radicals as Reaction Intermediates; CIDNP[171]

In a free-radical, interactions of the unpaired electron with neighbouring hydrogen nuclei affect nuclear relaxation processes profoundly and usually render n.m.r. spectra so broad as to be worthless. A recent technique has shown, however, that in reactions where there are free-radical intermediates en route to spin-paired products, the n.m.r. spectra of those products when initially formed may show characteristic effects. These chemically induced dynamic nuclear polarizations (CIDNP) promise to be an invaluable tool for the study of reactions which proceed via radical intermediates. The detailed mechanism of CIDNP is at the time of writing still a matter of some dispute, but we may illustrate a tentative explanation with the reaction between ethyl-lithium and isopropyl iodide. Around room temperature lithium iodide is rapidly formed accompanied by a number of organic products (Figure 50). If this reaction is carried out in an n.m.r. tube, then the spectra of both ethyl iodide and isopropyl iodide observed during the course of the reaction are strikingly different from their normal appearance, with signal enhancement and the presence of emission lines.[186]

The key to these observations is the radical cage intermediate 168. There are two electronic configurations relatively close in energy depending on whether the spin states of the radicals are parallel or antiparallel. Now the intensity of an n.m.r. signal depends on the relative populations of upper and lower nuclear spin quantum states, and of course these are normally in Boltzmann distribution. However, in the cage intermediate the interaction between electron and nuclear spin quantum levels perturbs this distribution. On its initial formation, therefore, the product will have a non-Boltzmann distribution of spin quantum states, thereby 'remembering' its radical precursor and giving rise to an enhanced adsorption or emission signal. If the proton signal is part of a multiplet, then the individual lines in the spectrum of initially

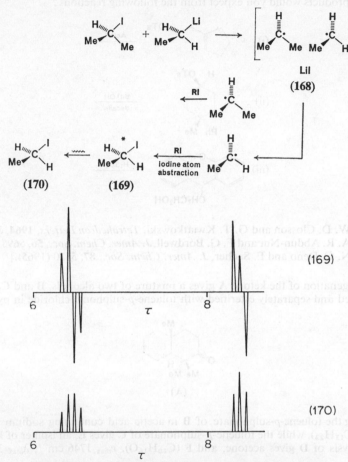

Figure 50. Chemically induced dynamic nuclear polarization. The n.m.r. spectrum of ethyl iodide produced by the interaction of ethyl-lithium and isopropyl iodide in benzene; **169** during formation and **170** after a few minutes. A similar effect is observed in the spectrum of isopropyl iodide during the reaction.

formed product are affected in different ways. It is not necessary to carry out the reaction in the cavity of an n.m.r. spectrometer to witness CIDNP effects, but the polarization dies away completely within a few minutes of the completion of reaction.

PROBLEMS

1. What products would you expect from the following reactions?

(i)

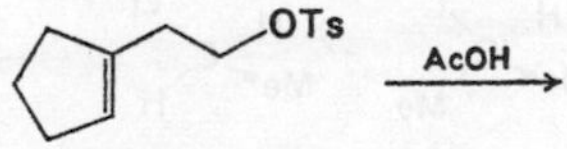

$\xrightarrow{\text{AcOH}}$

(ii)

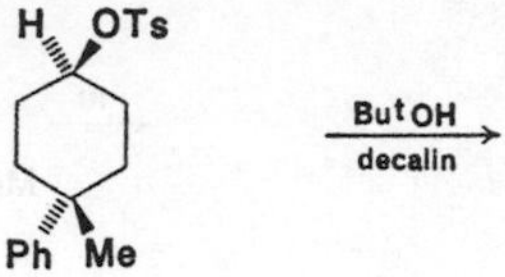

$\xrightarrow[\text{decalin}]{\text{Bu}^t\text{OH}}$

(iii)

$\xrightarrow{\text{H}_2\text{SO}_4}$

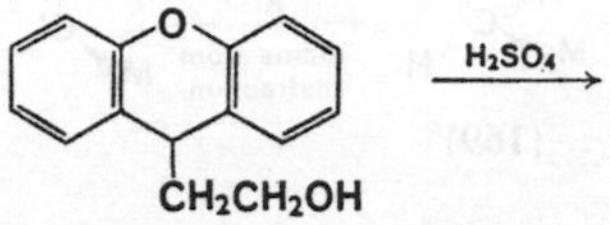

[see (i) W. D. Closson and G. T. Kwiatkowski, *Tetrahedron Letters*, **1964**, 3831.
(ii) A. R. Abdun-Nur and F. G. Bordwell, *J. Amer. Chem. Soc.*, **56**, 5695 (1964).
(iii) N. C. Deno and E. Sacher, *J. Amer. Chem. Soc.*, **87**, 5120 (1965).]

2. Hydrogenation of the ketone **A** gives a mixture of two alcohols, **B** and **C**, which are isolated and separately esterified with toluene-*p*-sulphonyl chloride in pyridine.

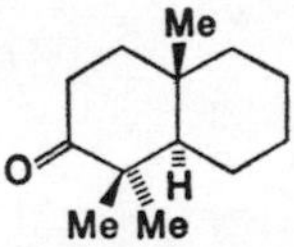

(A)

Heating the toluene-*p*-sulphonate of **B** in acetic acid containing sodium acetate gives **D** ($C_{13}H_{22}$), while the toluene-*p*-sulphonate of **C** gives **E**, an isomer of **D**.

Ozonolysis of **D** gives acetone, and **F** ($C_{10}H_{16}O$), ν_{max} 1746 cm^{-1}, λ_{max} 302 nm (ϵ 25).

Ozonolysis of **E** gives **G** ($C_{13}H_{22}O_2$), ν_{max} 1715 cm^{-1} (broad), λ_{max} 285 nm (ϵ 45).

Treatment of **G** with sodium hypochlorite followed by acidification gives **H** ($C_{11}H_{18}O_4$) whose calcium salt is converted by heat into **F**.

Give the graphic formulae of the compounds **A** to **F**, and explain the different behaviour of the toluene-*p*-sulphonates of **B** and **C** on solvolysis.

[See G. H. Whitham, *Alicyclic Chemistry*, Oldbourne Press, London, 1963, p. 79.]

3. Rates of solvolysis have been obtained for a series of 7-aryl-7-norbornyl and *syn*-7-aryl-*anti*-7-norbornenyl *p*-nitrobenzoates in 70% aqueous dioxan at 25°. For the former series, a linear correlation of log *k* against σ^+ was found with $\rho = -5.27$ but for the latter a sharp break occurred in the log k–σ^+ plot. A linear portion with

$\rho = -2.30$ was apparent when the substituent in the ring was 4-OMe, 4-H, and 4-CF₃, but when $X = NMe_2$ the rate constant differed considerably from that expected on extrapolation of this line. When $X = NMe_2$ and OMe the rate constants were almost identical in both series.

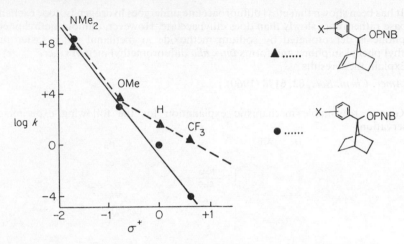

How would you rationalize these observations?

[See P. G. Gassman and A. F. Fentiman, Jr., *J. Amer. Chem. Soc.*, **92**, 2549 (1970).]

4. The 2-adamantyl system has been used as a model for solvolysis reactions since it was expected that solvent participation should be absent because a pentacovalent transition-state or intermediate would be strongly hindered by the axial hydrogens shown in the diagram.

(R = Me or H; X = Br)

How would you expect the $k_{R=Me}/k_{R=H}$ rate ratios to compare for the three systems, for the solvolysis of the bromides in 80% ethanol and, secondly, in acetic acid?

[See J. L. Fry, J. M. Harris, R. C. Bingham, and P. von R. Schleyer, *J. Amer. Chem. Soc.*, **92**, 2540 (1970).]

5. A solution of 1-methylcyclobutyl chloride in SbF₅–SO₂ClF at −80° gives an n.m.r. spectrum consisting of two singlets at τ 6.11 and 7.13 with an area ratio of 2/1. 1-(Trideuteriomethyl)cyclobutyl chloride in solution displays an n.m.r. spectrum with only one singlet at τ 6.07.

Give explanations which might fit these results, and consider a further experiment which might distinguish between the different possibilities.

[See M. Saunders and J. Rosenfeld, *J. Amer. Chem. Soc.*, **92**, 2548 (1970).]

6. It has been shown that ethyl difluoroacetate undergoes hydrogen isotope exchange in base rather more slowly than does ethyl acetate. However, difluoromethyl phenyl sulphone is deprotonated by sodium methoxide in methanol much faster than methyl phenyl sulphone and forms *inter alia* difluoromethyl methyl ether.

Explain these results.

[*J. Amer. Chem. Soc.*, **82**, 6178 (1960).]

7. Give a reasonable mechanistic explanation for the following experimental observation:

(23%)

[J. Hine and J. J. Porter, *J. Amer. Chem. Soc.*, **81**, 4719 (1959).]

8. Bearing in mind the discussion on solvent effects in Chapter 1, and the relative basicities of pyridine ($pK_b = 8.75$) and piperidine ($pK_b = 3.88$), explain why the equilibrium shown lies further to the right in pyridine than in piperidine.

[*Org. Reaction Mech.*, Wiley, London, **1969**, p. 145.]

9. The thermal rearrangement of optically active Feist's ester **A** to **B** gives a product which has maintained some degree of optical activity.

Suggest a pathway to explain this observation.

[E. F. Ullman, *J. Amer. Chem. Soc.*, **82**, 505 (1960).]

[*Hint*—see the discussion in *J. Amer. Chem. Soc.*, **89**, 715 (1967), and papers quoted therein.]

REFERENCES

Carbonium Ions

1. G. A. Olah and P. von R. Schleyer (ed.), *Carbonium Ions*, Vol. 1, Wiley-Interscience, New York, 1968.
2. Ref. 1, Vol. 2, 1970.
3. Ref. 1, Vol. 3,
   These three volumes provide an extensive study of carbonium ion chemistry.
4. D. Bethel and V. Gold, *Carbonium Ions*, Academic Press, London, 1967. Valuable text which does not rely on previous knowledge.
5. A. Streitwieser, *Solvolytic Displacement Reactions*, McGraw-Hill, New York, 1962. Most useful for wide survey of reactivity in tabulated form.
6. G. A. Olah and C. U. Pittman, *Adv. Phys. Org. Chem.*, 4, 305 (1966). Spectroscopic methods of studying carbonium ions (n.m.r., infrared, ultraviolet).
7. B. Capon, *Quart. Rev.*, 18, 45 (1964). Excellent review on neighbouring-group participation.
8. G. D. Sargent, *Quart. Rev.*, 20, 301 (1966). Critical discussion on non-classical carbonium ions.
9. R. C. Fort, Jr., and P. von R. Schleyer, *Adv. Alicyclic Chem.*, 1, 284 (1966). Review of reactivity of bridgehead carbonium ions, carbanions, and free-radicals.
10. A. Streitwieser, *Progr. Phys. Org. Chem.*, 1, 1 (1963); D. W. Turner, *Adv. Phys. Org. Chem.*, 4, 31 (1966). Reviews of ionization potentials.
11. N. C. Deno, *Progr. Phys. Org. Chem.*, 2, 129 (1964). Spectroscopic study of carbonium ions.
12. C. A. Bunton, *Nucleophilic Substitution at a Saturated Carbon Atom*, Elsevier, Amsterdam, 1963.
13. E. R. Thornton, *Solvolysis Mechanisms*, Ronald Press, New York, 1964.
14. R. A. More O'Ferrall, *Adv. Phys. Org. Chem.*, 5, 331 (1967). Formation of carbonium ions from diazoalkanes.
15. P. D. Bartlett, *Non-classical Ions*, Benjamin, New York, 1965.
16. J. L. Gleave, E. D. Hughes, and C. K. Ingold, *J. Chem. Soc.*, 1935, 236.
17. F. P. Lossing in *Mass Spectrometry* (ed. C. A. McDowell), McGraw-Hill, New York, 1963, p. 442.
18. M. Saunders and E. L. Hagan, *J. Amer. Chem. Soc.*, 90, 6881, 2436 (1968).
19. Ref. 1, E. M. Arnett and J. W. Larsen, p. 453.
20. L. C. Bateman, E. D. Hughes, and C. K. Ingold, *J. Chem. Soc.*, 1940, 974.
21. J. E. Williams, Jr., R. Sustmann, L. C. Allen, and P. von R. Schleyer, *J. Amer. Chem. Soc.*, 91, 1037 (1969).
22. W. G. Young, S. Winstein, and H. L. Goering, *J. Amer. Chem. Soc.*, 73, 1958 (1951).
23. S. Winstein, B. Appel, R. Baker, and A. Diaz in *Organic Reaction Mechanisms*, Chem. Soc. Special Publ. No. 19, 1965, p. 109; M. Szwarc, *Accounts Chem. Res.*, 2, 87 (1969).
24. H. L. Goering and E. F. Silversmith, *J. Amer. Chem. Soc.*, 77, 6249 (1955).
25. S. Winstein, P. E. Klinedinst, Jr., and G. C. Robinson, *J. Amer. Chem. Soc.*, 83, 885 (1961); S. Winstein, P. E. Klinedinst, Jr., and E. Clippinger, *ibid.*, p. 4986.
26. S. Winstein and G. C. Robinson, *J. Amer. Chem. Soc.*, 80, 169 (1958).

27. S. Winstein, R. Baker, and S. Smith, *J. Amer. Chem. Soc.*, **86**, 2072 (1964).
28. H. L. Goering, M. M. Pombo, and K. D. McMichael, *J. Amer. Chem. Soc.*, **85**, 965 (1963).
29. A. Streitwieser, Jr., and T. D. Walsh, *J. Amer. Chem. Soc.*, **87**, 3686 (1965).
30. R. A. Sneen and J. W. Larsen, *J. Amer. Chem. Soc.*, **91**, 362 (1969).
31. R. A. Sneen and J. W. Larsen, *J. Amer. Chem. Soc.*, **91**, 6031 (1969).
32. G. Baddeley, J. Chadwick, and H. T. Taylor, *J. Chem. Soc.*, **1954**, 2405.
33. H. C. Brown and J. D. Cleveland, *J. Amer. Chem. Soc.*, **88**, 2051 (1966).
34. H. C. Brown and R. S. Fletcher, *J. Amer. Chem. Soc.*, **71**, 1845 (1949).
35. S. Winstein and N. J. Holness, *J. Amer. Chem. Soc.*, **77**, 5562 (1955).
36. N. C. G. Campbell, D. M. Muir, R. R. Hill, J. H. Parish, R. M. Southam, and M. C. Whiting, *J. Chem. Soc.* (B), **1968**, 355.
37. V. J. Shiner and J. G. Jewett, *J. Amer. Chem. Soc.*, **86**, 945 (1964); *ibid.*, **87**, 1382, 1383 (1965).
38. P. von R. Schleyer, *J. Amer. Chem. Soc.*, **89**, 701 (1967).
39. C. J. Collins, Z. K. Cheema, R. G. Werth, and B. M. Benjamin, *J. Amer. Chem. Soc.*, **86**, 4913 (1964).
40. H. C. Brown, I. Rothberg, P. von R. Schleyer, M. M. Donaldson, and J. J. Harper, *Proc. Nat. Acad. Sci., U.S.A.*, **56**, 1653 (1966).
41. S. Winstein, E. Grunwald, R. E. Buckles, and C. Hanson, *J. Amer. Chem. Soc.*, **70**, 816 (1948).
42. S. Winstein and H. J. Lucas, *J. Amer. Chem. Soc.*, **61**, 1576, 2845 (1939).
43. S. Winstein, E. Allred, R. Heck, and R. Glick, *Tetrahedron*, **3**, 1 (1958).
44. H. C. Brown and R. S. Fletcher, *J. Amer. Chem. Soc.*, **72**, 1223 (1950).
45. H. C. Brown and M. Nakagawa, *J. Amer. Chem. Soc.*, **77**, 3610 (1955).
46. M. Cocivera and S. Winstein, *J. Amer. Chem. Soc.*, **85**, 1702 (1963).
47. R. Baker, J. Hudec, and K. L. Rabone, *J. Chem. Soc.* (B), **1970**, 1446.
48. C. A. Grob, *Angew. Chem. Internat. Edn. Engl.*, **8**, 535 (1969).
49. C. A. Grob, F. Ostermayer, and W. Raudenbusch, *Helv. Chim. Acta*, **45**, 1672 (1962).
50. C. A. Grob, R. M. Hoegerle, and M. Ohta, *Helv. Chim. Acta*, **45**, 1823 (1962).
51. C. A. Grob, V. Krasnobajew, and R. Frankhauser, *Helv. Chim. Acta*, **49**, 690 (1966).
52. C. W. Shoppee, *J. Chem. Soc.*, **1946**, 1147.
53. E. M. Kosower and S. Winstein, *J. Amer. Chem. Soc.*, **78**, 4347 (1956); S. Winstein and R. Adams, *ibid.*, **70**, 838 (1948).
54. S. Winstein, M. Shatavsky, C. Norton, and R. B. Woodward, *J. Amer. Chem. Soc.*, **77**, 4183 (1955).
55. J. B. Rogan, *J. Org. Chem.*, **27**, 3910 (1962).
56. R. G. Lawton, *J. Amer. Chem. Soc.*, **83**, 2399 (1961).
57. P. D. Bartlett and G. D. Sargent, *J. Amer. Chem. Soc.*, **87**, 1297 (1965).
58. W. D. Closson and S. A. Roman, *Tetrahedron Letters*, **1966**, 6015.
59. H. Tanida, T. Tsuji, and T. Irie, *J. Amer. Chem. Soc.*, **89**, 1953 (1967); M. A. Battiste, C. L. Deyrup, R. E. Pincock, and J. Haywood-Farmer, *ibid.*, p. 1954.
60. D. J. Cram and J. A. Thompson, *J. Amer. Chem. Soc.*, **89**, 6766 (1967).
61. C. J. Kim and H. C. Brown, *J. Amer. Chem. Soc.*, **91**, 4286, 4287, 4289 (1969).
62. D. J. Cram, *J. Amer. Chem. Soc.*, **86**, 3767 (1964).
63. S. Winstein and K. Schreiber, *J. Amer. Chem. Soc.*, **74**, 2165 (1952).
64. M. Goodwin Jones and J. L. Coke, *J. Amer. Chem. Soc.*, **91**, 4284 (1969).

65. C. J. Lancelot and P. von R. Schleyer, *J. Amer. Chem. Soc.*, **91**, 4291 (1969).
66. C. J. Lancelot, J. J. Harper, and P. von R. Schleyer, *J. Amer. Chem. Soc.*, **91**, 4294 (1969); C. J. Lancelot and P. von R. Schleyer, *ibid.*, p. 4296.
67. J. M. Harris, F. L. Schadt, P. von R. Schleyer, and C. J. Lancelot, *J. Amer. Chem. Soc.*, **91**, 7508 (1969).
68. P. von R. Schleyer and C. J. Lancelot, *J. Amer. Chem. Soc.*, **91**, 4297 (1969).
69. G. A. Olah, E. Namanworth, M. B. Comisarow, and B. Ramsey, *J. Amer. Chem. Soc.*, **89**, 711 (1967); G. A. Olah, M. B. Comisarow, and E. Namanworth, *ibid.*, **89**, 5259 (1967); L. Eberson, J. P. Petrovich, R. Baird, D. Dyckes, and S. Winstein, *ibid.*, **87**, 3504 (1965).
70. R. Baird and S. Winstein, *J. Amer. Chem. Soc.*, **85**, 567 (1963).
71a. W. G. Dauben and J. L. Chitwood, *J. Amer. Chem. Soc.*, **90**, 6876 (1968).
71b. G. M. Fraser and H. M. R. Hoffmann, *Chem. Comm.*, **1967**, 561.
72. E. D. Hughes, C. K. Ingold, and J. B. Rose, *J. Chem. Soc.*, **1953**, 3839.
73. D. J. Cram and J. Tadanier, *J. Amer. Chem. Soc.*, **81**, 2737 (1959).
74. T. P. Nevell, E. de Salas, and C. L. Wilson, *J. Chem. Soc.*, **1939**, 1188.
75. F. Brown, E. D. Hughes, C. K. Ingold, and J. F. Smith, *Nature*, **168**, 65 (1951).
76. S. Winstein and D. Trifan, *J. Amer. Chem. Soc.*, **74**, 1147, 1154 (1952).
77. H. C. Brown in *The Transition State*, Chem. Soc. Special Publ. No. 16, 1962, p. 140.
78. H. C. Brown, *Chem. in Britain*, **2**, 199 (1966).
79. H. C. Brown, F. J. Chloupek, and M-H Rei, *J. Amer. Chem. Soc.*, **86**, 1246 (1964).
80. P. von R. Schleyer, *J. Amer. Chem. Soc.*, **86**, 1854 (1964); C. S. Foote, *Tetrahedron Letters*, **1963**, 579; P. von R. Schleyer, *J. Amer. Chem. Soc.*, **86**, 1856 (1964).
81. H. L. Goering and C. B. Schewene, *J. Amer. Chem. Soc.*, **87**, 3516 (1965).
82. M. Saunders, P. von R. Schleyer, and G. A. Olah, *J. Amer. Chem. Soc.*, **86**, 5680 (1964).
83. G. A. Olah and A. M. White, *J. Amer. Chem. Soc.*, **91**, 3956 (1969).
84. G. A. Olah and A. M. White, *J. Amer. Chem. Soc.*, **91**, 6883 (1969); G. A. Olah, A. Commeyras, and C. Y. Lui, *ibid.*, **90**, 3882 (1968); G. A. Olah and A. M. White, *ibid.*, **91**, 3954 (1969); G. A. Olah, A. M. White, J. R. DeMember, A. Commeyras, and C. Y. Lui, *ibid.*, **92**, 4627 (1970).

Carbanions

85. D. J. Cram, *Fundamentals of Carbanion Chemistry*, Academic Press, New York, 1965. A broad discussion with particular emphasis on structure and reactivity.
86. H. Fischer and D. Rewicki, *Progr. Org. Chem.*, **7**, 116 (1968). The chemistry of acidic hydrocarbons.
87. A. Streitwieser, Jr., and J. H. Hammons, *Progr. Phys. Org. Chem.*, **3**, 41 (1965). Estimation of hydrocarbon acidity by kinetic and thermodynamic methods.
88. A. I. Shatenshtein, *Adv. Phys. Org. Chem.*, **1**, 155 (1963). Exchange reactions in liquid ammonia.
89. W. J. Albery, *Progr. Reaction Kinetics*, **4**, 353 (1967); M. Eigen, *Angew. Chem. Internat. Edn. Engl.*, **3**, 1 (1964); E. Grunwald, *Progr. Phys. Org. Chem.*, **3**, 317 (1965). Rapid proton-transfer reactions.

90. C. D. Broaddus, *Accounts Chem. Res.*, **1**, 231 (1968). Olefin exchange and equilibration in base.
91. A. W. Johnson, *Ylid Chemistry*, Academic Press, New York, 1966.
92. W. A. Burnett, *J. Chem. Educ.*, **44**, 17 (1967). Hybridization in cyclopropanes.
93. M. Schlosser, *Angew. Chem. Internat. Edn. Engl.*, **3**, 287, 362 (1964). Organosodium and organopotassium compounds.
94. M. Szwarc, *Carbanions, Living Polymers and Electron-transfer Processes*, Interscience, New York, 1968. Largely concerned with anionic polymerization but containing much work on ion-pairs in carbanion chemistry.
95. T. S. Stevens, *Progr. Org. Chem.*, **7**, 48 (1968). Wittig and related carbanion rearrangements.
96. E. C. Ashby, *Quart. Rev.*, **21**, 259 (1967). Structure of Grignard reagents.
97. A. Streitwieser, Jr., E. Ciuffarin, and J. H. Hammons, *J. Amer. Chem. Soc.*, **89**, 63 (1967), and preceding paper.
98. C. D. Ritchie and R. E. Uschold, *J. Amer. Chem. Soc.*, **89**, 1721, 2752 (1967).
99. R. E. Dessy, W. Kitching, T. Psarras, R. Salinger, A. Chen, and T. Chivers, *J. Amer. Chem. Soc.*, **88**, 460 (1966).
100. A. Streitwieser, Jr., and R. A. Caldwell, *J. Amer. Chem. Soc.*, **87**, 5394 (1965).
101. J. E. Hofmann, A. Schriesheim, and R. E. Nickols, *Tetrahedron Letters*, **1965**, 1745.
102. E.g. D. J. Cram and W. D. Kollmeyer, *J. Amer. Chem. Soc.*, **90**, 1791 (1968).
103. M. J. S. Dewar, *Hyperconjugation*, Ronald Press, New York, 1962, p. 53.
104. R. A. Alden and T. G. Traylor, *J. Amer. Chem. Soc.*, **90**, 74 (1968).
105. A. Streitwieser, R. A. Caldwell, and W. R. Young, *J. Amer. Chem. Soc.*, **91**, 527 (1969).
106. G. L. Closs and R. B. Larrabee, *Tetrahedron Letters*, **1965**, 287.
107. A. Streitwieser and W. R. Young, *J. Amer. Chem. Soc.*, **91**, 529 (1969).
108. R. Waack, M. A. Doran, E. B. Baker, and G. A. Olah, *J. Amer. Chem. Soc.*, **88**, 1272 (1966).
109. R. G. Pearson and R. L. Dillon, *J. Amer. Chem. Soc.*, **75**, 2439 (1953).
110. P. Belanger, J. G. Atkinson, and R. S. Stuart, *Chem. Comm.*, **1969**, 1067.
111. H. M. Walborsky, A. A. Youssef, and J. M. Motes, *J. Amer. Chem. Soc.*, **84**, 2465 (1962).
112. R. Breslow and M. Douek, *J. Amer. Chem. Soc.*, **90**, 2698 (1968).
113. H. M. Walborsky, F. J. Impastato, and A. E. Young, *J. Amer. Chem. Soc.*, **86**, 3283 (1964).
114. D. P. Craig, A. Maccoll, R. S. Nyholm, L. Orgel, and L. E. Sutton, *J. Chem. Soc.*, **1954**, 332.
115. J. C. J. Bart, *J. Chem. Soc.* (B), **1969**, 350.
116. E. J. Corey and T. H. Lowry, *Tetrahedron Letters*, **1965**, 793.
117. S. Wolfe, A. Rauk, and I. G. Csizmadia, *J. Amer. Chem. Soc.*, **91**, 1567 (1969); A. Rauk, S. Wolfe, and I. G. Csizmadia, *Can. J. Chem.*, **47**, 113 (1969).
118. C. Rappe, *Acta Chem. Scand.*, **23**, 2305 (1969).
119. A. Streitwieser, Jr., A. P. Marchand, and A. H. Pudjaatmaka, *J. Amer. Chem. Soc.*, **89**, 693 (1967).
120. A. Streitwieser and F. Mares, *J. Amer. Chem. Soc.* **90**, 2444 (1968).
121. J. Hine, L. G. Malone, and C. L. Liotta, *J. Amer. Chem. Soc.*, **89**, 5911 (1967).
122. H. G. Viehe, R. Merenyi, J. F. M. Oth, J. R. Senders, and P. Valange, *Angew. Chem. Internat. Edn. Engl.*, **3**, 755 (1964).

123. R. B. Bates, R. H. Carnighan, and C. E. Staples, *J. Amer. Chem. Soc.*, **85**, 3030 (1963).
124. H. E. Zimmerman, *Tetrahedron*, **16**, 169 (1961).
125. R. A. Dytham and B. C. L. Weedon, *Tetrahedron*, **8**, 246 (1960), and preceding papers.
126a. C. C. Price and W. H. Snyder, *J. Amer. Chem. Soc.*, **83**, 1773 (1961).
126b. S. Bank, A. Schriesheim, and C. A. Rowe, Jr., *J. Amer. Chem. Soc.*, **87**, 3245 (1965).
127a. R. J. Bushby, *Quart. Rev.*, **24**, 585 (1970).
127b. F. Sondheimer and Y. Gaoni, *J. Amer. Chem. Soc.*, **82**, 5765 (1960).
128. U. Schöllkopf and H. Schaefer, *Ann. Chem.* **663**, 22 (1963).
129. U. Schöllkopf, *Angew. Chem. Internat. Edn. Engl.*, **9**, 763 (1970).
130. E. Grovenstein and G. Wentworth, *J. Amer. Chem. Soc.*, **89**, 1852, 2348 (1967).
131. J. K. Brown, D. R. Burnham, and N. A. J. Rogers, *J. Chem. Soc.* (B), **1969**, 1149.
132. A. J. Birch, *Quart. Rev.*, **4**, 69 (1950); A. J. Birch and H. Smith, *ibid.*, **12**, 17 (1958).
133a. R. Gompper, *Angew. Chem. Internat. Edn. Engl.*, **3**, 560 (1964).
133b. N. Kornblum, R. Seltzer, and P. Haberfield, *J. Amer. Chem. Soc.*, **85**, 1148 (1963), and earlier papers.
134. T. E. Hogen-Esch and J. Smid, *J. Amer. Chem. Soc.*, **88**, 307, 318 (1966).
135. P. West, J. I. Purmont, and S. V. McKinley, *J. Amer. Chem. Soc.*, **90**, 797 (1968).
136. G. M. Whitesides, J. E. Nordlander, and J. D. Roberts, *Discuss. Faraday Soc.*, **34**, 185 (1962).
137. G. Fraenkel, D. T. Dix, and M. Carlson, *Tetrahedron Letters*, **1968**, 579.
138. M. B. Smith and W. E. Becker, *Tetrahedron*, **23**, 4215 (1967); F. W. Walker and E. C. Ashby, *J. Amer. Chem. Soc.*, **91**, 3845 (1969); and earlier papers by these groups.

Carbenes

139. T. L. Gilchrist and C. W. Rees, *Carbenes, Nitrenes and Arynes*, Nelson, London, 1969.
139a. W. Kirmse, *Carbene Chemistry*, Academic Press, New York, 1964.
140. J. Hine, *Divalent Carbon*, Ronald Press, New York, 1964.
 These three books provide a broad background to the field.
141. D. Bethell, *Adv. Phys. Org. Chem.*, **7**, 153 (1969). A good modern review on structural and mechanistic aspects of carbenes.
142. G. Köbrich, *Angew. Chem. Internat. Edn. Engl.*, **6**, 41 (1967). The chemistry of α-halogenolithium compounds.
143. G. L. Closs, *Topics Stereochem.*, **3**, 193 (1968). Stereochemical course of carbene addition to olefins.
144. W. Kirmse, *Progr. Org. Chem.*, **6**, 164 (1964).
145. H. M. Frey, *Progr. Reaction Kinetics*, **2**, 131 (1964); W. B. DeMore and S. W. Benson, *Adv. Photochem.*, **2**, 219 (1964); J. A. Bell, *Progr. Phys. Org. Chem.*, **2**, 1 (1963). Reactions of methylene.
146. J. Hine and S. J. Ehrenson, *J. Amer. Chem. Soc.*, **80**, 824 (1958).
147. R. Hoffmann, *J. Amer. Chem. Soc.*, **90**, 1475 (1968).
148. A. M. Trozzolo, *Accounts Chem. Res.*, **1**, 329 (1968).

149. V. Franzen, *Chem. Ber.*, **95**, 1964 (1962).
150. H. W. Wanzlick, *Angew. Chem. Internat. Edn. Engl.*, **1**, 75 (1962).
151. C. S. Elliott and H. M. Frey, *Trans. Faraday Soc.*, **64**, 2352 (1968).
152. W. Kirmse, *Angew. Chem.*, **74**, 183 (1962); J. A. Smith, H. Schechter, and L. Friedman, *J. Amer. Chem. Soc.*, **87**, 659 (1965); S. J. Cristol and J. K. Harrington, *J. Org. Chem.*, **28**, 1413 (1963).
153. A. M. Mansoor and I. D. R. Stevens, *Tetrahedron Letters*, **1966**, 1733.
154. J. W. Powell and M. C. Whiting, *Tetrahedron*, **7**, 305 (1959); L. Friedman and H. Schechter, *J. Amer. Chem. Soc.*, **81**, 5512 (1959); R. H. Shapiro, J. H. Duncan, and J. C. Clopton, *ibid.*, **89**, 471, 1442 (1967).
155. W. von E. Doering and seven co-authors, *Tetrahedron*, **23**, 3943 (1967).
156. W. von E. Doering and A. K. Hoffmann, *J. Amer. Chem. Soc.*, **76**, 6162 (1954).
157. P. S. Skell and R. C. Woodworth, *J. Amer. Chem. Soc.*, **78**, 4496 (1956); P. S. Skell and A. Y. Garner, *ibid.*, p. 3409.
158. H. E. Simmons, E. P. Blanchard, and R. D. Smith, *J. Amer. Chem. Soc.*, **86**, 1347 (1964), and preceding paper.
159. A. Ledwith and L. Phillips, *J. Chem. Soc.*, **1962**, 3796.
160. G. L. Closs and J. J. Coyle, *J. Amer. Chem. Soc.*, **87**, 4270 (1965); R. A. Moss and R. Gerstl, *Tetrahedron*, **22**, 2637 (1966).
161. G. Köbrich, H. Büttner, and E. Wagner, *Angew. Chem. Internat. Edn. Engl.*, **9**, 169 (1970).
162. P. S. Skell and M. S. Cholod, *J. Amer. Chem. Soc.*, **91**, 7131 (1969).

Free Radicals

163. C. Walling, *Free-Radicals in Solution*, Wiley, New York, 1957. A classic text in the field.
164. W. A. Pryor, *Free-Radicals*, McGraw-Hill, New York, 1966. A lucid general account of free-radical chemistry.
165. A. R. Forrester, J. M. Hay, and R. H. Thomson, *The Organic Chemistry of Stable Free-Radicals*, Academic Press, New York, 1968.
166. *Free-Radicals in Solution*, Butterworths, 1967. The Proceedings of an international symposium, also appearing as *Pure Appl. Chem.* (I.U.P.A.C.), **15**, 1 ff (1967).
167. A. F. Trotman-Dickenson, *Adv. Free-Radical Chem.*, **1**, 1 (1965); R. S. Davidson, *Quart. Rev.*, **21**, 249 (1967). Hydrogen-atom abstraction reactions.
168. H. G. Kuivila, *Accounts Chem. Res.*, **1**, 299 (1968). Radical generation by trialkyltin hydrides.
169. J. A. Kerr and A. F. Trotman-Dickenson, *Progr. Reaction Kinetics*, **1**, 105 (1961). Reactions of alkyl radicals.
170. P. B. Ayscough, *Electron-Spin Resonance in Chemistry*, Methuen, London, 1967; R. O. C. Norman and B. C. Gilbert, *Adv. Phys. Org. Chem.*, **5**, 53 (1967); M. C. R. Symons, *ibid.*, **1**, 283 (1963). The latter two reviews on e.s.r. deal with long-lived and short-lived radicals and largely complement one another.
171. CIDNP is reviewed by H. Fischer and J. Bargon, *Accounts Chem. Res.*, **2**, 110 (1969).
172a. P. Jacobson, *Chem. Ber.*, **38**, 196 (1905).
172b. H. Lankamp, W. T. Nauta, and C. Maclean, *Tetrahedron Letters*, **1968**, 249; H. A. Staab, H. Brettshnieder, and H. Brunner, *Chem. Ber.*, **103**, 1101 (1970); R. D. Guthrie and G. R. Weisman, *Chem. Comm.*, **1969**, 1316.

173. D. J. Carlsson and K. U. Ingold, *J. Amer. Chem. Soc.*, **90**, 7047 (1968).
174. P. D. Bartlett and T. Funahashi, *J. Amer. Chem. Soc.*, **84**, 2596 (1962).
175. T. Koenig and R. Wolf, *J. Amer. Chem. Soc.*, **91**, 2574 (1969).
176. P. D. Bartlett and R. R. Hiatt, *J. Amer. Chem. Soc.*, **80**, 1398 (1958).
177. P. Lorenz, C. Rüchardt, and E. Schacht, *Tetrahedron Letters*, **1969**, 2787.
178. Described in most textbooks of reaction kinetics, e.g. A. A. Frost and R. G. Pearson, *Kinetics and Mechanism*, pp. 275–6.
179. J. M. Tedder, *Quart. Rev.*, **14**, 336 (1960).
180. R. E. Pearson and J. C. Martin, *J. Amer. Chem. Soc.*, **85**, 3142 (1963).
181. J. K. Kochi and A. Bemis, *J. Amer. Chem. Soc.*, **90**, 4038 (1968).
182. E. J. Corey and J. Casanova, Jr., *J. Amer. Chem. Soc.*, **85**, 165 (1963).
183. R. H. Hesse, *Adv. Free-Radical Chem.*, **3**, 83 (1969).
184. L. K. Montgomery and J. W. Matt, *J. Amer. Chem. Soc.*, **89**, 934 (1967).
185. J. K. Kochi and P. J. Krusic, *J. Amer. Chem. Soc.*, **90**, 7157 (1968).
186. See *Chem. and Eng. News*, March 9th, 1970, p. 36; many other examples are reported including the series of papers by G. L. Closs and co-workers, *J. Amer. Chem. Soc.*, **91**, 4549 (1969) and references quoted therein.

Chapter 3

Synchronous Reactions

We now discuss an important class of reactions where the timings of bond-making and bond-breaking at carbon may be interlinked. The most important examples are heterolytic reactions which may be subdivided into three broad classes: bimolecular electrophilic substitution (S_E2 reactions), bimolecular nucleophilic substitution (S_N2 reactions), and bimolecular elimination ($E2$ reactions). Considering substitution reactions first of all, the essential difference between the two types is that electrophilic attack on R_3C—X with X as electrophilic leaving group will presumably occur at the σ-bond, directly, while nucleophilic attack on R_3C—X' with X' as nucleophilic leaving group is likely to occur at the carbon atom. This arises because of the general principle that an electrophile will attack at a centre of high electron-density whereas a nucleophile will prefer to attack at a point of lower electron-density, and has important stereochemical consequences (Figure 1).

(a)   (b)

Figure 1. Possible transition-state arrays for (a) electrophilic attack by X and (b) nucleophilic attack by X, with substitution of Y.

It is well known that in S_N2 reactions, where the geometry of the saturated carbon atom undergoing substitution is defined, reaction always proceeds with complete inversion of configuration. The stereochemistry of S_E2 reactions is less clear-cut, but retention of configuration is more frequently encountered than inversion. A rationalization may be made by considering the electrons involved in bond-making and bond-breaking; at the S_N2 transition-state four electrons must be distributed over three atoms, and at the

180

S_E2 transition-state two electrons over three atoms. If the carbon atom undergoing substitution and the entering and leaving groups are considered in isolation, a rationalization may be made in terms of simple Hückel molecular-orbital theory.[1] The energy-levels obtained for a 2-electron 3-centre system very strongly favour a triangular arrangement (corresponding to a retention transition-state) but for a 4-electron 3-centre system a linear arrangement (corresponding to an inversion transition-state) is much preferred (Figure 2).

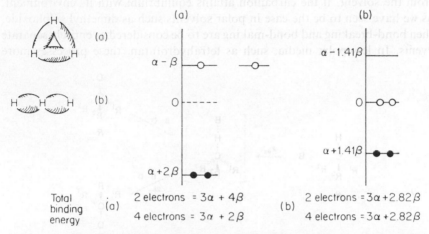

Total binding energy	(a)	(b)
	2 electrons $= 3\alpha + 4\beta$	2 electrons $= 3\alpha + 2.82\beta$
	4 electrons $= 3\alpha + 2\beta$	4 electrons $= 3\alpha + 2.82\beta$

Figure 2. Hückel molecular orbital energies for linear and triangular three-atom systems in terms of Coulomb integrals (α) and resonance integrals (β). Note the preference for a triangular array for two electrons (corresponding to retention of configuration in an S_E2 reaction) and a preferred linear array with four electrons (corresponding to inversion of configuration in an S_N2 reaction).

BIMOLECULAR ELECTROPHILIC SUBSTITUTION REACTIONS

Reactions which may be broadly classified under this heading are encountered in a wide variety of situations in organic and organometallic chemistry. There has been rather less mechanistic work here than in bimolecular nucleophilic substitution, and we shall therefore concentrate more on presenting typical examples than on undertaking a detailed study of transition-state structure. The electrophilic reagent is most frequently a proton or metal ion capable of covalent bonding to carbon, although positive halogen displacements may be considered in this category. The leaving group is most commonly another proton or metal-ion, but certain electrophilic reactions in which carbon–carbon bonds are cleaved may be considered in this context.

Protons as Electrophiles

Proton Exchange in Basic Media

Consider, first of all, reactions in which both the entering and leaving groups are hydrogen ions; these may be conveniently followed when the net result is an isotopic hydrogen exchange. At one mechanistic extreme is the dissociative pathway discussed in Chapter 2 where attack of base on a C—H bond results in formation of a carbanion which subsequently accepts a proton from the solvent. If the carbanion attains equilibrium with its environment, as we have seen to be the case in polar solvents such as dimethyl sulphoxide, then bond-breaking and bond-making are to be considered as entirely separate events. In less polar media, such as tetrahydrofuran, these processes more

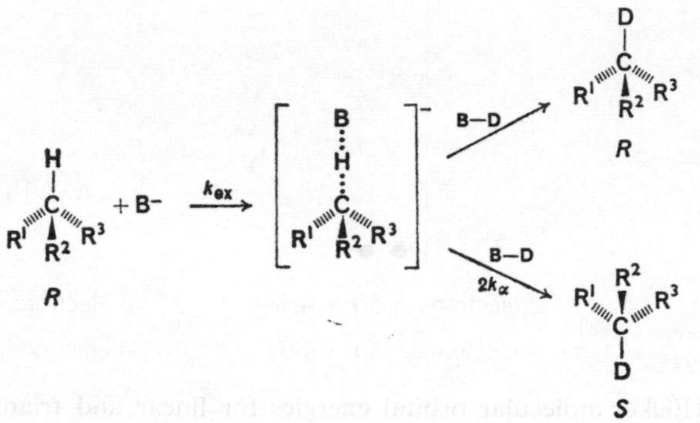

Figure 3. Pathways for exchange at an optically active site.

often occur in concert. Cram and coworkers particularly have studied this reaction,[2] and a reasonably clear picture of its stereochemical course is now available. If we consider the various pathways for exchange at an optically active site (Figure 3), we may distinguish between possible courses of reaction by concurrently measuring the rate of isotope exchange (k_{ex}) (following, for example, uptake of deuterium by mass-spectrometric methods), and the rate of racemization (k_α). These measurements may be related to the overall stereochemical path; thus a value of $k_{ex}/k_\alpha = 10$ means that, for every twenty exchange events, nineteen occur with retention of configuration and one with inversion. A value $k_{ex}/k_\alpha = 1.0$ would require complete racemization during exchange, and $k_{ex}/k_\alpha = 0.6$ implies that five out of every six exchanges occur with inversion. Cram's studies on exchange reactions of hydrocarbons show that, in non-polar media, exchange generally occurs with retention of configuration, even in situations where the intermediate carbanion might be expected to be planar and sp^2-hybridized. A particularly clear example is afforded by the optically active fluorene (1). With 0.3 M ammonia in tetra-hydrofuran,[3] where the solvent cannot donate protons, exchange probably occurs within the confines of a tightly bound contact ion-pair (2) with strong hydrogen-bonding between the counter-ions (Figure 4). To produce exchange with retention, the propylammonium ion must rotate through 109°, and to produce inversion the fluorenide anion partner must rotate through 180°, with complete loss of hydrogen-bonding at some stage. As might be expected, the former process occurs with much greater facility, and exchange was shown to be much more rapid than racemization under these conditions.

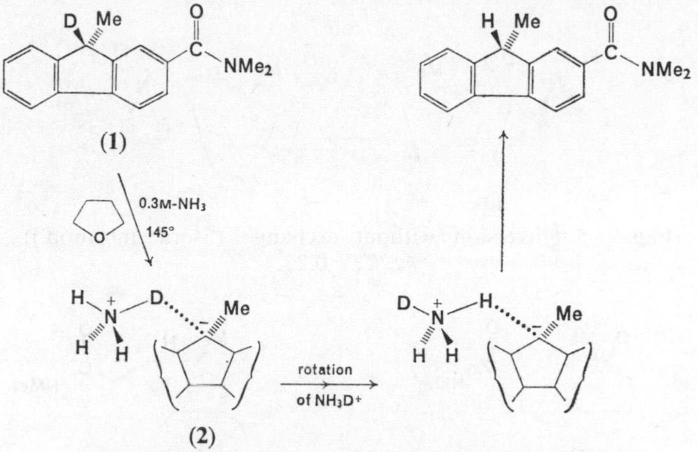

Figure 4. Exchange with retention; $k_{ex}/k_\alpha = 148$.

Where the amine does not supply protons in the intermediate, and the solvent is a poor proton-donor, a rather different situation ensues.[4] The ion-pair (3) when formed undergoes rotation faster than external proton-transfer to the carbanion partner, with the overall effect that inversion is very much faster than exchange; this situation is termed 'isoracemization' by Cram (Figure 5).

In polar protic solvents such as methanol or ethylene glycol, exchange involving predominant *inversion* is commonly observed.[5] This is in keeping with a mechanism in which ionization produces an ion-pair (4) with the ionizing proton still strongly hydrogen-bonded to the carbanion (Figure 6). This strong single hydrogen-bond is displaced preferentially from the rear-side by electrophilic attack of a molecule of solvent, resulting in simultaneous isotopic exchange and inversion of configuration within the ion-pair, which may then collapse to product.

7

A strongly hydrogen-bonded ion-pair analogous to **2** may mediate in the base-catalysed isomerization of olefins. Substituted indenes, e.g. **5**, are particularly attractive substrates for this type of study since they may be

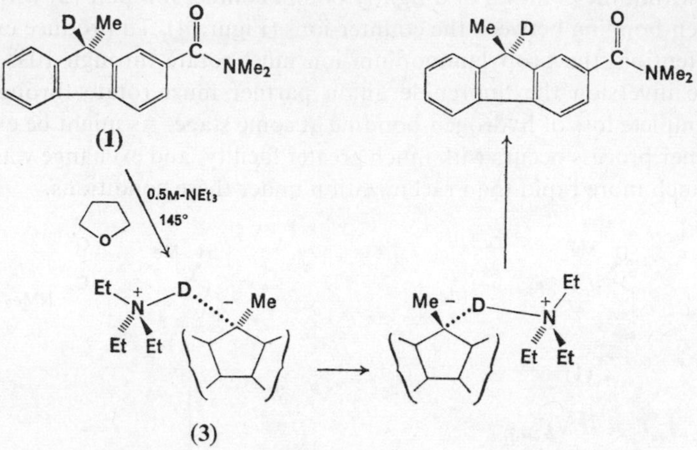

(1)

0.5M-NEt₃

145°

(3)

Figure 5. Inversion without exchange ('isoracemization');
$k_{ex}/k_\alpha = 0.23$.

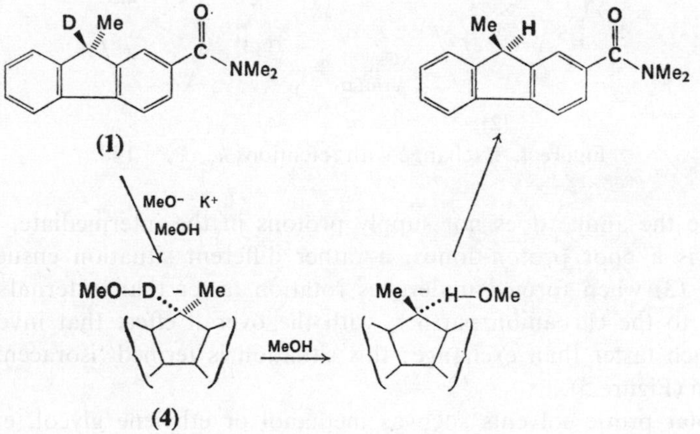

(1)

MeO⁻ K⁺

MeOH

MeOH

(4)

Figure 6. Exchange with inversion; $k_{ex}/k_\alpha = 0.69$. A component of isoracemization is probably involved.

deprotonated under mildly basic conditions, and are readily available as individual enantiomers. A number of systems of this type have been studied by Bergson[6] and Cram[7] and their respective coworkers. In a typical experiment, isotopic exchange and isomerization of **5** to **6** were followed using either propyl[$^2H_2$]amine in tetrahydrofuran or the tertiary amine diazabicyclo-[2.2.2]octane in deuteriated t-butyl alcohol. These experiments (Figure 7) suggested that collapse and rotation within an intermediate ion-pair occurred at similar rates, and that there was little isotope exchange between ion-pair and external solvent.

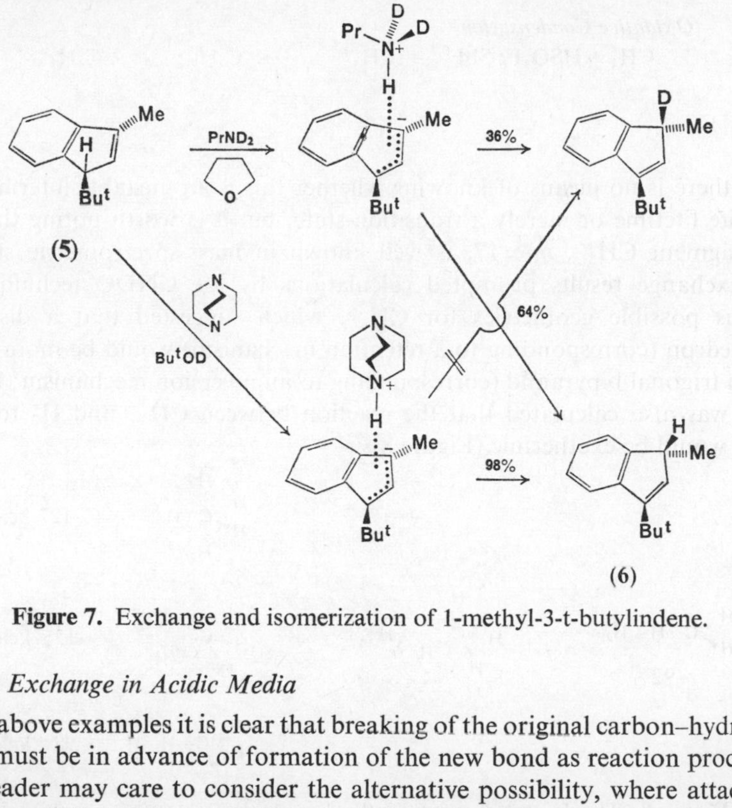

Figure 7. Exchange and isomerization of 1-methyl-3-t-butylindene.

Proton Exchange in Acidic Media

In the above examples it is clear that breaking of the original carbon–hydrogen bond must be in advance of formation of the new bond as reaction proceeds. The reader may care to consider the alternative possibility, where attack by an electrophile proton precedes C—H bond breaking; such a mechanism would require acidic rather than basic reaction conditions. This acid-catalysed reaction has only been recognized relatively recently, for in 1968 the schools of Hogeveen[8] and Olah[9] both recorded that [$^2H_4$]methane in very strongly acidic media such as HF–SbF$_5$ underwent isotopic exchange with the solvent. Many 'super-acid' media are also oxidizing agents, and in HSO$_3$F–SO$_2$ClF–SbF$_5$ the exchange reaction **7** is complicated by competing oxidative condensation reactions, e.g. **8**, in which polymers and stable carbonium-ions are formed.

The observation of an exchange reaction suggests the involvement of a protonated methane, CH$_5^+$, on the reaction coordinate. Without further

Exchange

$$CD_4 + HSO_3F; SbF_5 \ \rightleftharpoons \ \overset{+}{C}D_4H \ (\ddagger, ?) \ \rightleftharpoons \ CD_3H \qquad (7)$$

$$\longrightarrow\!\!\!\!/\!\!\!\!\longrightarrow \ CD_3^+ + HD$$

Oxidative Condensation

$$CH_4 + HSO_3F; SbF_5 \ \rightarrow \ CH_3^+ \ \xrightarrow[CH_4]{(a)} \ C_2H_7^+ \rightarrow \ \rightarrow \ C_4H_9^+ \qquad (8)$$

work there is no means of knowing whether this is an unstable intermediate of finite lifetime or merely a transition-state, but it is worth noting that the ion-fragment CH_5^+, m/e 17, is well known in mass-spectrometric studies. The exchange results prompted calculations by the CNDO technique on various possible geometries for CH_5^+, which suggested that a distorted tetrahedron (corresponding to a retention mechanism) would be more stable than a trigonal bipyramid (corresponding to an inversion mechanism; Figure 1). It was also calculated that the reaction between CH_3^+ and H_2 to form CH_5^+ would be exothermic (Figure 8).[9]

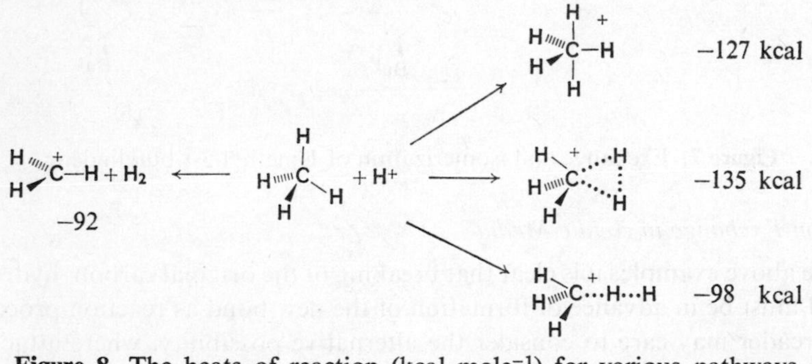

Figure 8. The heats of reaction (kcal mole^{-1}) for various pathways of methane protonation, obtained by an all valence electron calculation.

Reaction (a) in the oxidative condensation reaction **8** suggests that a carbonium ion may act as an electrophile in S_E2 reactions. The reverse of this reaction, which is fragmentation of hydrocarbons in strongly acidic media with carbonium ion as leaving group, might therefore be expected to occur under appropriate conditions. Only one or two examples are known, and these involve loss of a stable carbonium ion. Thus hexamethylethane reacts rapidly in HF–SbF$_5$ to produce isobutane and t-butyl cation.[8] The reaction may be regarded as occurring through protonation of the central carbon–carbon bond, and it is interesting to note that other work has established exchange between the methine proton of isobutane and t-butyl cation with a very low energy of activation.[10] The reaction profiles for these two processes are very probably interconnected (Figure 9).

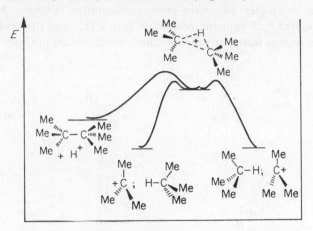

Figure 9. Reaction profiles for the production of isobutane and t-butyl cation from hexamethylethane in HF–SbF₅.

Two factors will determine the facility of cleavage of carbon–carbon bonds; firstly the stability of the carbonium-ion product, and secondly the amount of strain built into the carbon–carbon bond. Highly strained hydrocarbons such as **9** and **10** react very readily with acids and other electrophiles, and simple cyclopropanes are cleaved by strong acids. The reaction involves initial transfer of a proton to cyclopropane, which could occur in one of three different modes. The 'face-protonated' cyclopropane (**11**) is generally considered to be unimportant, but it is rather more difficult to choose between 'edge-protonated' (**12**) and 'corner-protonated' (**13**) species, which CNDO calculations suggest to be relatively close in energy.[10a] The intricacy of the system is revealed in the reaction of cyclopropane with DCl or D₂SO₄ when n-propyl products are formed with deuterium at every position.[11] Only one atom of deuterium is incorporated in each molecule of product so that this

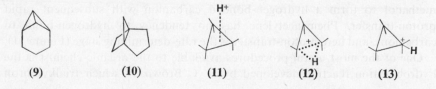

(9)   **(10)**   **(11)**   **(12)**   **(13)**

isotopic scrambling must be occurring in an intermediate faster than that intermediate proceeds to product. Although the detailed mechanism is not absolutely clear, an interconversion of edge-protonated cyclopropanes (Figure 10) may very well be involved. It is worthwhile to note that an edge-protonated cyclopropane resembles an olefin–proton π-complex (Chapter 5), and a corner-protonated cyclopropane resembles both CH_5^+ and the bridged non-classical ions such as norbornyl cation which are discussed in Chapter 2.

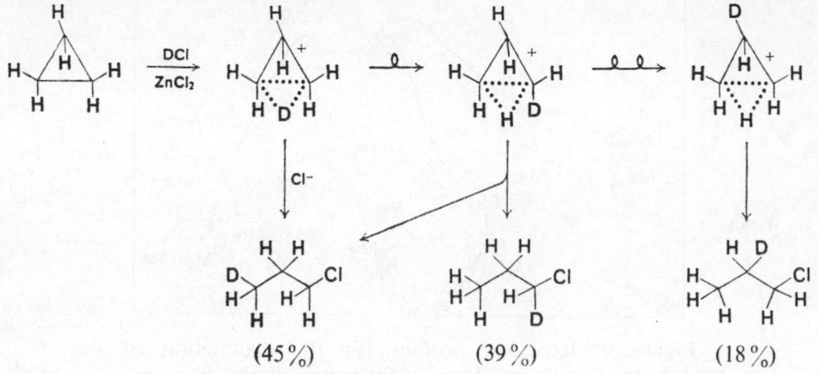

Figure 10. Deuterium scrambling in the addition of DCl to cyclopropane.

Proton Cleavage of Carbon–Metal Bonds

The ease of cleavage of bonds between carbon and electropositive elements by proton sources varies very markedly with the metal. Systems containing electropositive metals, such as butylmagnesium bromide and butyl-lithium, react exothermically with water, but di-n-butylmercury reacts relatively slowly with acetic acid at 25° and tri-n-butylboron requires heating in acetic acid to 100° in order to effect electrophilic cleavage to butane. In a general study of proton-transfer to reactive organometallic compounds, Pocker and Exner[12] showed that there were two fairly distinct mechanisms depending on whether an oxygen or carbon 'acid' was involved in delivering the proton. Thus the rates of proton-transfer by methanol and by methan[²H]ol to benzylmagnesium chloride (**14**) were in ratio 1.15/1.0 but PhC≡C—D reacted 3.4 times more slowly than PhC≡C—H with **14**. It appears that in the first reaction the rate-determining stage is replacement of the metal by methanol to form a hydrogen-bonded carbanion with subsequent rapid proton-transfer. Phenylacetylene has no tendency to hydrogen-bond to carbanions, and here proton-transfer is the rate-determining stage (Figure 11).

One of the most useful procedures available to the organic chemist is the hydroboration reaction developed by H. C Brown, in which trialkylboron compounds which are produced as intermediates may be converted into useful products by electrophilic displacement of boron.[13] Suitable electrophiles include hydrogen peroxide, electrophilic olefins such as acrylonitrile, and proton acids; Brown noticed in his initial studies that carboxylic acids cleaved the carbon–carbon bond more effectively than did aqueous mineral acids. A later kinetic study of the reaction between triethylboron and various

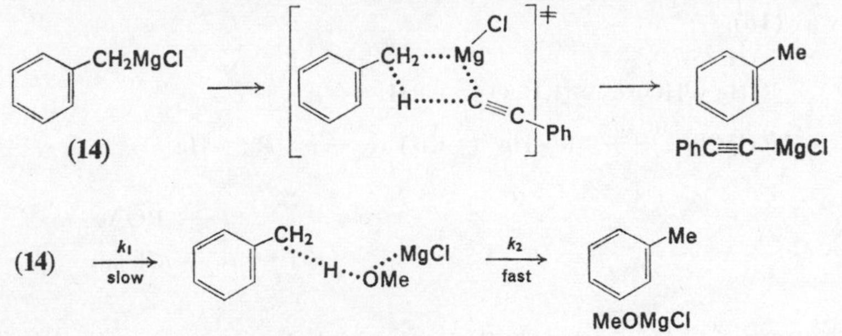

Figure 11. Variation in mechanism in the protonolysis of benzyl-magnesium chloride in ether.

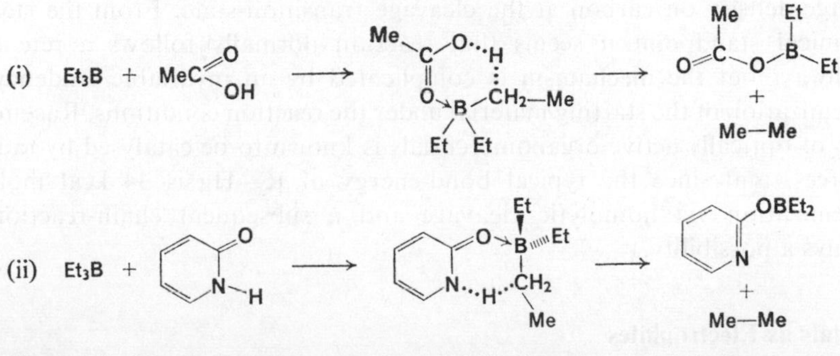

Figure 12. The first stage in the electrophilic cleavage of triethylboron by (i) acetic acid and (ii) 2-pyridone.

organic acids showed an *inverse* relationship between rate of cleavage and acidity.[14] This seemed to be in accordance with an earlier suggestion by H. C. Brown that the reaction pathway involved synchronous electrophilic attack at carbon and nucleophilic attack by carboxy oxygen at boron with a cyclic six-membered transition-state. Particularly good evidence for this pathway stems from the observation[14] that 2-pyridone ($pK_a = 9.7$) is as effective an acid catalyst as acetic acid, but 4-pyridone ($pK_a = 9.4$) is unreactive (Figure 12).

There have probably been more organometallic studies on mercury compounds than on those of any other metal since these species are generally stable, relatively easy to handle, and capable of resolution into enantiomers. Organomercurials[15] are reactive towards acids, but the course of reaction depends on the substrate. For example, dialkyl compounds R_2Hg undergo carbon–mercury bond cleavage in acetic acid by an S_E2 mechanism (15) but under the same conditions alkylmercuric halides react by a carbonium-ion pathway (16).

$$R_2Hg + HOAc \rightarrow RHgOAc + RH \qquad\qquad (15)$$

$$RHgCl \xrightarrow{\text{HOAc}} R\text{---}Hg^+ (+\,Cl^-) \xrightarrow{\text{rapid}} R^+ + Hg \qquad\qquad (16)$$

$$\qquad\qquad\qquad\qquad\qquad\qquad\qquad\qquad \begin{array}{l} \longrightarrow ROAc \\ \longrightarrow \text{Olefins} \end{array}$$

The S_E2 acid cleavage of dialkyl mercurials has been examined in some detail both by the Russian coworkers of Reutov and the American group of Jensen.[15] In general it is apparent that the ease of cleavage of a given R—Hg bond is related to the carbanion stability of R^-, suggesting some negative charge-density on carbon at the cleavage transition-state. From the stereochemical standpoint it seems that reaction normally follows a retention pathway, but the mechanism is complicated by an invariable tendency to racemization of the starting material under the reaction conditions. Racemization of optically active organomercurials is known to be catalysed by radical sources, and since the typical bond-energy of R—Hg is 34 kcal mole^{-1}, racemization via homolytic cleavage and a subsequent chain-reaction is always a possibility.

Metals as Electrophiles

One common property of metal alkyls is their tendency to undergo redistribution reactions. For example, a mixture of diisopropylmagnesium and diphenylmercury in ether is converted, essentially quantitatively, into a mixture of diphenylmagnesium and diisopropylmercury. The tendency of the alkyl or aryl group which is more stable as a carbanion to bond to the more electropositive metal in this reaction has been used as a measure of relative carbanion stabilities (Chapter 2, page 121). We may consider this exchange reaction as an electrophilic substitution at the carbon of the carbon–metal bond. In many cases, alkyl transfers of this type are sufficiently fast to be studied by nuclear magnetic resonance,[16] and the spectral appearance of a solution of a mixture of metal alkyls will be temperature-dependent. It is instructive to consider the behaviour of Group III metal derivatives. Trimethylindium and trimethylgallium both show singlet ¹H n.m.r. spectra in toluene solution, and an equimolar mixture of the two solutions shows a singlet at the mid-point of the chemical shifts for the pure components, even at −65°. This behaviour is consistent with rapid intermolecular exchange between

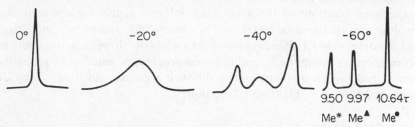

0°   −20°   −40°   −60°

9.50 9.97 10.64τ

Me* Meᐟ Meᐧ

Figure 13. N.m.r. spectra of a mixture of GaMe₃ and Al₂Me₆ in cyclopentane at various temperatures.

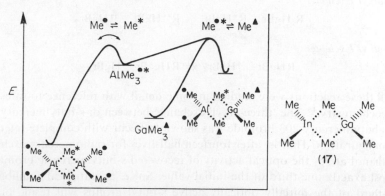

Figure 14. Energy diagram showing the pathways available for methyl exchange in GaMe₃ and Al₂Me₆.

Me₃Ga and Me₃In (both of which are known to be monomeric in solution), and a reasonable transition-state structure is **17** which resembles the known ground-state structure of dimeric trimethylaluminium.* The two types of methyl group in the latter show distinct n.m.r. resonances in toluene solution at −65°, which gradually broaden and coalesce as the temperature is raised, becoming one sharp singlet at room temperature. A ready rationalization is

* Recent X-ray diffraction studies[17] on the structure of Al₂Me₆ suggest that the bridging methyl groups may bond to aluminium through both carbon and hydrogen.

that Al_2Me_6 is in equilibrium with a very small amount of $AlMe_3$, and when equilibrium is sufficiently fast on an n.m.r. time-scale only one methyl signal may be discerned. A mixture of Al_2Me_6 and $GaMe_3$ in toluene shows three methyl resonances at $-60°$, as expected for a slowly exchanging system, which all broaden as the temperature rises and coalesce around $-20°$. The various transition-states involved in these processes may be represented in terms of an energy diagram (Figure 14).

Exchange reactions of this type using different metals, or non-equivalent bonding sites on the same metal, may be successfully studied by n.m.r. Ingold and coworkers used a different approach to study the slower reactions between organomercury compounds, and followed progress of reaction by redistribution of radioactive mercury among different mercury substrates. They distinguished three possible types of reaction:

Three-alkyl Exchange

$$RHgR + R^*HgBr \; \rightleftharpoons \; R^*HgR + RHgBr \tag{18}$$

Two-alkyl Exchange

$$R'HgBr + R^*HgBr \; \rightleftharpoons \; R'^*HgBr + RHgBr \tag{19}$$

One-alkyl Exchange

$$RHgBr + {}^*HgBr_2 \; \rightleftharpoons \; RHg^*Br + HgBr_2 \tag{20}$$

All these reactions were studied in some detail with reference to rates and stereochemistry.[18] The 'three-alkyl' exchange between di-s-butylmercury and (L)-s-butylmercury-203 bromide was shown to occur with complete retention of configuration. That is, after fourteen half-lives for radiomercury exchange in ethanol at $35°$ the optical activity of recovered s-butylmercuric bromide is almost exactly one-third of the initial value. Since, at statistical equilibrium, one-third of the initially optically active s-butyl groups will reside on the mercuribromide residue, this corresponds to 100% retention in every exchange event. The reaction was found to be accurately first-order in each component.

A common feature of many S_E2 reactions of this type is their sensitivity to nucleophilic catalysis, which is particularly apparent in one-alkyl exchanges. The kinetic results of Ingold and Hughes on this reaction in ethanol compelled them to suggest that three simultaneous processes were occurring in the presence of added halide ion: an uncatalysed reaction, a reaction catalysed by one halide ion, and a reaction catalysed by two halide ions. It is reasonable to ascribe transition-states to these processes in which the catalytic halide ions are bonded to mercury (Figure 15). The extent of catalysis can be quite dramatic; for example, addition of lithium bromide in 0.1M concentration to the exchange reaction between MeHgBr and *HgBr$_2$ increases the rate by a factor of 100.

Among many contributions by Reutov and other Russian workers to the field[19] there is a study of the self-immolation of benzylmercuric halides such as **21**. The reaction is promoted by ammonia, which may serve to complex with mercuric bromide and thereby displace the equilibrium. Variation of ring-substituent shows that the transformation obeys a Hammett relationship with

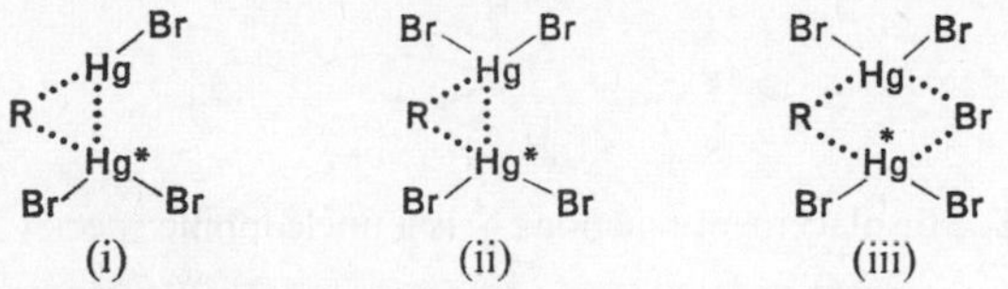

(i)   (ii)   (iii)

Figure 15. Possible transition-state structures in the reaction:

$$RHgBr + *HgBr_2 \rightarrow R*HgBr + HgBr_2$$

where (i) is uncatalysed, (ii) involves 'one-anion' catalysis, and (iii) 'two-anion' catalysis.

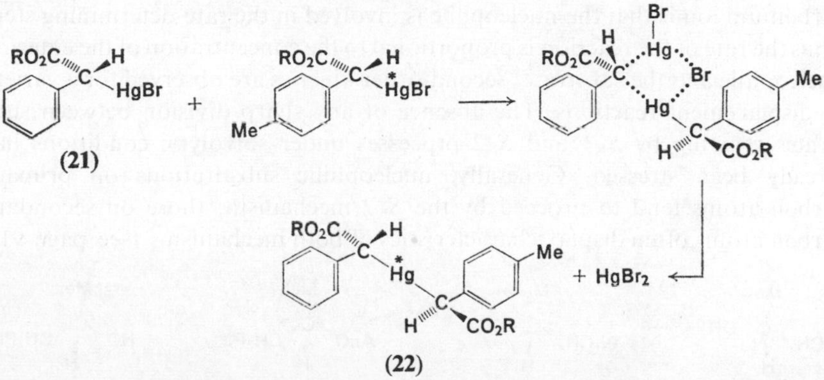

Figure 16. Metathesis of benzylmercuric halides.

$\rho = 2.85$; this demonstrates that reaction is accelerated by electron-with-drawing substances, and implies carbanionic character on the atom under-going substitution at the transition-state. In a reaction mixture containing two different substrates, the 'crossed' process (Figure 16) was found to be dominant. This observation was explained on the basis that electron-donation by Me would facilitate cleavage of the mercury–bromine bond, and radio-mercury labelling of the Me-component confirmed that mercury in the product (**22**) was derived from this.

S_N2 DISPLACEMENT REACTIONS[20-25]

Substitution at a saturated carbon atom through a two-step process and intermediate carbonium ion was discussed in an earlier chapter. Hughes, Ingold, and coworkers also recognized the existence of another important mechanism for substitution, the S_N2 mechanism (substitution, nucleophilic, bimolecular).[26]

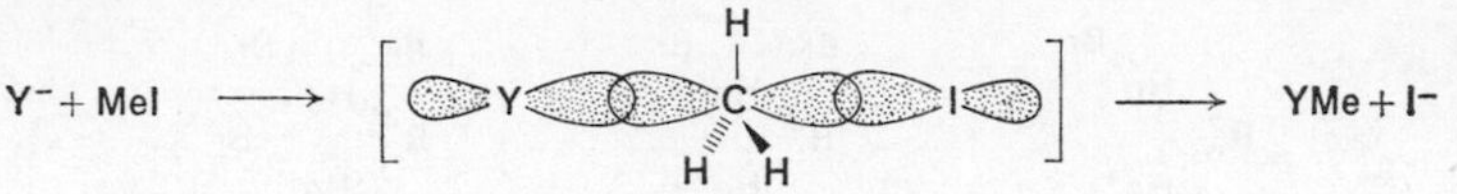

S_N2 displacement reaction; Y is a nucleophilic species

This is a one-step process in which a nucleophile becomes associated with the central carbon atom, whilst the bond between the carbon and the leaving group is breaking. No compelling evidence has been found that 5-coordinate intermediates of finite life are formed in S_N2 processes (but see page 186 for discussion of CH_5^+).

The fundamental difference between this process and that involving a carbonium ion is that the nucleophile is involved in the rate-determining step. Thus the rate of the reaction is proportional to the concentration of the entering nucleophile and the substrate; second-order kinetics are observed for a variety of displacement reactions. The absence of any sharp division between substrates reacting by S_N1 and S_N2 processes under solvolytic conditions has already been stressed. Generally, nucleophilic substitutions on primary carbon atoms tend to proceed by the S_N2 mechanism; those on secondary carbon atoms often display characteristics of both mechanisms (see page 91).

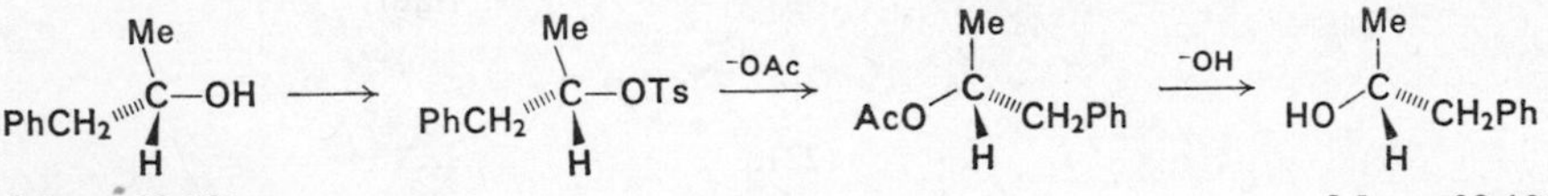

Figure 17. Conversion of an optically active alcohol into its enantiomer; the α-values designate the specific rotation for each of the compounds (5461 Å, 23°).

It was realized from an early stage that an S_N2 displacement at carbon always gives complete inversion of configuration; this is now known as the Walden Inversion. The experimental proof of this was obtained from a group of experiments by Kenyon, Phillips, and coworkers.[27] A series of reactions were conducted in which an optically active alcohol was converted into its enantiomer, and it was unequivocally demonstrated that inversion of configuration occurred in nucleophilic displacement reactions (Figure 17).

The resulting alcohol has a rotation opposite in sign but nearly equal in magnitude to that of the original alcohol. An inversion of configuration must have occurred in the displacement of the tosylate by acetate, since in both the formation of the tosylate and hydrolysis of the acetate the bond from the oxygen to the central carbon remains unchanged.

Radioactive iodide ion was also employed to good advantage and demonstrated strikingly the utility of isotopes in investigations of reaction mechanisms.[28] If optically active (+)-2-octyl iodide is added to a solution of iodide ion in acetone, racemization of the organic iodide takes place (Figure 18). The rate of racemization is proportional to the concentration of iodide ion. When half the (+)-iodide molecules are converted into (−)-iodide, the solution will have become optically inactive since subsequent displacements will bring about no change in specific rotation. When radioactive iodide was used, the iodide exchange resulted in incorporation of radioactive iodine atoms in the organic iodide. The fraction of radioactive iodine in the organic iodide increased until it was equal to the fraction of the iodide in solution. The results showed that the rate of racemization was twice the rate of iodide exchange. Every displacement of iodide therefore results in inversion since (+)-2-octyl

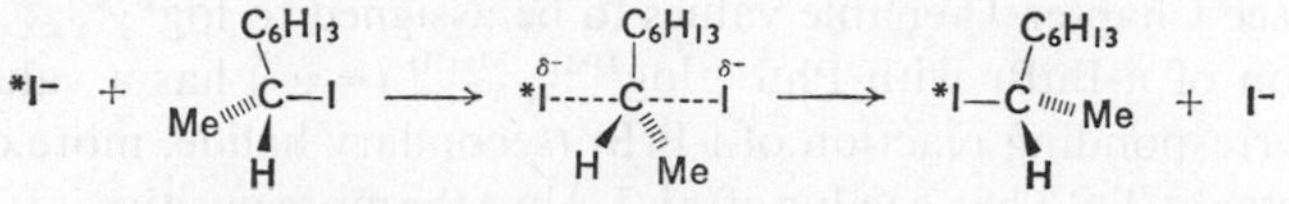

Figure 18. Displacement of 2-octyl iodide by radioactive iodide.

iodide will be completely racemized when half has been inverted to (−)-2-octyl iodide and exchange has proceeded half-way to completion. The consequence of inversion in nucleophilic displacement is readily understandable in molecular-orbital terms. On approach of the nucleophile to the sp^3-orbital a rehybridization to an sp^2-carbon and a p-orbital takes place. Effective overlap is then possible between the leaving group, nucleophile, and the p-orbital of the central carbon. As the leaving group departs, a further rehybridization to an sp^3-carbon takes place. This has already been seen to be the most favourable stereochemical pathway on molecular orbital considerations (see Fig. 2).

Electronic, Solvent, and Steric Effects

Extensive evidence is available that the nature of the transition-state in S_N2 reactions varies with the leaving group, attacking nucleophile, solvent, and the groups attached to the reaction centre. Furthermore, these reactions are extremely sensitive to steric effects and these frequently outweigh electronic effects. Owing to these factors only a few generalizations can be made about electronic effects in these reactions.

Parker suggests that comparison of solvent activity coefficients of the transition-states for S_N2 reactions might give some idea of their structure and charge distribution.[20] Solvent activity coefficients are defined such that $^o\gamma^s$ is proportional to the change in standard chemical potential, μ of a solute, on transfer from a reference solvent, o, to another solvent, s, at temperature T:

$$\mu^s = \mu^o + RT \ln {}^o\gamma^s$$

methanol being chosen as reference solvent. It can be derived from this that the rate constant for a bimolecular reaction in a solvent s is related to the rate constant in a reference solvent o through the appropriate solvent activity coefficients: i.e., for

$$Y^- + RX \rightarrow [YRX^-]^\ddagger \rightarrow products$$

$$\log\left(\frac{k_s}{k_o}\right) = \log {}^o\gamma^s_{Y^-} + \log {}^o\gamma^s_{RX} - \log {}^o\gamma^s_{YRX^-\ddagger}$$

Measurement of rate constants k_s and k_o, and estimates of $\log{}^o\gamma^s_{Y^-}$ and $\log{}^o\gamma^s_{RX}$ (see Chapter 1) enable values to be assigned to $\log{}^o\gamma^s_{YRX^\ddagger}$. Now for the reaction of n-BuBr with PhS$^-$, $\log{}^{DMF}\gamma^{MeOH}_{YRX\ddagger}$ ($= \gamma^\ddagger$) has a value of $+0.8$, and the corresponding reaction of i-PrBr (secondary halide, more carbonium ion character at T.S.) has a value of $+1.7$. Thus the more positive is $\gamma^\ddagger$, the more positive charge is localized at carbon in the transition-state.

The values obtained were considered to be consistent with a spectrum of S_N2 transitions (**23—25**) between the extremes of a completely synchronous process (**23**) and an S_N2 process with substantial carbonium ion character (**25**). From **23** to **25** bond-breaking moves increasingly ahead of bond-formation so that negative charge becomes increasingly localized on Y and X. **23** was termed a 'tight' transition-state and **25** a 'loose' transition-state. The S_N1 transition-state, where bond-breaking proceeds completely ahead of bond-forming is, of course, the extreme form of **25**. Similar conclusions on the nature of these transition-states have been drawn from the measurement of deuterium[29] and chlorine[30a] isotope effects in S_N2 reactions, and also the examination of the k_{OTs}/k_{Br} ratio for nucleophilic displacement reactions of a number of systems.[30b] From this approach it was shown that the ratio k_{OTs}/k_{Br} is small if the nucleophile is powerful and the substrate has a poor ionizing tendency, but this ratio varied over a large range (0.30—5000) corresponding to a continuous spectrum of transition-states. The variation of ratio is due to the potential for p-toluenesulphonate, unlike bromide, to delocalize negative charge over its three oxygens at the transition-state. Consequently the k_{OTs}/k_{Br} ratio depends largely on the degree of charge-separation between carbon and the leaving group in the transition-state; large ratios indicate very ionic transition-states, and small ratios little charge development in the transition-state.

It is instructive to consider the detailed effect of a change of solvent from methanol to a dipolar aprotic solvent on the S_N2 transition-state $[YRX]^{\neq}$ *vis à vis* solvent activity coefficients. The solvation changes depend upon the nature of both Y and X. For example, small anions which are good hydrogen-bond acceptors (Cl^-, N_3^-, $MeCOO^-$) are more effectively solvated by protic than by dipolar aprotic solvents. Large anions, which are polarizable (see page 204) but poor hydrogen-bond acceptors, are better solvated by aprotic solvents. One of the main factors causing this is that polar solvents such as water and formamide have highly developed structures, held together by strong hydrogen-bonds. Dipolar aprotic solvents have 'looser' structures and although dipole–dipole interactions are important there is no hydrogen-bonding. Thus, in general, large polar species, anions or cations or polar molecules, are more solvated by dipolar aprotic solvents. Solvation is also extremely sensitive to the amount of charge on the species, and in the spectrum

of transition-states both in **23** and **24** considerable negative charge is localized on the entering group (Y), but more charge is localized on the leaving group (X) in **24** compared to **23**. Similarly a different amount of positive charge is developed at the carbon atom in the spectrum of transition-states. In **24** and **25** bond-breaking is ahead of bond-forming and the central carbon atom has some carbonium ion character. Although many difficulties exist, it should be apparent to the reader that detailed considerations of solvent change from protic to aprotic solvents can be of value in the investigation of the ionic character of S_N2 transition-states.

Increasing values of γ^+ imply increasingly loose S_N2 transition-states; thus for the reaction $R—Br + N_3^-$, γ^+ becomes more positive as we go from methyl (0.7) to *n*-butyl (1.5) to *i*-propyl (2.1), and likewise from 4-nitrobenzyl (0.2) to 4-methoxybenzyl (+3.0). Clearly the transition-state for 4-nitrobenzyl looks like **23** and for 4-methoxybenzyl like **25**.

This suggests that, for both the reaction of the secondary bromide and 4-methoxybenzyl bromide, the transition-state is a better hydrogen-bond acceptor than the transition-state for the reaction of the primary bromide and 4-nitrobenzyl bromide respectively. In terms of the spectrum of transition-states 23 and 24, that for reaction of the secondary bromide has more negative charge localized on PhS^- and Br^- than in the transition-state for the primary bromide (or for the secondary bromide the transition-state has more of the character found in 25 than that for the primary bromide). A similar conclusion can be drawn for the transition-state of 4-methoxybenzyl bromide compared to that for 4-nitrobenzyl bromide.

Parker and others have pointed out the origin of these differences in transition-state character. The position of the S_N2 transition-state in the spectrum of 23 and 24 is determined by a fine balance between four variables: solvation,

Table 1. Relative reactivities of RCl toward potassium iodide in acetone at 50°.[31]

R	Relative rate[a]	R	Relative rate
Me	200[b]	$PhCH_2CH_2$	1.2
$CH_2{=}CHCH_2$	79	Ph	$<10^{-3}$
$PhCH_2$	200	$ClCH_2$[c]	0.2
$PhCOCH_2$	100,000	$PhOCH_2CH_2$	0.3
$MeCOCH_2$	36,000	$HOCH_2CH_2$	1.7
$NCCH_2$	3000	$EtO_2CCH_2CH_2$	1.6
$MeOCH_2$	920		

[a] Relative to $Bu^nCl = 1$.
[b] Assuming that the reactivity of MeBr relative to Pr^nBr is the same as that of MeCl relative to Pr^nCl.
[c] Ref. 32.

charge development, non-bonded interactions, and bond-energies. The free-energy of the transition-state is lower for stronger contributions from partial C_α—Y and C_α—X bonds, but is raised by non-bonded interactions between X, Y, and the three R groups. Charge development on X, C_α, and Y raises the energy of 24 but a positive charge on C_α can be stabilized by electron-donating R groups, and a negative charge on X can be stabilized by being distributed over a number of atoms (e.g. X = tosylate). Since solvation favours charge production, transition-state 25 is more favoured in protic solvents. For the change from primary alkyl bromide to secondary alkyl bromide, non-bonding interactions in 24 could increase unless a loosening of the C_α—Y and C_α—X bonds takes place. This produces more positive charge on C_α but this is more readily accommodated than at the primary carbon atom, because of the electron-donating methyl groups. The overall balance is for a transition-state more like 25 for the secondary alkyl compound.

In the benzyl system, non-bonding interactions are constant because changes occur only at the *para*-position. The 'looser' or more carbonium ion-like character of the transition-state for the reaction of 4-methoxybenzyl bromides compared to that for the 4-nitrobenzyl bromide is due to the greater electronic stabilization of positive charge on C_α by the 4-methoxybenzyl group.

Some generalizations of electronic effects on nucleophilic substituent reactions can be made. For saturated halides, electron-withdrawing substituents usually decrease S_N2 reactivity and electron-donating substituents produce rate increases, but both effects are fairly small. Thus, 1,2-dibromoethane

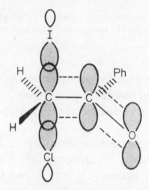

Figure 19. The reaction of phenacyl chloride with iodide ion:

$I^- + PhCOCH_2Cl \rightarrow PhCOCH_2I + Cl^-$

indicating the overlap of the orbital on the carbon atom at which displacement takes place with the π-orbital of the carbonyl group.

reacts about 8 times slower than ethyl bromide with sodium thiophenoxide in methanol. Stabilization of S_N2 transition-states by delocalization produces marked increases in rates. Table 1 is a summary of this effect for the reactivity of allyl, benzyl, and phenacyl chlorides and other compounds towards potassium iodide in acetone. Phenacyl chloride reacts 10^5 times as rapidly as n-propyl chloride. This striking rate increase is almost certainly caused by overlap of the orbital on the carbon atom, at which displacement takes place, with the π-orbital of the carbonyl group (Figure 19).

Streitwieser has assessed some average values for the effect of substituents for a series of nucleophilic displacement reactions of alkyl halides under a variety of conditions (Table 2). A fall in the rate from the methyl halides to ethyl halides is partly attributed to steric hindrance. A further replacement

Table 2. Average relative rates[a] for alkyl halides in S_N2 displacements.[b]

Alkyl group	Relative rate	Alkyl group	Relative rate
Me	3.0	Me_2CH	0.025
$MeCH_2$	1.0	Me_2CHCH_2	0.03
$MeCH_2CH_2$	0.4	$CH_2{=}CHCH_2$	40
$MeCH_2CH_2CH_2$	0.4	$PhCH_2$	120
		Me_3CCH_2	0.00001

[a] From ref. 22.
[b] The average of the displacement reaction of alkyl iodides and bromides with a variety of nucleophiles in acetone, ethanol, methanol, benzene, and nitrobenzene.

of an α-hydrogen atom by an alkyl group produces a further fall in reactivity. The reactivity of neopentyl halides is particularly striking; steric effects cause these derivatives to undergo displacement reactions 10^5 times slower than ethyl derivatives.

Steric hindrance in the transition-state for nucleophilic displacement of neopentyl derivatives (Figure 20) affects both Arrhenius parameters. Substantial steric interactions between Y and X and the three β-methyl groups are present in the transition-state, and these compressions increase the activation energy ($\Delta H^{\ddagger}$). There is a decrease in the entropy of activation ($\Delta S^{\ddagger}$) since interference between the β-methyl groups and X and Y prevent rotation about the C_α—C_β bond; the number of possible configurations of the transition-state, therefore, decrease relative to those of the initial state. Entropy

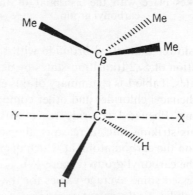

Figure 20. Transition-state for the S_N2 reaction between Y and Me_3C—CH_2X.

effects also contribute to the rate differences for displacement of other alkyl halides, as indicated by the successful calculations of enthalpy and entropy differences in halogen-exchange reactions by Hughes, Ingold, and coworkers.[33]

The reactivity of cycloalkyl halides with nucleophiles varies with the size of the ring. Three- and four-membered ring compounds are much less reactive than the cyclopentyl derivatives. Since the transition-state for an S_N2 reaction has sp^2-hybridized carbon, in the ideal case, the bond angles at carbon should change from 109°28′ in the starting material to 120° in the transition-state (see page 84). In cyclopropyl bromide the bond angle is only 60°, so that the molecule is considerably strained (page 11). Formation of the transition-state for displacement increases this strain in a similar way to that encountered for the S_N1 process, and this results in an increased activation energy and a

Table 3. The reactivity of cycloalkyl bromides towards iodide ion in acetone.[34]

Alkyl group	$10^7 k$ at 70° (l mole^{-1} sec^{-1})
Cyclopropyl	<0.01
Cyclobutyl	0.98
Cyclopentyl	208
Cyclohexyl	1.29
Cycloheptyl	127
Isopropyl	130

slower reaction. The same effect, to a lesser degree, is found for cyclobutyl bromide. Part of the origin of the low reactivity of these compounds also arises out of the substantial non-bonded interactions between the attacking nucleophile and the carbon–hydrogen bonds; this effect also explains the low reactivity of the cyclohexyl halides. Displacement reactions on cyclo-pentyl bromide are about as fast as on isopropyl bromide.

The conformation of the cyclohexyl derivative is important and it could have been anticipated that the axial derivatives would have been the more reactive towards S_N2 displacement owing to their higher ground-state energy and smaller steric hindrance to the approach of the nucleophile. cis-4-t-Butylcyclohexyl bromide (26) undergoes substitution 58 times faster than trans-4-t-butylcyclohexyl bromide (27) with sodium thiophenoxide in 87% aqueous ethanol.[35] Substantial amounts of olefins are also produced in these reactions. Factors affecting the competition between substitution and elimination reactions are discussed later (page 227).

Bridgehead halides are extremely unreactive to S_N2 attack and apocamphyl chloride (28) and 1-iodotriptycene (29) are completely inert to concentrated alcoholic potassium hydroxide.[36, 37] Approach of the nucleophilic reagent from the 'rear side' is, of course, not possible in these cases and thus S_N2 reaction is prevented.

Nucleophilicity

Although nucleophilic reactivity and basicity are governed by many of the same factors, some reagents are frequently more reactive in displacement reactions than could be predicted from their basicity. Thus, both the iodide and hydrosulphide ions are considerably more reactive than the hydroxyl ion although their conjugate acids are stronger acids than water. Basicity is a measure of the affinity of the nucleophile to bond to hydrogen, and some differences from its affinity towards carbon might well be expected particularly since the former is determined by thermodynamic and the latter by kinetic factors. A series of nucleophilic constants were derived by Swain and Scott (Table 4) who suggested the use of the linear free-energy relationship:[38]

$$\log (k/k_0) = ns$$

where k is the rate constant for the reaction of a given anion with an organic substrate, k_0 is the rate constant for the reaction of the same substrate with water, n is the nucleophilicity constant, and s is a parameter characteristic of the substrate that measures the susceptibility of the reaction rate to changes in the nucleophilic activity. The scale was defined by setting $s = 1.0$ for methyl bromide but, in general, this approach has limited application. Attempts to relate nucleophilicity to other experimental values such as oxidation–reduction potentials[39] and molar refractivities[40] have also been made.

An examination of the reactivity of halide ions towards methyl bromide in water immediately illustrates the problem in predicting nucleophilic reactivity.[24] The reactivity sequence $I^- > Br^- > Cl^-$ is opposite to that expected from either the sequence of the strength of the bonds to be formed (C—Cl > C—Br > C—I) or the order expected on the basis of the size of the ion; the smaller ions would have a more concentrated charge and should be more effective nucleophiles ($Cl^- > Br^- > I^-$). The dominant effect is certainly the solvation of ions but ion-aggregation in solution is frequently important in modifying nucleophilic reactivity.

Solvation would be expected to be more extensive with small ions, since the charge is more concentrated, compared to that with large ions. For attack in a nucleophilic displacement reaction, therefore, small ions must lose a greater amount of solvation energy than larger ions, and this is reflected in the reactivity order $I^- > Br^- > Cl^-$ in water for displacement of methyl iodide.[41] The observation that the nucleophilic reactivity of halide ions, in acetone or acetonitrile, is greatly reduced by the addition of hydroxylic

Table 4. Nucleophilic constants, n, of various nucleophilic reagents.[38]

Reagent	n	Reagent	n
H_2O	0	Br^-	3.89
$p\text{-}MeC_6H_4SO_2O^-$	<1.0	N_3^-	4.00
NO_3^-	1.03	$(NH_2)_2CS$	4.1
Picrate	1.9	OH^-	4.20
F^-	2.0	$PhNH_2$	4.49
SO_4^{2-}	2.5	SCN^-	4.77
$MeCOO^-$	2.72	I^-	5.04
Cl^-	3.04	CN^-	5.1
	3.6	SH^-	5.1
HCO_3^-	3.8	SO_3^{2-}	5.1
HPO_4^{2-}	3.8	$S_2O_3^{2-}$	6.36
		$HPSO_3^{2-}$	6.6

solvents, also indicates the importance of solvation. This observation has found successful application to preparative work in the use of anionic nucleophiles in dipolar aprotic solvents. For example, the formation of primary and secondary alkyl cyanides from alkyl halides and cyanide ion is most advantageously performed in dimethylformamide.[23] This solvent effect had been considered to be due to enhanced solvation of the nucleophile in the protic solvent but evidence has been presented that the effect is the result of greater solvation of the transition-states in the dipolar aprotic solvents; this and other medium-effects were considered in more detail on page 40.

The assignment of solvation of ions as a major effect in determining nucleophilic reactivity is confirmed by the observation of the reactivity order $R_4N^+Cl^- > R_4N^+Br^- > R_4N^+I^-$ for the reaction of tetrabutylammonium salts with n-butyl brosylate and isobutyl tosylate in acetone.[42] There is no doubt, however, that a second effect is important since a reactivity order of LiI > LiBr > LiCl is found for the reaction of the above substrates with lithium salts in acetone. The explanation was found in the measurement of the dissociation constants of the salts in solution. From conductivity measurements in acetone and other solvents of low dielectric constant and low ion-solvating ability, it was found that lithium chloride exists predominantly as ion-aggregates. Dissociation to ions depends upon (a) the nature of the counterion and (b) the solvent, and the amount of ion-aggregation decreases from LiCl to LiI. Since these aggregates are very unreactive as nucleophiles the reactivity of the lithium salts increases from chloride to iodide. Ion aggregation with the tetrabutylammonium salts is minimal so that the discrepancy in the reactivity order for lithium and tetrabutylammonium salts, in acetone, is understandable.

Polarizability is also considered to be an important factor in determining the nucleophilicity of ions. In simple terms this has been viewed as the ease

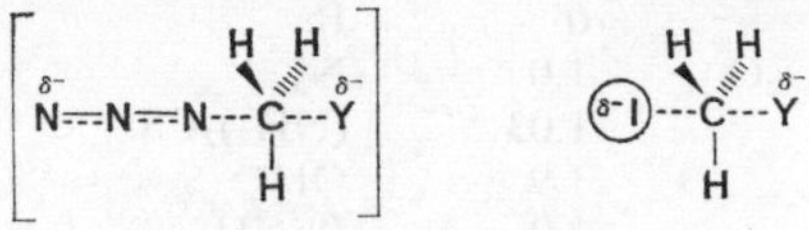

Figure 21. Polarizability in nucleophilic displacement reactions.

with which the electrons of a large atom can move in response to a demand since they are less firmly held by the nucleus than those of a smaller atom. Alternatively, the greater ease of distortion of the outer shell has the effect of decreasing the coulombic repulsion-energy term of the two negative charges in the transition-state. This is also possible with ions which can distribute their charge over several atoms such as the azide ion (Figure 21).

The steric requirement of the attacking reagent has some significance in determining reactivity and this factor becomes more important as steric interactions in the substrate increase. Thus quinuclidine, in which the substituents on the N are held back allowing less steric interference with the substance, reacts over 50 times as rapidly with methyl iodide as triethylamine, but this factor is increased to 700 if the more hindered isopropyl iodide is used.[43] Similarly the relative rates for the reaction of pyridine, 2-methylpyridine, 4-methylpyridine, and 2,6-dimethylpyridine and methyl iodide in nitrobenzene are 1.0, 0.47, 2.27, and 0.042 respectively.[44]

In general, since the leaving group must accommodate a pair of electrons, the facility with which a group departs from a substrate in a displacement reaction is related to its basicity. Streitwieser has summarized average leaving-group ability for a series of reactions (Table 5). However, the effectiveness of a leaving group ion in displacement reactions is interrelated with the attacking nucleophile and reaction conditions employed.

Table 5. Average relative leaving-group[a] ability for S_N2 displacement reactions.

Leaving group	Relative rate
OSO_2F	10^5
OSO_2Ph	6
I	3
Br	1
$^+OH_2$	1^b
$^+SMe_2$	0.5^b
Cl	0.02
ONO_2	0.01
F	0.0001

[a] From ref. 22.
[b] Solvent-dependent; the values listed are for ethanol.

Hard and Soft Acids and Bases

Recently, in an attempt to generalize reactivity effects in nucleophilic and other reactions, the principle of hard and soft acids and bases (HSAB) was proposed.[45] These terms were defined as follows.

Soft Base. Donor atom is of high polarizability, low electronegativity, easily oxidized, and associated with empty, low-lying orbitals.

Hard Base. Donor atom is of low polarizability, high electronegativity, difficult to oxidize, and associated with empty orbitals of high energy.

Soft Acid. Acceptor is of low positive charge, large size, and has several easily excited outer electrons.

Hard Acid. Acceptor atom is of highly positive charge, small size, and does not have easily excited outer electrons.

From equilibrium measurements on the reaction (where B is the base):

$$MeHg^+(aq.) + BH^+(aq.) \rightleftharpoons MeHgB^+(aq.) + H^+(aq.)$$

the following order of decreasing softness of a series of bases was obtained:

$$I^- > Br^- > Cl^- > S^{2-} > RS^- > CN^- > H_2O > NH_3 > F^- > OH^-$$

The proton is the simplest hard acid and the methylmercury cation is one of the simplest soft acids. Table 6 summarizes the nature of bases from the evidence of these and other equilibrium measurements.

It should be appreciated that the terms 'hard' and 'soft' do not mean the same as 'strong' and 'weak'; acids and bases are associated with two properties, their strength and their hardness or softness. Generalized acid–base exchange reactions of the type $AB' + A'B \rightleftharpoons AB + A'B'$, are expected to proceed in such a way that the strongest acid A is found co-ordinated to the strongest base B. The HSAB principle then states, however, that there is an extra stabilization in AB if both the acid and bases are hard or if both are soft.

Table 6. Classification of bases.

Hard	Soft	Borderline
H_2O, OH^-, F^-, $MeCOO^-$, ClO_4^-, NO_3^-, ROH, RO^-, R_2O, NH_3, RNH_2, N_2H_4	R_2S, RSH, RS^-, I^-, SCN^-, $S_2O_3^{2-}$, Br^-, CN^-, H^-, R^-	$PhNH_2$, C_5H_5N, N_3^-, Cl^-, NO_2^-, SO_3^{2-}

In an S_N2 reaction the rate depends on the difference in free-energy of the transition-state and the reactants:

$$B' + A\text{—}B \rightleftharpoons B'\text{----}A\text{----}B \rightleftharpoons B'\text{—}A + B$$

The HSAB principle can be utilized in the consideration of S_N2 displacement reactions since the transition state B'---A---B can be regarded as an acid–base complex. Applied to nucleophilic substitution reactions the general rule can be stated: hard electrophilic centres (acids) react rapidly with hard nucleophiles (bases) and soft electrophilic centres react rapidly with soft nucleophiles. These principles appear to be useful in predicting the mode of reaction of nucleophiles and organic compounds with multiple electrophilic centres. For example, the following reactions of isopropyl bromide are found:

$$Me_2CHBr \xrightarrow[\text{EtONa}]{\text{EtOH}} MeCH{=}CH_2$$

$$Me_2CHBr + H_2C(CO_2Et)_2 \xrightarrow[\text{EtONa}]{\text{EtOH}} Me_2CHCH(CO_2Et)_2$$

in both cases 80 % of the product shown can be isolated but a small amount of substitution and elimination take place in the first and second reaction, respectively. These reactions can be rationalized by attack of ethoxide ion (hard) at the proton (hard) giving elimination and malonate anion (soft) at the tetrahedral carbon (soft) with displacement of bromide ion. The rates of reactions are dependent upon the nature of the leaving group and attacking nucleophiles. As a general pattern the rates of substitution involving soft leaving groups increase with increasing softness of the nucleophile. Thus, with hard bases such as alcohols, alkoxides, and amines in methanol, methyl iodide reacts about 100 times faster than methyl chloride. When softer nucleophiles are employed such as SCN^- ions in methanol, however, methyl iodide reacts ca. 10^4 times faster than methyl chloride as a result of the greater softness of the iodide leaving group. Pearson and Sonntag stress that the HSAB

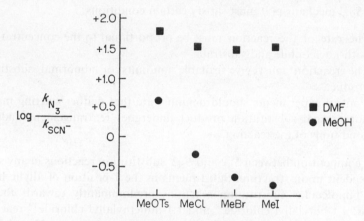

Figure 22. Rate ratios for the reaction of two nucleophiles, N_3^- and SCN^-, with substrates having different leaving groups, in methanol and dimethylformamide (at 25°).

principle is not a theory but a statement about experimental facts. Although the approach appears to have some value, little account is taken of solvent effects. Illustrating this, Figure 22 summarizes the rate ratios of reaction of two nucleophiles with a series of substrates with leaving groups of differing softness. N_3^- has been classified as on the borderline between hard and soft, whilst SCN^- is a soft nucleophile. Some differences would be expected in the rate ratio $\log(k_{N_3^-}/k_{SCN^-})$ with substrate variation. The general trend predicted from the hard–soft principles is found in methanol; methyl iodide, the substrate with the softest leaving group, reacts faster with SCN^- than with N_3^-. However, a marked difference is found in dimethylformamide and no general trend is found in this solvent. It is apparent that detailed solvent studies are essential for any discussion of nucleophilic reactivity.

S_N2' Reactions

For allylic halides a further mechanism is possible in nucleophilic substitution reactions (30).[46] One of the earliest examples of the S_N2' process was the reaction of 1-ethylallyl chloride with the sodium salt of diethyl malonate in

$$X \frown CH_2 = CH - CHR \frown Y \qquad (30)$$

$$MeCH_2CHClCH = CH_2 + {}^-CH(CO_2Et)_2 \rightarrow MeCH_2CH = CHCH_2CH(CO_2Et)_2$$
$$(31)$$

ethanol (equation 31).[47] Generally, for a reaction to be classified as an example of the S_N2' mechanism it must satisfy certain conditions:

1. The rate of the reaction must be proportional to the concentration of both nucleophile and substrate.
2. The reaction must give isolable amounts of abnormal substitution product.
3. Control experiments should demonstrate that neither starting material nor normal substitution product undergoes rearrangement under the conditions of the reaction.

The competition between S_N2 and S_N2' substitution reactions in any system is dependent upon structure. Substituents in the 1-position of allylic halides direct bimolecular substitution reactions predominantly towards the S_N2' process; 1-t-butylallyl chloride[48] and 1,1-dimethylallyl chloride[49] react with nucleophiles to yield almost entirely rearranged product (e.g. 32 and 33). In

$$CH_2 = CHCHBu^tCl \xrightarrow[Cl^-]{EtO^-} EtOCH_2CH = CHBu^t \qquad (32)$$

$$CH_2 = CHCMe_2Cl \xrightarrow{PhS^-} PhSCH_2CH = CMe_2 \qquad (33)$$

both cases steric effects inhibit the formation of the product of direct substitution. Substituents in the 3- and 2-positions, however, inhibit the S_N2' process, and in the reaction of 3-methylallyl bromide with bromide ion in acetone[50] the S_N2 process was at least 10^4 more favoured than the reaction involving rearrangement.

Quantum-mechanical calculations indicated that, in a similar way to substitution at a saturated carbon atom, the S_N2' process should be associated with a distinctive stereochemistry.[51] The stereoelectronic factor in this case is a requirement that the entering and leaving groups are *cis* to each other. This was clearly demonstrated by Stork and White in the reactions of *trans*-6-alkylcyclohex-2-enyl 2,6-dichlorobenzoates (equation 34) with piperidine.[52]

The entering group was shown to take up the position *cis* to that formerly occupied by the displaced group. Similar results were obtained with the diethyl malonate ion but S_N2 reaction accompanied S_N2' for the 6-methyl compound; with larger alkyl groups, however, no S_N2 displacement occurred. Reductive debromination of the allylic bromide by lithium aluminium hydride or deuteride (equation **35**)[53] has also been demonstrated to be an S_N2' process, and it was established that the entering deuterium and leaving bromine were *cis* to each other.

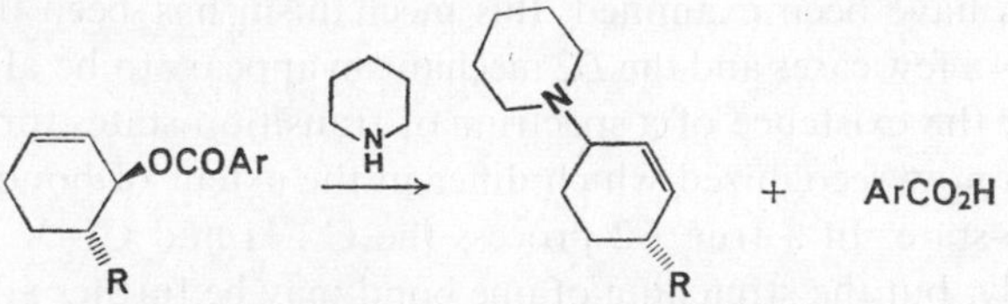

$$\text{(34)}$$

(Ar = 2,6-dichlorophenyl; R = Me, Pr^i, Bu^t)

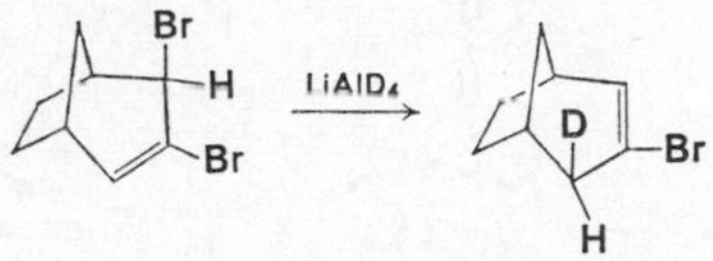

$$\text{(35)}$$

BIMOLECULAR ELIMINATION REACTIONS[54-58]

Accompanying many S_N2 displacement reactions is another process in which more than one bond is undergoing change simultaneously. This is the formation of olefins by a 1,2- or β-elimination reactions. Elimination reactions have already been encountered in the loss of a proton from carbonium ions (*E1*, page 99). Synthetically, however, the most important process for olefin formation is in basic media. Under these conditions a mechanism termed the *E2* process (elimination, bimolecular) operates for a wide variety of leaving groups and bases (equation **36**). Provided that the base and substrate are of comparable concentration, second-order kinetics should be observed for the *E2* process and this has been demonstrated for a large number of combinations of base and leaving group.

E1

$$R_2CHCXR_2 \rightleftharpoons R_2CHC^+R_2 \rightarrow R_2C{=}CR_2$$

E2

$$B + R_2CHCXR_2 \rightarrow BH^+ + R_2C{=}CR_2 + X^- \qquad \text{(36)}$$

($X = {}^+NR_3, {}^+PR_3, SO_2R, OCOR, F, Cl, Br, I;$
$B = NR_3, {}^-OH, {}^-OAc, {}^-OR, {}^-OAr, {}^-NH_2, I^-, {}^-CN,$ *and organic carbanions*)

E1cB

$$B + R_2CHCXR_2 \underset{k_{-1}}{\overset{k_1}{\rightleftharpoons}} BH^+ + R_2\bar{C}CXR_2$$

$$R_2\bar{C}CXR_2 \xrightarrow{k_2} R_2C{=}CR_2 + X^-$$

(37)

A third mechanism, also consistent with second-order kinetics is the *E1cB* (elimination, conjugate base) or carbanion mechanism (equation 37).[56] This is a two-step process, in which proton-abstraction is followed by unimolecular loss of the leaving group from the conjugate base of the substrate. Although many systems have been examined, this mechanism has been demonstrated to exist in only a few cases and the *E2* mechanism appears to be almost universal. However, the existence of a spectrum of transition-states for bimolecular elimination is now recognized which differ in the extent of bond-breaking in the transition-state. In a true *E2* process the C—H and C—X bonds break simultaneously, but the stretching of one bond may be further advanced than the other in the transition-state. The transition-state may resemble a carbonium

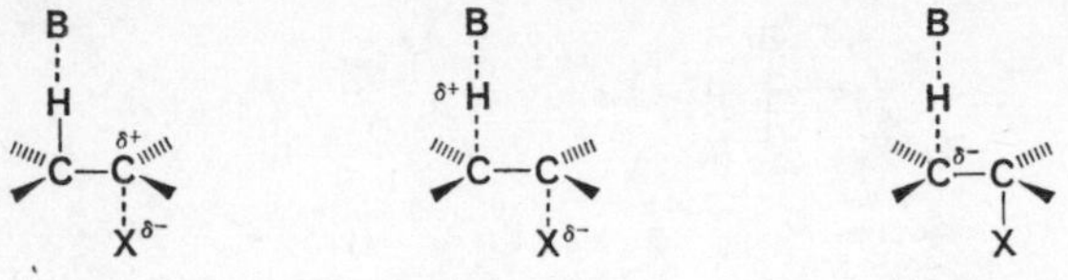

(a) *E1*-like   (b) Central   (c) *E1cB*-like

Figure 23. Spectrum of *E2* transition-states.

ion (a), the olefin product (b), or a carbanion (c) depending on the timing of bond-breaking processes (Figure 23). The carbonium ion and carbanion intermediates of the pure *E1* and *E1cB* mechanisms are thus limiting structures in the spectrum. Evidence for the *E1cB* mechanism will be discussed first since, although much less important than the *E2* process, details of the variable *E2* transition-state will be more fully appreciated when the properties of the extreme mechanisms have been considered.

E1cB Mechanism

Applying the steady-state theory to equation **37**, the following relationship is obtained:

$$Rate = \frac{k_1 k_2 [Substrate][B]}{k_{-1}[BH^+] + k_2}$$

so that second-order kinetics:

$$Rate = k_{obs}[Substrate][B]$$

will be observed for two conditions:

(i) The first step is rate-limiting, and the second is the relatively rapid ejection of the leaving group from the α-carbon ($k_2 \gg k_{-1}[BH^+]$).

(ii) The first step is a rapidly attained equilibrium and the second is the rate-determining, unimolecular decomposition of the carbanion ($k_{-1}[BH^+] \gg k_2$).

For the two mechanistic categories of $E1cB$ eliminations, the first is called the 'irreversible' type and the second, in which the carbanion and substrate are in equilibrium, has been termed the 'pre-equilibrium mechanism'. A number of experimental tests have been applied to assign the mechanistic type of numerous elimination reactions. Positive information has been obtained from hydrogen-exchange experiments, isotope effects, and a detailed study of the kinetics of the reactions.

Hydrogen-exchange experiments have been widely used as a probe for the $E1cB$ mechanism. If a pre-equilibrium carbanionic elimination is proceeding in a protic solvent, β-hydrogen atoms of the substrate will exchange with the solvent protons at a rate which is rapid compared to that of olefin formation. This exchange can be demonstrated by using either the substrate labelled with deuterium or tritium in the β-position, or solvent suitably labelled (e.g. EtOD). If exchange is absent, then an irreversible-type carbanion or $E2$ mechanism is indicated. 1,1,1-Trifluoro-2,2-dihalogenoethanes (38) undergo base-catalysed exchange considerably faster than dehydrofluorination.[59] In these cases, the electron-withdrawing effect of the halogens and the poor leaving ability of fluoride are the important factors favouring carbanion formation before elimination. Trichloroethylene (39) undergoes base-catalysed

$$CF_3CHX_2 \underset{EtONa}{\overset{EtOD}{\rightleftharpoons}} CF_3\bar{C}X_2 \rightleftharpoons CF_3CDX_2$$
$$\text{(38)} \qquad \qquad \downarrow$$
$$CF_2{=}CX_2 + HF(DF)$$

$$CDCl{=}CCl_2 \underset{MeOH}{\overset{^-OMe}{\rightleftharpoons}} \bar{C}Cl{=}CCl_2 \overset{MeOH}{\rightleftharpoons} CHCl{=}CCl_2$$
$$\text{(39)} \qquad \qquad \downarrow$$
$$ClC{\equiv}CCl + HCl(DCl)$$

deuterium exchange considerably faster than dehydrohalogenation.[60] Breslow has raised the point that the demonstration of the existence of a carbanion does not necessarily indicate that this is an intermediate in the elimination reaction, since the carbanion may be an intermediate on a reaction pathway other than the elimination process.[61] Although formally correct, this point can probably be neglected since it could be anticipated that loss of a leaving group from a carbanion, if formed, would be more likely than from the neutral substrate.

The measurement of isotope effects has also been used as a probe for the $E1cB$ mechanism. A substrate can be labelled in the β-position with deuterium or tritium, and the kinetic isotope effects, k_H/k_D, measured for the elimination reaction (equation **40**). For an $E2$ elimination which passes through a central-type transition-state, the theoretical maximum (about 7) for the deuterium

$$
\left.
\begin{array}{l}
CF_3CDCl_2 \xrightarrow{\;k_D\;} \\[2em]
CF_3CHCl_2 \xrightarrow{\;k_H\;}
\end{array}
\right\} \longrightarrow CF_2{=}CCl_2
\tag{40}
$$

isotope effect is predicted. As the character of the transition-state becomes increasingly carbanionic the isotope effect should decrease, and be 1 in the limiting $E1cB$ mechanism since any isotope label would be exchanged rapidly with the medium. Thus the isotope effects k_H/k_D in the dehydrofluorination of CF_3CHCl_2 and $CF_3CHBrCl$ are 1.26 and 1.41 respectively.[62] The observed isotope effects supplement the observation of β-hydrogen exchange as evidence for the assigned pre-equilibrium carbanion mechanism for the elimination reactions of these substrates.

Strong evidence for an $E1cB$ mechanism has recently been obtained for the elimination reactions of the phenyl ethers **41** and **42**.[63] Again it should be

$$
MeSOCH_2CH_2OPh \qquad\qquad Me_2S^+CH_2CH_2OPh\;I^-
$$

$$
\textbf{(41)} \qquad\qquad\qquad\qquad \textbf{(42)}
$$

$$
XCH_2CH_2OPh \underset{k_{-1}}{\overset{k_1}{\rightleftharpoons}} X\bar{C}HCH_2OPh \overset{k_2}{\rightleftharpoons} XCH{=}CH_2 + {}^-OPh
$$

observed that these systems contain a poor leaving group (^-OPh) and a substituent capable of stabilizing a carbanion. Proof of the character of the elimination mechanism was obtained by measurement of the elimination rates of **41** and **42** in aqueous sodium hydroxide, which were compared with the rates of elimination of the corresponding β-deuteriated compounds, XCD_2CH_2OPh, with NaOD in D_2O. For an $E1cB$ mechanism, isotope effects k_H/k_D of about unity would be expected but, since deuteroxide is 1.6 times as strong a base as hydroxide, this would be modified to a $k_H/k_D \approx 0.7$, since the equilibrium formation of a carbanion should be favoured in D_2O over H_2O. The experimental values obtained for the sulphonium salt and the sulphoxide were 0.66 and 0.78 respectively, so the $E1cB$ mechanism was considered to be operating. It should be emphasized again that the experimental evidence of hydrogen-exchange and a k_H/k_D of unity can only indicate the operation of the 'pre-equilibrium' $E1cB$ process. Evidence for the irreversible-type mechanism, which is one extreme of the $E2$ process, will be discussed in the next section.

Transition States in $E2$ Eliminations

Considerable evidence is now available for a spectrum of $E2$ transition-states. Isotope effects have already been mentioned in connection with the $E1cB$ mechanism. Substitution of a deuterium into the β-position of a substrate

Table 7. β-Deuterium isotope effects in the elimination reactions of $PhCD_2CH_2X$ at 30°.[58]

Leaving group (X)	Solvent[a]	k_H/k_D
Br	EtOH	7.1
Br	Bu^{t}OH	7.9
OTs	EtOH	5.7
OTs$^+$	Bu^{t}OH	8.0
SMe$_2{}^+$	EtOH	5.1
SMe$_2{}^+$	H$_2$O	5.9[b]
NMe$_3{}^+$	EtOH	3.0[c]
NMe$_3{}^+$	50% EtOH–H$_2$O	3.0[e]

[a] The base was the conjugate base of solvent.
[b] Extrapolated from 60°, 70°, and 80°.
[c] Determined at 50°.

produces a distinctive effect on the rate of elimination which is dependent upon the nature of the transition-state. The magnitude of the isotope effect for an elimination reaction depends on the bonding of the β-hydrogen in the transition-state. If the hydrogen lies midway between the base and C_β, the value of k_H/k_D will be a maximum. Either more, or less, C—H bond stretching will result in a lower isotope effect. Other evidence must, therefore, also be taken into account when isotope effects are considered for information on the nature of elimination reactions. Measured isotope effects for a number of 2-phenylethyl derivatives are summarized in Table 7. β-Deuterium effects fall in the order Br > OTs > $^+$SMe$_2$ > $^+$NMe$_3$, and this is consistent with the reverse order of increasing proton-transfer to the base in the transition-state; thus the elimination reaction of $PhCH_2CH_2NMe_3{}^+$ passes through a more $E1cB$-like transition-state than $PhCH_2CH_2Br$. Isotope effects associated with the leaving group have also been studied. The elimination reaction of $MeCH_2NMe_3^+$ with EtONa–EtOH showed a nitrogen isotope effect (k^{14}/k^{15}) of 1.017, whilst $PhCH_2CH_2NMe_3^+$ gave a corresponding value of 1.009.[64] This illustrates the ability of the 2-phenyl group to enhance the $E1cB$ character of the elimination of the latter case which allows a smaller amount of C—H bond stretching in the transition-state (and thus a smaller isotope effect). Similar results were found by the use of $^{32}S/^{34}S$ isotope effects.[65]

Further information on the nature of the transition-state of these reactions has been provided by the application of the Hammett equation. The rates of elimination of a series of m- and p-substituted 2-phenylethyl derivatives have been correlated against σ-values. The ρ-values for the reaction are summarized in Table 8. In all cases, a positive ρ-value indicates that the reactions

Table 8. ρ-Values for reaction of $RC_6H_4CH_2CH_2X$ in EtO⁻–EtOH at 30—50°.[58]

X	I	Br	OTs	Cl	$^+SMe_2$	F	$^+NMe_3$
$+\rho$	2.07	2.14	2.27	2.61	2.75	3.12	3.77

are accelerated by electron-withdrawing substituents and that the 2-carbon must be electron-rich in the transition-state. The increasing size of ρ with change in leaving group indicates an increase in the extent of C—H bond stretching in the transition-state and increasing $E1cB$ character. Some idea of the degree of carbanion character in the reactions can be made since DePuy has suggested that the ρ-value of +5 found for the rates of carbanion polymerization of styrenes should be used as a guide for the value expected for an $E1cB$ mechanism.[66]

Orientation in $E2$ Eliminations

Two facets of orientation arise when olefin is produced from an unsymmetrical substrate. One type of orientation (geometrical) is concerned with the proportion of *cis*- and *trans*-olefins which are formed (Figure 24a). Secondly, positional orientation is concerned with the direction of elimination reactions, and for alkyl derivatives two general statements relating leaving group with orientation can be made (Figure 24b). For groups such as halogen or arenesulphonates the elimination proceeds by the preferential removal of hydrogen from the carbon that has the smaller number of hydrogen atoms. This is called the Saytzeff rule and the olefin is formed with the most alkyl groups substituted in the double bond. A second generalization, the Hofmann rule,

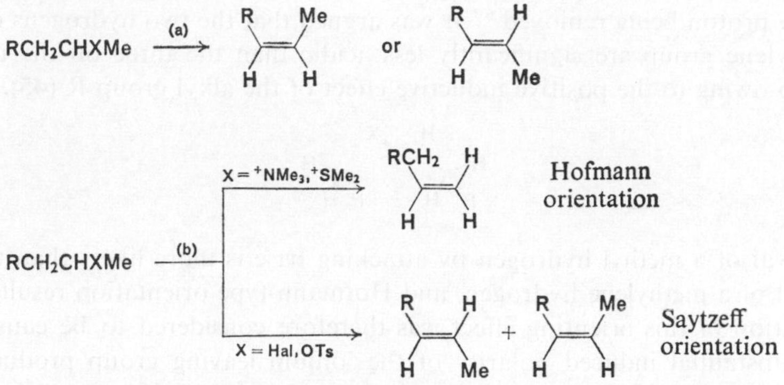

Figure 24. Geometrical and positional orientation in elimination reactions.

states that, in the elimination reactions of tetra-alkylammonium and tri-alkylsulphonium salts, the olefin predominantly obtained is that which contains the smallest number of alkyl groups situated about the double bond. The orientation rules apply only to alkyl structures with no polar substituents, and it should be emphasized that Hofmann or Saytzeff orientation for any substrate may be predominant but not exclusive, and the alternative olefin is produced in varying amounts (equations **43** and **44**). However, the subject

$$MeCH_2CHBrMe \xrightarrow{-OH} McCH{=}CHMe + MeCH_2CH{=}CH_2 \quad (43)$$
$$(81\%) \qquad\qquad (19\%)$$

$$MeCH_2CHMeS^+Me_2 \xrightarrow{-OEt} MeCH_2CH{=}CH_2 + MeCH{=}CHMe \quad (44)$$
$$(74\%) \qquad\qquad (26\%)$$

is discussed in detail in this text since it illustrates the difficulties in mechanistic interpretations due to the interplay of more than one factor. Saytzeff orientation has been already discussed in relation to $E1$ eliminations where the most stable olefin is formed, this stability probably being due to the mesomeric effect of the groups linked α and β about the double bond (page 100). The origin of Hofmann orientation has provoked substantial controversy, and two theories have been advanced to rationalize the experimental facts: one, that it is determined by the polarity of the leaving group and its effect on the acidity of the β-hydrogens, and secondly that Hofmann orientation is determined by steric interactions.

The first hypothesis was that an elimination reaction would proceed according to the Hofmann rule if its direction was determined by the acidity

8

of the proton being removed.[67] It was argued that the two hydrogens on the methylene group are significantly less acidic than the three on the methyl group owing to the positive inductive effect of the alkyl group R (45). Thus,

$$
\begin{array}{c}
\text{H}_{\text{\tiny///}}\quad\text{X} \\
\text{H}\!-\!\underset{\underset{\text{R}\quad\text{H}}{\beta}}{\overset{\alpha}{\diagup}}\!\!-\!\!\underset{\underset{\text{H}\quad\text{H}}{\beta}}{\diagdown}\!\!-\!\text{H}
\end{array}
\qquad (45)
$$

removal of a methyl hydrogen by attacking base is more likely than the removal of a methylene hydrogen, and Hofmann-type orientation results. The operation of this orienting effect was therefore considered to be caused by the substantial induced polarity of the 'onium leaving group producing a transition-state with more carbanion character and less double-bond character than that leading to Saytzeff elimination. Although these suggestions form part of the now well accepted 'variable $E2$ transition state theory', they met considerable criticism from H. C. Brown and others who considered that the major cause of Hofmann orientation was steric interaction.[68] It was noted that leaving groups of substrates which undergo Hofmann elimination are larger than halide ions which are associated with Saytzeff orientation. The fraction of 'Hofmann product' in the mixtures of olefins resulting from $E2$ reactions of 2-pentyl derivatives increases with the size of leaving group (46).[69]

$$
\text{MeCH}_2\text{CH}_2\text{CHMe} \longrightarrow \text{MeCH}_2\text{CH}_2\text{CH}=\text{CH}_2 \;+\; \underset{\text{H}}{\overset{\text{MeCH}_2}{\diagdown}}\text{C}=\text{C}\underset{\text{Me}}{\overset{\text{H}}{\diagup}} \;+\; \underset{\text{MeCH}_2}{\overset{\text{H}}{\diagdown}}\text{C}=\text{C}\underset{\text{Me}}{\overset{\text{H}}{\diagup}}
$$

$$
\overset{|}{\text{X}}
$$

(46)

	X =	Br	$^+\text{SMe}_2$	OSO_2Me	$^+\text{NMe}_3$
$\dfrac{\text{Pent-1-ene}}{\text{Pent-2-ene}}$		0.45	6.7	7.7	ca. 50

Other evidence was also quoted to show that in the $E2$ reactions of alkyl halides the proportion of Hofmann orientation increases with the degree of branching in the substrate or the steric requirements of the attacking base. The hypothesis that these trends were caused by steric interactions was based on a detailed consideration of the transition-states. Assuming an *anti*-periplanar arrangement of attacking base and leaving group (page 220), the conformation leading to the transition-states for Hofmann (47) and Saytzeff products (48) can be written as Newman projections (49 and 50). For 50 it was considered that steric interactions might arise between the groups R and X if either were large. If this was the case conformation 49 would be favoured and Hofmann-type elimination would be observed. Although steric effects must play some part in determining orientation, no allowance is made in this hypothesis for a change in the nature of the transition-state with type of substrate.

A detailed and definitive study of the factors which control the orientation

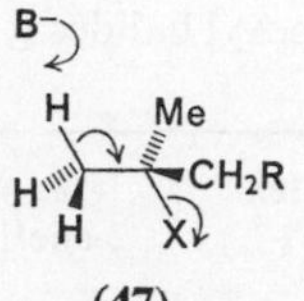

B⁻

H Me
H⟋⟍CH₂R
H   X

(47)

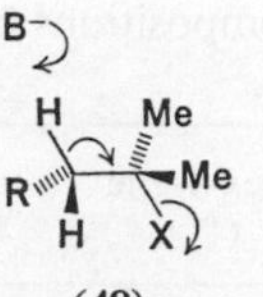

B⁻

H Me
R⟋⟍Me
H   X

(48)

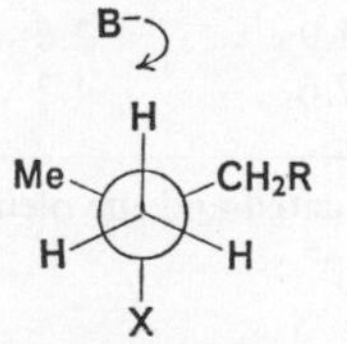

B⁻

H
Me   CH₂R
H   H
X

(49)

B⁻

H
Me   Me
R   H
X

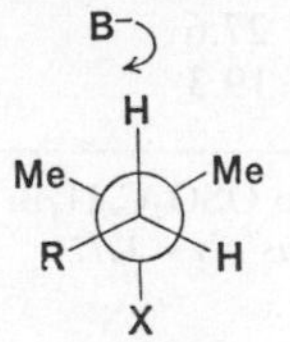

(50)

of $E2$ reactions for halides has been made by Bartsch and Bunnett.[70] Products and rates of reaction of a series of 2-hexyl halides with sodium methoxide in methanol were determined (51). The 2-ene/1-ene ratio and the *trans/cis* ratio increased regularly with change in leaving group $F < Cl < Br < I$ (Table 9). A similar order for the 2-ene/1-ene ratio was also obtained from studies on 2-pentyl halides. It is evident that the elimination of 2-fluorohexane follows the Hofmann rule. A number of observations indicate strongly that all these eliminations occur by the $E2$ mechanism; the $E1$ mechanism is immediately ruled out by the second-order kinetics. Hydrogen-isotope exchange in 2-fluoropentane does not accompany its elimination reaction with sodium ethoxide in ethan[²H]ol, demonstrating that reaction does not occur through carbanion in equilibrium with reactants ($E1cB$).[71] The irreversible formation of a carbanion was also excluded on the basis that 2-fluorohexane, which has

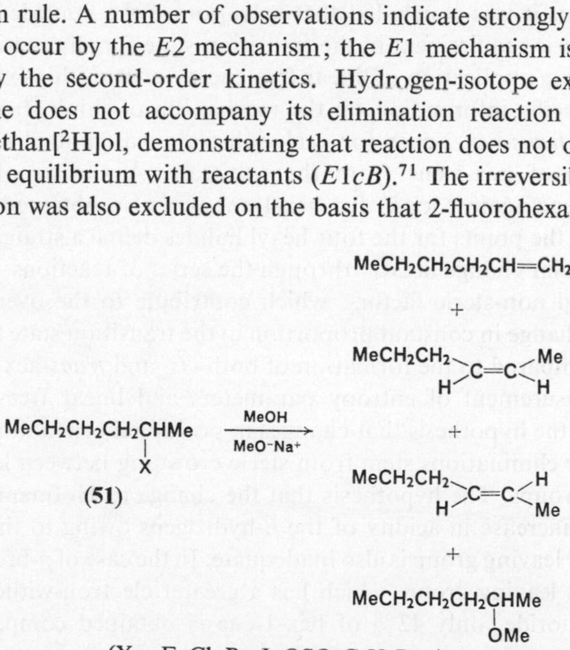

$$MeCH_2CH_2CH_2CH=CH_2$$

+

$$MeCH_2CH_2\!\!\diagdown\!\!{}_{H}C\!\!=\!\!C\!\!\diagup\!\!{}^{Me}\!\!\diagdown\!\!H$$

$$MeCH_2CH_2CH_2CHMe \xrightarrow[\text{MeO-Na}^+]{\text{MeOH}}$$
|
X

(51)

+

$$MeCH_2CH_2\!\!\diagdown\!\!{}_{H}C\!\!=\!\!C\!\!\diagup\!\!{}^{H}\!\!\diagdown\!\!Me$$

+

$$MeCH_2CH_2CH_2CHMe$$
|
OMe

$$(X = F, Cl, Br, I, OSO_2C_6H_4Br\text{-}p)$$

Table 9. Olefin compositions from reaction of 2-hexyl halides with NaOMe–MeOH at 100°.[70]

Leaving group	Hex-1-ene (%)	*trans*-Hex-2-ene (%)	*cis*-Hex-2-ene (%)	2-ene/1-ene	*trans/cis*
F	69.9	21.0	9.1	0.43	2.3
Cl	33.3	49.5	17.1	2.0	2.9
Br	27.6	54.5	17.9	2.6	3.0
I	19.3	63.0	17.6	4.2	3.6

The leaving group $OSO_2C_6H_4Br$-*p* was also investigated and the olefin ratios were 2-ene/1-ene = 1.4 and *trans/cis* = 1.7.

the most strongly electron-withdrawing halogen, would react faster than other halides, whereas it reacts slower.

The observation that 2-fluorohexane yields more hex-1-ene than hex-2-enes would seem completely to rule out the hypothesis that Hofmann orientation is always due to the large steric requirement of the leaving group. However, there is the possibility of solvation of a leaving group and the steric requirement of a group in a reaction is the most important consideration, rather than the size of the group itself. From considerations of the entropy of activation of the reactions it was concluded that solvation of the fluoride ion does not increase its steric requirement in the transition-state for elimination.

An interesting application of linear free-energy relationships was employed to provide specific information on the mechanism of elimination. $\log k_2$ for formation of *trans*-hex-2-enes for each substrate was plotted against $\log k_2$ for formation of hex-1-ene from the substrate. A similar plot of $\log k_2$ for *cis*-hex-2-ene against $\log k_2$ for hex-1-ene was also obtained (Figure 25). In both plots, the points for the four hexyl halides define a straight line which signifies a regular change in $\Delta G^{\ddagger}$ through the series of reactions. It is unlikely that steric and non-steric factors, which contribute to the overall observed $\Delta G^{\ddagger}$, would change in constant proportion in the transition state for hex-1-ene formation compared to the formation of both *cis*- and *trans*-hex-2-ene. Thus, both the measurement of entropy parameters and linear free-energy plots argue against the hypothesis that changes in positional orientation among the 2-hexyl halide eliminations stem from steric crowding between leaving group and β-alkyl groups. The hypothesis that the change to Hofmann orientation is due to the increase in acidity of the β-hydrogens owing to the increase in polarity of the leaving group is also inadequate. In the case of *p*-bromobenzene-sulphonate as leaving group, which has a greater electron-withdrawing property than fluoride, only 42% of hex-1-ene is obtained compared to 70% with 2-fluorohexane.

An explanation for both the positional and geometrical orientation is found in the 'Theory of the variable $E2$ transition state'.[57] The character of the transition-state is determined by a number of factors: the nature of the leaving group which determined the energy required to break the C—X bond, α- and β-substituents, and the nature of the medium. According to this theory Hofmann or Saytzeff orientation is determined by the kinetic effect of a β-alkyl substituent and this is determined by the character of the transition-state. For elimination reactions with central-type transition-states a β-alkyl group lowers the energy of the transition-state by electron-release to the developing double

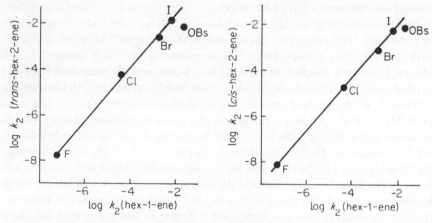

Figure 25. Linear free-energy relationships between logarithms of rates of formation of *trans*-hex-2-ene against hex-1-ene and of *cis*-hex-2-ene against hex-1-ene.

bond so that Saytzeff elimination is favoured. For carbanion-like transition-states the degree of double-bond character is small, but a β-alkyl group interacts unfavourably with the negative charge on C_β and Hofmann orientation predominates. In the reactions of 2-hexyl halides, the nature of the transition-state would be expected to change through the series from iodide to fluoride. As the leaving group becomes poorer the carbanion-like character of the transition-state is increased. The C—X bond becomes harder to break and is augmented by C_β—H breakage in the transition-state. The change to a more carbanion-like transition-state with decreasing atomic number of halogen is also reinforced by an increase in the electron-withdrawing effect of the halogens. The experimentally observed order of increasing amounts of Hofmann orientation from iodide to fluoride conforms to that predicted by this analysis.

Geometrical orientation is also determined by the character of this transition-state. The degree of double-bond character in the C_β—C_α bond is small at both extremes of the spectrum of $E2$ transition-states but relatively large in the centre. Thus, for 'central transition states', the formation of *cis*-olefin is unfavourable owing to eclipsing effects between alkyl groups. A decrease in the *trans/cis* ratio would be predicted with increasing carbanion-like character of the transition-state. For 2-hexyl halides the *trans/cis* ratio does decrease through the series from iodide to fluoride (Table 9) in line with the hypothesis.

Both positional and geometrical orientation, therefore, appear to be predominantly controlled by the nature of the transition-state. While steric interactions appear to have little effect for determining positional orientation in the products from the reaction of 2-hexyl halides with sodium methoxide in methanol, they cannot be discounted in more highly branched structures with larger leaving groups. In fact, predominantly hex-1-ene (about 90%) is obtained from the reaction of all the 2-hexyl halides with potassium butoxide in t-butanol.[72] Positional orientation is therefore seen to vary with the medium employed. While the character of the transition-state for elimination reactions probably changes to more carbanion-like with increasing basicity of the medium, steric effects were considered to be the origin of the substantially greater amount of Hofmann orientation in t-butanol. In the transition-state for formation of hex-2-ene, it has been suggested that steric interactions exist between the large butoxide ion and the n-propyl group on C_β which are much less important when the base (methoxide) or β-alkyl group is smaller. Evidence that the large alkyl group is important is found in the result that the olefin mixture from the reaction of 2-bromobutane with potassium t-butoxide in t-butanol contains only 54% of but-1-ene[73] whereas that from 2-bromo-hexane contains 87% of hex-1-ene. These steric interactions would be expected to be even greater with larger leaving groups. In conclusion, both the character of the transition-state and steric effects must be considered in making prediction of orientation in elimination reactions.

Stereochemistry in $E2$ Eliminations

Stereospecificity is a feature of many olefin-forming reactions. Generally β-eliminations proceed most readily if the leaving groups are in an *anti*-periplanar conformation (52) and the torsional angle between C—H and C—X is 180°. Thus the H—C_β and C_α—X bonds are coplanar and the process is termed an *anti*-elimination. One of the factors governing the stereospecificity is believed to originate from the necessity to minimize the repulsion energy between the migrating electron-pairs, and a theoretical explanation has been proposed by Fukui and Fujimoto.[74] In *syn*-eliminations torsional effects favour a staggered arrangement (53) but this is opposed by the stereoelectronic demand for coplanarity. It is important to notice that the terms *anti-* and *syn*-elimination do not refer to the stereochemistry of the olefin produced but to the arrangement of leaving groups in the conformation leading to reaction.

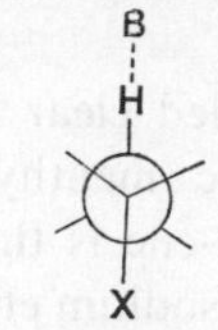

(52) 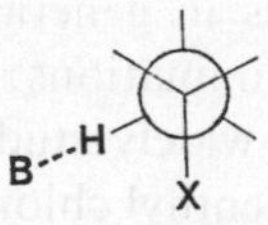   (53)

Numerous examples are available illustrating the preference for *anti*-eliminations. Thus, dehydrobromination of the two diastereoisomeric dibromostilbenes yields two different α-bromostilbenes. The *meso*-dibromide (54) gives the *cis*-olefin whereas *trans*-olefin is produced from (±)-dibromide

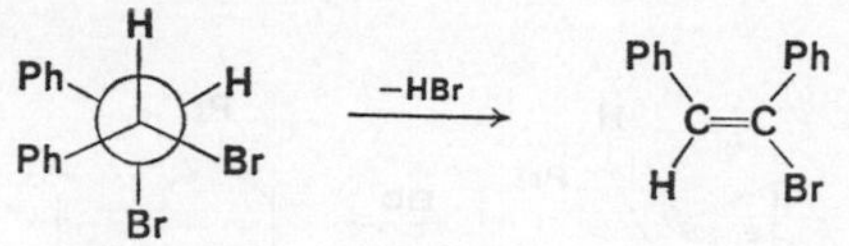

meso (54)

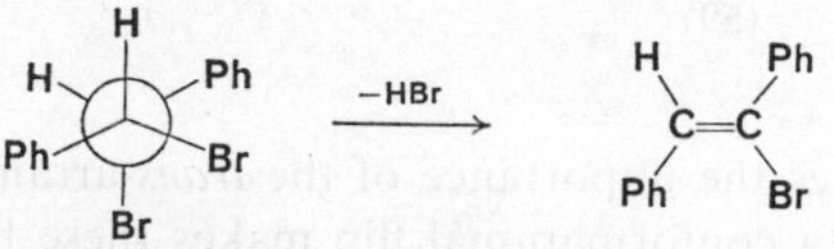

(±) (55)

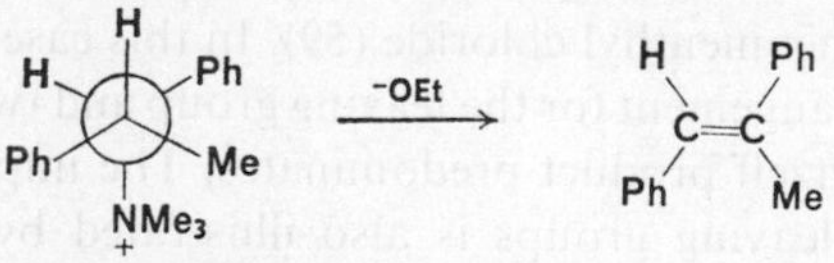

threo (56)

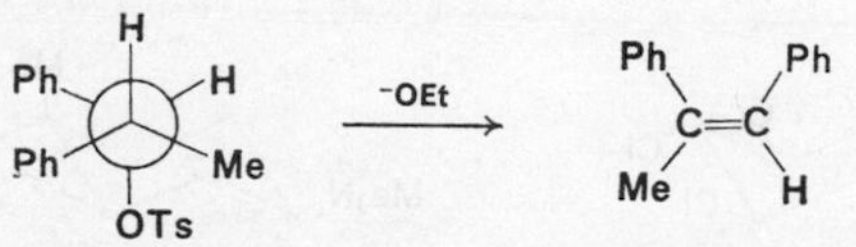

erythro (57)

(55). Similarly, the *erythro*-compound (57) yields *cis*-olefin, and *trans*-olefin is produced by the reaction of the *threo*-compound (56) with sodium ethoxide in ethanol. Furthermore, 56 reacts over 50 times as rapidly as the *erythro*-form, corresponding to a difference in ΔG^{+} of 2.3 kcal mole^{-1}.[75] There is little doubt that larger steric interaction in the transition-state leading to *cis*-olefins, than in that leading to *trans*-olefins, is largely responsible for this energy difference.

Alicyclic compounds in general have provided clear information on the stereochemistry of eliminations; the isomeric menthyl and neomenthyl derivatives have been widely studied. Menth-2-ene is the exclusive product from the reaction of menthyl chloride (58) with sodium ethoxide in ethanol.[76]

This result illustrates the importance of the *trans*-arrangement of the C—H and C—Cl bonds (a conformational flip makes these bonds axial) since, in contrast to accepted orientation rules, the *anti*-Saytzeff product is formed. A mixture of menth-2-ene (25%) and menth-3-ene (75%) is formed from a similar reaction of neomenthyl chloride (59). In this case, the structure allows a *trans*-coplanar arrangement for the leaving group and two β-axial hydrogens: as expected the Saytzeff product predominates. The importance of the *trans*-orientation of the leaving groups is also illustrated by the extremely slow dehydrochlorination of 60, in which all the chlorine atoms are equatorial. Elimination of HCl from this compound is 10^4 times slower than from any other isomer of 1,2,3,4,5,6-hexachlorocyclohexanes.[77] Similarly 3β-chol-estanyltrimethylammonium salts **(61)** do not show $E2$ reactions, but only react by S_N2 substitution (other factors determining elimination/substitution ratio are discussed later). This steroid derivative has the leaving group in an equatorial position and can only take up a *trans*-coplanar arrangement with the C—H bond in a highly unfavourable boat conformation.

Although the above discussion has illustrated the preferential *anti*-stereo-chemistry for elimination reactions, numerous examples have been documented where the eliminated groups are *syn* and coplanar. DePuy and coworkers[78] showed that the elimination reaction of *trans*-2-phenylcyclopentyl tosylate **(62)** with potassium t-butoxide in t-butyl alcohol was only 9 times slower than elimination from *cis*-2-phenylcyclopentyl tosylate **(63)** (Table 10). 1-Phenyl-cyclopentene is formed in both cases so that a *syn*-elimination must be occurring

Table 10. Second-order rate constants for elimination re-actions with potassium t-butoxide in t-butyl alcohol at 50°.[78]

Toluene-*p*-sulphonate	Type of elimination	10^4k (l mole^{-1} sec^{-1})	k_{anti}/k_{syn}
63	*anti*	26.4	9.1
62	*syn*	2.9	
65	*anti*	1.92	} $>10^4$
64	*syn*	no reaction }	

from the former compound and an *anti*-elimination from the latter. Similar measurements for the analogous 2-phenylcyclohexyl compounds **(64** and **65)** show that *anti*-elimination occurs at least 10^4 times faster than the *syn*-elimination. This difference was attributed to the fact that the eliminating groups can take up a *cis*-coplanar arrangement easily in the cyclopentyl but not in the cyclohexyl system. A boat conformation would be necessary for this arrangement in the latter system. Earlier workers had considered that *syn*-eliminations might proceed by an $E1cB$ mechanism. Evidence that this is not the case, and that the *syn*-elimination is concerted and does not involve an intermediate carbanion, was obtained for the cyclopentyl system. Intro-duction of substituents into the phenyl ring enabled the ρ-value for the *syn*-and *anti*-elimination of toluene-*p*-sulphonic acid to be measured. For the reaction of *trans*-2-arylcyclopentyl tosylate with potassium t-butoxide in t-butyl alcohol, the elimination (*syn*) has a ρ-value of 2.8 which is only a little larger than the ρ-value of 1.5, obtained for the *anti*-elimination of *cis*-2-arylcyclopentyl tosylate. It is apparent that, although the transition-state for

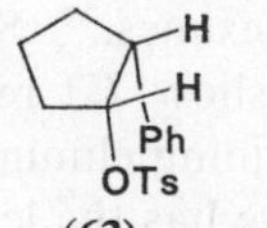

(63)

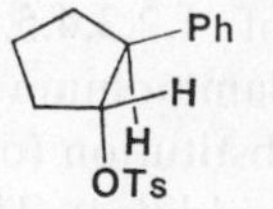

(62)

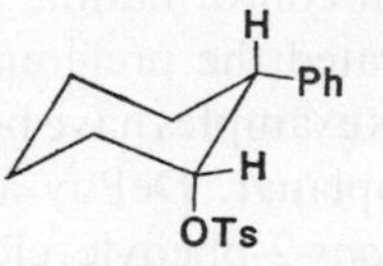

(65)

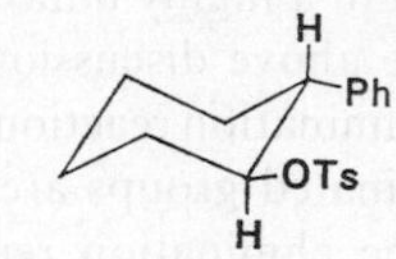

(64)

the *syn*-elimination has slightly more carbanionic character than the *anti*-elimination, both are essentially concerted. Further evidence for this was found in that the kinetic isotope effect k_H/k_D is 5.6 at 50° for the elimination reaction of **62**.

The above discussion has illustrated that the ease of *syn*-eliminations can be comparable with that of *anti*-eliminations for some substrates. In fact a *syn*-elimination can also become more favourable in certain structures. In some substrates, although the leaving groups are *trans*, a dihedral angle of only about 120° or 150° can be achieved due to restrictions imposed by ring systems (**66**). This effect of non-coplanarity is substantial; thus it is found that *syn*-elimination from *trans*-2,3-dihalogenonorbornane (**67**) is faster than *anti*-elimination from *cis,endo*-2,3-dihalogenonorbornane (**68**), since co-planarity can be achieved in the transition-state for *syn*-elimination but not for the *trans*-dehydrohalogenation owing to the geometric rigidity of the system.[79] The same reactivity order *syn* > *anti* has also been found for the dehydrochlorination of the 11,12-dichloro-9,10-ethanoanthracenes (**69** and **70**).[80]

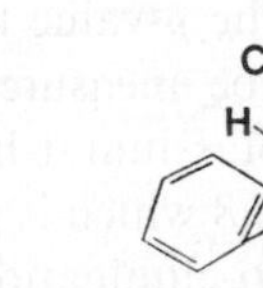

(66)

(67)

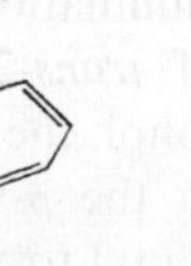

(68)

(69)

(70)

The factors determining whether a *syn*-elimination takes preference over *anti*-elimination are not yet clearly defined. Certainly the existence of a *cis*-coplanar arrangement of leaving groups is essential, and it also appears that a *syn*-elimination is more favoured when the transition-state has more carbanion character. Sicher and coworkers have conducted and inspired numerous studies on the stereochemistry of elimination reactions.[81] They suggested that the elimination reactions of some cycloalkyltrimethylammonium ions to yield *trans*-cycloalkenes proceeded by a *syn*-mechanism.[82] Of course, the formation of a *trans*-double bond is only possible for rings containing eight or more carbon atoms. The rates of formation of *cis*- and *trans*-olefins from a series

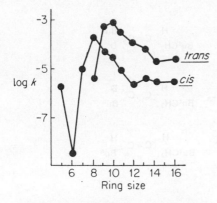

Figure 26. Effect of ring size on rates of formation of *cis*- and *trans*-cycloalkenes from cycloalkyldimethylamine *N*-oxides in t-butyl alcohol at 70.6°.

Figure 27. Effect of ring size on rates of formation of *cis*- and *trans*-cycloalkenes from cycloalkyltrimethylammonium chloride and potassium t-butoxide in t-butyl alcohol at 55°.

of cycloalkyldimethylamine *N*-oxides in t-butyl alcohol were measured. The rates for the two series showed a maxima for the eight- and ten-membered rings for formation of *cis*-olefin and *trans*-olefins respectively (Figure 26). Since it is known that amine *N*-oxides undergo elimination by a *syn*-mechanism (page 230), the dependence of the rate of elimination of cycloalkyldimethylammonium ions on ring size was taken to be characteristic of *syn*-eliminations. A similar treatment was also applied to the rates of elimination of cycloalkyltrimethylammonium chlorides with potassium t-butoxide in t-butyl alcohol for the formation of *cis*- and *trans*-olefins (Figure 27). Comparison of Figures 26 and 27 shows that the formation of *trans*-olefin from the trimethylammonium ions and amine oxides changes in a similar manner with ring size, but that formation of *cis*-olefins differs considerably. From this evidence it was suggested that the $E2$ elimination of cycloalkylammonium ions to form *trans*-olefin proceeds by a *syn*-mechanism, and that *cis*-olefins are formed by an *anti*-mechanism. These studies were extended to other solvent–base systems, and it was found that the proportion of *trans*-olefin formed from all the cycloalkylammonium salts studied increased as the solvent–base system varied in the series $HOCH_2CH_2OH-HOCH_2CH_2OK < MeOH-MeOK < EtOH-EtOK < Pr^iOH-Pr^iOK < Bu^tOH-Bu^tOK$. The increasing *trans*-olefin production was considered to be associated with an increase in the $E1cB$-like

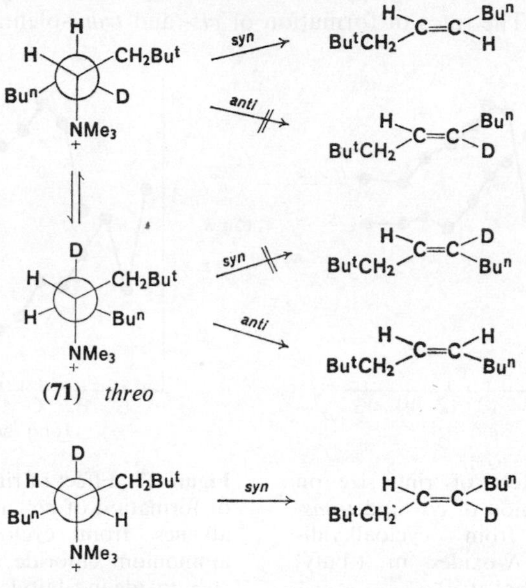

Figure 28. *syn*- and *anti*-Elimination in the elimination reaction of *threo*- and *erythro*-quaternary ammonium salts.

character of the transition-state resulting from an increase in the proton affinity of the base and a decrease in the solvating power of the solvent. These results were taken as evidence that the transition-states of $E2$ *syn*-elimination have $E1cB$-like character.

Further extensive studies have also demonstrated a *syn*-elimination mechanism for the formation of *trans*-olefin from the acyclic quaternary ammonium salts **71** (*threo*) and **72** (*erythro*) in the media MeOH–MeOK, Bu^tOH-Bu^tOK and Bu^tOK-Me_2SO.[80] Formation of *trans*-olefin from **71** showed an isotope effect k_H/k_D in the region 2.3–4.2 (depending upon the medium). This indicated a *syn*-elimination mechanism, and consistent with this assignment the *trans*-olefin was found to be formed with almost complete deuterium loss (Figure 28). The formation of *cis*-olefin from **71** was associated with an isotope effect k_H/k_D of 3.1—4.7 and loss of deuterium from the original quaternary salt. These results point emphatically to the formation of *trans*-olefin by a *syn*-mechanism and the formation of a *cis*-olefin by an *anti*-elimination process. Consistent with the hypothesis, formation of both *cis*- and *trans*-olefins from the *erythro*-isomer showed virtually no isotope effect (k_H/k_D= 0.9—1.2), and both were formed without loss of deuterium by *anti*- and *syn*-mechanisms respectively.

These results are seen to parallel those described previously for the cycloalkyl derivatives. Similar results have also been found for other leaving groups. Formation of *trans*-olefin from medium- and large-ring cycloalkyl bromides with potassium t-butoxide in t-butyl alcohol also shows characteristics of a *syn*-mechanism, but reactions to produce both *cis*- and *trans*-olefin in potassium ethoxide in ethanol have all the characteristics of an *anti*-elimination. Whilst it is clear that the transition-state for *syn*-elimination has some carbanion character, ion-pairing may well be of great importance. Thus, in **73** the anion may attack close to the 'onium ion preferentially, and with the cycloalkyl bromides the *syn*-elimination in t-butyl alcohol, but not in ethyl alcohol, may be due to the more favoured formation of ion-pairs in the former solvent (**74**).

(73)   (74)

Further studies with alkyl ammonium salts, bromides, and toluene-*p*-sulphonates support the view that ion-pairing is an important factor in the *syn*-mechanism since the amount of this process increases with decreasing dissociating powers of the solvent.

Competition between Elimination and Substitution

The attack of a nucleophilic reagent on a substrate RX may result in either an S_N2 or an $E2$ reaction, so that the yield of olefin is determined by the relative rates of the reactions. In some structures one of these reactions is frequently favoured almost exclusively because of the nature of the structure or the choice of reaction conditions. A knowledge of the factors affecting the ratio $E2/S_N2$ can be of considerable advantage in the synthetic applications of these reactions.

The nature of the alkyl group has a considerable effect on the competition between elimination and substitution, and Table 11 summarizes results of the reaction of alkyl bromides with sodium ethoxide. Branching at either the α- or β-carbon generally retards S_N2 reactions as a result of steric effects. At the same time branching at the α-carbon accelerates the elimination reactions owing to stabilization of the transition-state by hyperconjugative effects. The effect of β-substitution on the rate of elimination varies with the type of $E2$ reaction. For these reactions with central-like $E2$ transition-states, such as

Table 11. Rates of elimination and substitution in the reaction of alkyl bromides with sodium ethoxide.[83]

Alkyl bromide	Temp. (°C)	$10^5 k_2$ (l mole^{-1} sec^{-1})			Olefin (%)
		Total	S_N2	$E2$	
EtBr	55	174	172	1.6	0.9
Pr^{n}Br	55	60	54.7	5.3	8.9
Bu^{n}Br	55	43.9	39.6	4.3	9.8
Pe^{n}Br	55	39.2	35.7	3.5	8.9
Bu^{i}Br	55	14.3	5.8	8.5	59.5
PhCH$_2$CH$_2$Br	55	593	32	561	94.6
MeCHBrMe	25	0.295	0.058	0.237	80.3
EtCHBrMe	25	0.422	0.075	0.347	82.2[a]
Pr^{n}CHBrMe	25	0.343	0.067	0.276	80.7[b]
EtCHBrEt	25	0.455	0.054	0.401	88.1
MeCBrMe$_2$	25	4.17	<0.1	4.17	>97
EtCBrMe$_2$	25	9.44	<0.2	9.44	>97[c]

[a] 81% But-2-ene, 19% but-1-ene.
[b] 71% Pent-2-ene, 29% pent-1-ene.
[c] 72% 2-Methylbut-1-ene, 28% 2-methylbut-1-ene.

halides, β-substitution aids the elimination process similarly to α-alkyl substitution. Elimination reaction with $E1cB$-like transition-states will be retarded by β-alkyl substitution owing to the inductive effect of the alkyl group destabilizing the transition-state. Even in these cases, however, the S_N2 process is retarded more by steric effects so that the net effect of β-substitution is a greater proportion of olefin product. Generally more olefin is produced when 'onium ions are decomposed than with halogen employed as the leaving group. Thus, the reaction of ethyl bromide with alkali in ethanol yields only 1% of ethylene but the reaction of the triethylsulphonium ion with alkali in 80% ethanol produces 100% of ethylene.[84] It has also been observed that toluene-p-sulphonates give a smaller fraction of elimination than the corresponding halides.[85]

The favoured requirement of a *trans*-coplanar arrangement for an $E2$ elimination is also further exhibited by the olefin-substitution ratio of a number of reactions. Thus substitution occurs exclusively for the reaction of the *trans*-4-t-butylcyclohexyl derivative **75** in potassium t-butoxide in t-butyl alcohol whilst 92% of olefin is produced from the *cis*-isomer (**76**) under the same conditions.[86] Both substituents are equatorial in **75** and the conformation

100% S_N2	92% $E2 + 8\%$ S_N2
(75)	(76)

with both groups in the axial positions is energetically unfavourable. When a *trans*-coplanar arrangement of leaving groups exists, as in **76**, then the elimination process predominates. A further example of this effect is the exclusive S_N2 reaction of 3β-cholestanyl-'onium ion in basic solution (page 222).

The $E2/S_N2$ ratio is affected by the nature of the nucleophilic reagent. An increase in the basicity of the reagent is usually accompanied by an increase in the percentage of elimination. Good nucleophiles, but poor bases, such as PhS^- and N_3^-, are effective for bimolecular substitution and lower the $E2/S_N2$ ratio. Strong bases favour the elimination reaction, and alkoxides and hydroxide ion are used widely if olefin is the desired product; alkali metals, amides, KNH_2, and $NaNH_2$ can also be used. Since S_N2 reactions are very susceptible to steric effects, large bases such as t-butoxide and ethyldiisopropylamine can be used to effect almost exclusive elimination (equation 77).

$$MeCH_2OCHClMe \xrightarrow{\text{Pr}^i_2NEt} MeCH_2OCH=CH_2 \qquad (77)$$

The elimination/substitution ratio associated with a given substrate is also influenced by the polarity of the solvent. An examination of the transition-states **78** and **79** for the two reactions results in the prediction that, since

(78)	(79)

charge is dispersed, both reactions should be favoured by a decrease in solvent polarity. There is greater dispersal for the elimination reaction, however, so that low solvent polarity will favour elimination more than substitution. In line with these predictions the $E2/S_N2$ ratio for the reaction of alkyl halides with hydroxide or alkoxide ion increases as the solvent is made less polar. The reaction of 2-bromopropane with sodium hydroxide results in an $E2/S_N2$ ratio of 1.17 when carried out in 60% aqueous ethanol, 1.44 in 80% ethanol, and 2.45 in dry ethanol.

Of the two types of reactions, eliminations have a less negative $\Delta S^{\ddagger}$ and a higher $\Delta H^{\ddagger}$. The consequence is that, although both substitution and elimination are accelerated by an increase in temperature, the elimination process is accelerated more. A typical example is that 2-bromopropane yields 53% of olefin on reaction with alkali in 60% aqueous ethanol at 45° but 64% at 100°.[87]

Thermal *syn*-Eliminations[88, 89]

Olefin is produced when a number of types of organic compounds are heated in the liquid or vapour phase or, in some cases, in inert solvent. Those reactions which have proved the most synthetically useful are the pyrolysis of carboxylate esters, xanthates (Chugaev reaction), and especially amine oxides (Cope reaction), which occur at a relatively low temperature. The decompositions

$$Me_3CCH(Me)OCOMe \xrightarrow{500°} Me_3CCH=CH_2 + MeCO_2H \qquad (80)$$

$$Me_3CCH(Me)OC(S)SMe \xrightarrow{180-200°} Me_3CCH=CH_2 + MeSC(S)OH \qquad (81)$$

$$PhCH_2CH_2N^+Me_2O^- \xrightarrow{85-120°} PhCH=CH_2 + Me_2NOH \qquad (82)$$

obey first-order rate laws and exhibit negative entropies of activation. Extensive experimental evidence exists that these reactions (i.e. retro-ene reactions; see Chapter 4) occur predominantly by cyclic, concerted mechanisms (Figure 29). The *syn*-character of the elimination from esters was demonstrated by Curtin and Kellom[90] who prepared the esters **83** and **84** and showed the almost complete retention of deuterium in the former and the loss of most

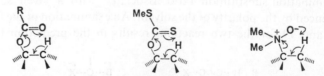

Figure 29. Pyrolytic decomposition of esters, xanthates, and dialkylamine oxides.

of the deuterium from the latter. Similar stereochemical evidence has been obtained for the other pyrolysis reactions.

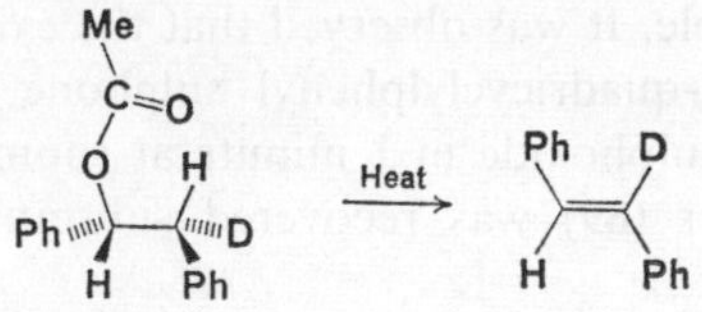

erythro (83)

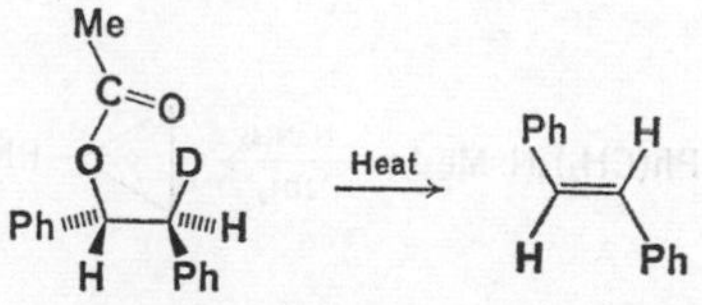

threo (84)

Although the cyclic transition-states are now generally accepted, numerous experiments were conducted to investigate the polarity of the pyrolysis reactions. Thus, for ester decomposition, either the carbon–oxygen bond or carbon–hydrogen bond could, in principle, break slightly ahead of any other process in the transition-state. Rate measurements on 1-arylethyl acetates, for which a ρ-value of -0.66 was found, and on other systems, indicate the very small charge development in the transition-state.[91] The transition-state for ester pyrolysis is therefore considered to be highly concerted, with only a small amount of charge separation.

Olefins can also be formed by the thermal decomposition of other groups such as sulphoxides, amides, vinyl esters, alcohols, and alkyl halides.[89] All proceed essentially by a *syn*-mechanism but, for some, varying amounts of concurrent radical processes may also occur. The decomposition of alkyl halides in the gas phase has been extensively studied. In the presence of inhibitors for the competing radical-chain processes, a unimolecular elimination is observed which is considered to pass through a four-centred transition-state which has a small amount of carbonium ion character.[89]

γ-Elimination

A class of reactions, termed γ- or 1,3-eliminations, is known in which a three-membered ring is formed.[92] In basic media mechanisms ranging from concerted E2 reactions to E1cB type can be envisaged. In acyclic systems cyclopropane formation has been found to occur with a number of different activating groups (equations **85** and **86**).[93] An activating group at the 3-position is required to labilize the hydrogen atom and to stabilize the carbanion if the mechanism is stepwise. The stereochemical requirements for 1,3-eliminations with *E1cB* character are predictable. It was observed that the *exo*-bromo-compound **87** was converted into 2-quadricyclylphenyl sulphone (**88**) by potassium t-butoxide in dimethyl sulphoxide in 1 minute at room temperature, whereas the *endo*-bromo-epimer (**89**) was recovered substantially unchanged after

$$Cl(CH_2)_3CN \quad \xrightarrow[NH_3]{NaNH_2} \quad \triangleright\!\!-CN \qquad\qquad (85)$$

$$Ph(CH_2)_3N^+Me_3I^- \quad \xrightarrow[NH_3]{NaNH_2} \quad \triangleright\!\!-Ph \qquad\qquad (86)$$

(87)   **(88)**   **(89)**

treatment for 7 hours.[94] Thus the formation of cyclopropane from the intermediate carbanion has the normally expected requirement for inversion in nucleophilic attack; inversion is not possible with the *endo*-bromo-compound.

1,3-Eliminations are also involved in a number of base-catalysed rearrangements in which a three-membered ring is formed as an intermediate but frequently not isolated. In the Favorskii rearrangement, an α-halogeno-ketone, when treated with base, is transformed into a carboxylic acid derivative via a cyclopropanone.[95] The Ramberg–Bäcklund reaction is the base-catalysed conversion of α-bromo-sulphones into alkenes, and this involves a 1,3-elimination reaction step to form an intermediate episulphone.[96,97] Experiments showed that the decomposition of episulphones is stereospecific so that a mixture of two intermediates was formed in the cyclization step (equation **90**).

$$MeCH_2SO_2CH(Cl)Me + {}^-OH \quad \underset{fast}{\rightleftharpoons} \quad Me\bar{C}HSO_2CH(Cl)Me + H_2O$$

$$\xrightarrow{fast} \quad MeCH{=}CHMe \qquad\qquad (90)$$
79% *cis*
21% *trans*

PROBLEMS

1. The ether **A** on reaction in strongly acid media forms an aromatic hydrocarbon $C_{19}H_{30}$.

(A)

Suggest a possible structure, and mechanism of formation.

[L. R. C. Barclay and M. C. Macdonald, *Tetrahedron Letters*, **1968**, 881.]

2. Write down two possible reaction schemes by which $MeCHBrCH_2Br$ undergoes dehalogenation by iodide ion in methanol. Which of these mechanisms would fully account for the observation that dehalogenations leading to terminal olefins are considerably faster than those leading to olefins of the type $RCH{=}CHR$?

[See J. Hine and W. H. Brader, Jr., *J. Amer. Chem. Soc.*, **77**, 361 (1955).]

3. Use the predictions involving hard and soft acids and bases to decide the expected products from the reactions of methyl iodide and acetyl chloride with thiocyanate ion in acetone.

[See ref. 45.]

4. The polarizability and solvation of the attacking nucleophile are two factors which govern the rates of substitution reactions. Indicate how the measurement of chlorine kinetic isotope effects (when chloride is the leaving group) can assist in distinguishing between the importance of these two factors.

[See ref. 30.]

5. The reaction of *cis*-dibromoethylene and triethylamine in dimethylformamide to yield monobromoacetylene has a deuterium isotope effect $k_H/k_D = 1.00$. In the presence of triethylammonium deuteriobromide in dimethylformamide the unreacted dibromoethylene did not become deuteriated although in methanol hydrogen exchange was observed.

Comment on the mechanism of these reactions in dimethylformamide and methanol. (*Hint*—consider ion pairs.)

[See W. K. Kwok, W. G. Lee, and S. I. Miller, *J. Amer. Chem. Soc.*, **91**, 468 (1969).]

6. What products would you expect from the reactions of the *cis*- and *trans*-8-D-derivatives of 1,1,4,4-tetramethylcyclodecyl-7-trimethylammonium chloride in $Bu^tOK{-}Me_2SO$?

trans-8-D   *cis*-8-D

Indicate the magnitude of the isotope effects which would be expected to be associated with these processes. Show how the energy differences between the elimination modes might be estimated. (*Hint*—the methyl groups have no significance.)

[See J. Závada, M. Svoboda, and J. Sicher, *Coll. Czech. Chem. Comm.*, **33**, 4027 (1968).]

REFERENCES

1. See J. D. Roberts, *Molecular-Orbital Calculations*, Benjamin, New York, 1961, p. 118.
2. Much of the earlier work is reviewed in the text by D. J. Cram, *Fundamentals of Carbanion Chemistry*, Academic Press, New York, 1965.
3. D. J. Cram, W. T. Ford, and L. Gosser, *J. Amer. Chem. Soc.*, **90**, 2598 (1968); D. J. Cram and L. Gosser, *ibid.*, **86**, 5445 (1964).
4. D. J. Cram and L. Gosser, *J. Amer. Chem. Soc.*, **86**, 5457 (1964).
5. W. T. Ford, E. W. Graham, and D. J. Cram, *J. Amer. Chem. Soc.*, **89**, 4661 (1967).
6. L. Ohlsson, S. Wold, and G. Bergson, *Arkiv Kemi*, **29**, 351 (1968); G. Bergson and L. Ohlsson, *Acta Chem. Scand.*, **21**, 1393 (1967); and earlier papers quoted therein.
7. J. Almy and D. J. Cram, *J. Amer. Chem. Soc.*, **91**, 4459, 4468 (1969), and earlier papers quoted therein.
8. H. Hogeveen and C. J. Gaasbeck, *Rec. Trav. Chim.*, **87**, 319 (1968); H. Hogeveen and A. F. Bickel, *ibid.*, **88**, 371 (1969).
9. G. A. Olah and R. Schlosberg, *J. Amer. Chem. Soc.*, **90**, 2726 (1968); G. A. Olah, G. Klopman, and R. Schlosberg, *ibid.*, **91**, 3261 (1969).
10. G. M. Kramer, B. E. Hudson, and M. T. Melchior, *J. Phys. Chem.*, **71**, 1525 (1967); see also ref. 8 and D. M. Bronwer and J. M. Oelderik, *Rec. Trav. Chim.*, **87**, 721 (1968).
10a. H. Fischer, H. Kollmar, H. O. Smith, and K. Miller, *Tetrahedron Letters*, **1968**, 5821.
11. Reviewed by C. J. Collins, *Chem. Rev.*, **69**, 543 (1969).
12. Y. Pocker and J. H. Exner, *J. Amer. Chem. Soc.*, **90**, 6764 (1968).
13. For a summary see H. C. Brown, *Hydroboration*, Benjamin, New York, 1962.
14. L. H. Toporcer, R. E. Dessy, and S. I. E. Green, *J. Amer. Chem. Soc.*, **87**, 1236 (1965).
15. Discussed in detail in the book by F. R. Jensen and B. Rickborn, *Electrophilic Substitution of Organomercurials*, McGraw-Hill, New York, 1968.
16. A review, with examples mainly from organolithium chemistry, by T. L. Brown, *Accounts Chem. Res.*, **1**, 23 (1968); see also E. A. Jeffrey and T. Mole, *Austral. J. Chem.*, **22**, 1129 (1969), and K. C. Williams and T. L. Brown, *J. Amer. Chem. Soc.*, **88**, 5460 (1966).
17. S. K. Byram, J. K. Fawcett, S. C. Nyburg, and R. J. O'Brien, *Chem. Comm.*, **1970**, 16.
18. H. B. Charman, E. D. Hughes, C. K. Ingold, and H. C. Volger, *J. Chem. Soc.*, **1961**, 1142, and earlier papers in the series.
19. O. A. Reutov, *Russ. Chem. Rev.* (Chem. Soc. English Translation), **36**, 163 (1967).
20. A. J. Parker, *Chem. Rev.*, **69**, 1 (1969). Protic and dipolar aprotic solvent effects on rates of bimolecular reactions.

21. C. A. Bunton, *Nucleophilic Substitution at a Saturated Carbon Atom*, Elsevier, London, 1963.
22. A. Streitwieser, Jr., *Solvolytic Displacement Reactions*, McGraw-Hill, New York, 1962.
23. A. J. Parker, *Adv. Phys. Org. Chem.*, **5**, 173 (1967). Bimolecular reactions in dipolar aprotic solvents.
24. J. F. Bunnett, *Ann. Rev. Phys. Chem.*, **14**, 271 (1963); J. O. Edwards and R. G. Pearson, *J. Amer. Chem. Soc.*, **84**, 16 (1962). Factors governing nucleophilic activity.
25. S. I. Miller, *Adv. Phys. Org. Chem.*, **6**, 185 (1968). Excellent review on factors governing stereochemical pathways in organic reactions.
26. J. L. Gleave, E. D. Hughes, and C. K. Ingold, *J. Chem. Soc.*, **1935**, 236.
27. C. M. Bean, J. Kenyon, and H. Phillips, *J. Chem. Soc.*, **1936**, 303, and preceding papers.
28. E. D. Hughes, F. Juliusburger, S. Masterman, B. Topley, and J. Weiss, *J. Chem. Soc.*, **1935**, 1525.
29. G. J. Frisone and E. R. Thornton, *J. Amer. Chem. Soc.*, **90**, 1211 (1968).
30a. E. P. Grimsrud and J. W. Taylor, *J. Amer. Chem. Soc.*, **92**, 739 (1970).
30b. H. M. R. Hoffmann, *J. Chem. Soc.*, **1965**, 6748, 6753, 6762.
31. W. R. Kirner, *J. Amer. Chem. Soc.*, **50**, 2446 (1928).
32. J. Hine, C. H. Thomas, and S. J. Ehrenson, *J. Amer. Chem. Soc.*, **77**, 3886 (1955).
33. P. B. D. de la Mare, L. Fowden, E. D. Hughes, C. K. Ingold, and J. D. H. Mackie, *J. Chem. Soc.*, **1955**, 3200.
34. P. J. C. Fierens and P. Verschelden, *Bull. Soc. Chim. Belges*, **61**, 427 (1952).
35. E. L. Eliel and R. G. Haber, *J. Amer. Chem. Soc.*, **81**, 1249 (1959).
36. P. D. Bartlett and F. D. Greene, *J. Amer. Chem. Soc.*, **76**, 1088 (1954).
37. P. D. Bartlett and L. H. Knox, *J. Amer. Chem. Soc.*, **61**, 3184 (1939).
38. C. G. Swain and C. B. Scott, *J. Amer. Chem. Soc.*, **75**, 141 (1953).
39. J. O. Edwards, *J. Amer. Chem. Soc.*, **76**, 1540 (1954).
40. J. O. Edwards, *J. Amer. Chem. Soc.*, **78**, 1819 (1956).
41. R. H. Bathgate and E. A. Moelwyn-Hughes, *J. Chem. Soc.*, **1959**, 2642.
42. S. Winstein, L. G. Savedoff, S. Smith, I. D. R. Stevens, and J. S. Gall, *Tetrahedron Letters*, **1960**, 24.
43. H. C. Brown and N. R. Eldred, *J. Amer. Chem. Soc.*, **71**, 445 (1949).
44. H. C. Brown, D. Girtis, and H. Podall, *J. Amer. Chem. Soc.*, **78**, 5375 (1956).
45. R. G. Pearson and J. Songstad, *J. Amer. Chem. Soc.*, **89**, 1827 (1967).
46. R. H. DeWolfe and W. G. Young, *Chem. Rev.*, **56**, 753 (1956). Substitution and rearrangement reactions of allylic compounds; see also F. G. Bordwell, *Accts. of Chem. Res.*, **3**, 281 (1970).
47. R. E. Kepner, S. Winstein, and W. G. Young, *J. Amer. Chem. Soc.*, **71**, 115 (1949).
48. P. B. D. de la Mare, E. D. Hughes, P. C. Merriman, L. Pichat, and C. A. Vernon, *J. Chem. Soc.*, **1958**, 2563.
49. P. B. D. de la Mare and C. A. Vernon, *J. Chem. Soc.*, **1953**, 3555.
50. B. D. England and E. D. Hughes, *Nature*, **168**, 1002 (1951).
51. W. Drenth, *Rec. Trav. Chim.*, **86**, 318 (1967).
52. G. Stork and W. N. White, *J. Amer. Chem. Soc.*, **78**, 4609 (1956).
53. C. W. Jefford, S. N. Mahajan, and J. Gunsher, *Tetrahedron*, **24**, 2921 (1968).

54. D. V. Banthorpe, *Elimination Reactions*, Elsevier, London, 1963.
55. J. F. Bunnett, *Angew. Chem. Internat. Edn. Engl.*, **1**, 225 (1962). Mechanism of elimination reactions.
56. D. J. McLennan, *Quart. Rev.*, **21**, 490 (1967). *E1cB* mechanism.
57. J. F. Bunnett, *Survey Progr. Chem.*, **5**, 53 (1969). Mechanism of elimination reactions.
58. D. V. Banthorpe, *Studies on Chemical Structure and Reactivity* (ed. J. H. Ridd), Methuen, London, 1966. The nature of the transition-state in *E*2 reactions.
59. S. I. Miller and W. G. Lee, *J. Amer. Chem. Soc.*, **81**, 6313 (1959).
60. T. J. Houser, R. B. Bernstein, R. G. Miekka, and J. C. Angus, *J. Amer. Chem. Soc.*, **77**, 6201 (1955).
61. R. Breslow, *Tetrahedron Letters*, **1964**, 399.
62. J. Hine, R. Wiesboeck, and R. G. Ghirardelli, *J. Amer. Chem. Soc.*, **83**, 1219 (1961).
63. J. Crosby and C. J. M. Stirling, *J. Amer. Chem. Soc.*, **90**, 6869 (1968).
64. G. Ayrey, A. N. Bourns, and V. A. Vyas, *Can. J. Chem.*, **41**, 1759 (1963).
65. W. H. Saunders, Jr., A. F. Cockerill, S. Ašperger, L. Klasinc, and D. Stefanović, *J. Amer. Chem. Soc.*, **88**, 848 (1966).
66. C. H. DePuy, G. F. Morris, J. S. Smith, and R. J. Smat, *J. Amer. Chem. Soc.*, **87**, 2421 (1965).
67. C. K. Ingold, *Proc. Chem. Soc.*, **1962**, 265.
68. H. C. Brown and R. L. Klimisch, *J. Amer. Chem. Soc.*, **88**, 1425 (1966).
69. H. C. Brown and O. H. Wheeler, *J. Amer. Chem. Soc.*, **78**, 2199 (1956).
70. R. A. Bartsch and J. F. Bunnett, *J. Amer. Chem. Soc.*, **90**, 408 (1968).
71. W. H. Saunders, Jr., and M. R. Schreiber, *Chem. Comm.*, **1966**, 145.
72. R. A. Bartsch and J. F. Bunnett, *J. Amer. Chem. Soc.*, **91**, 1376 (1969).
73. H. C. Brown and R. L. Klimisch, *J. Amer. Chem. Soc.*, **88**, 1425 (1966).
74. K. Fukui and H. Fujimoto, *Tetrahedron Letters*, **1965**, 4303.
75. A. Elhafez and D. J. Cram, *J. Amer. Chem. Soc.*, **75**, 339 (1953).
76. W. Hückel, W. Tappe, and G. Legutke, *Ann. Chem.*, **543**, 191 (1940).
77. S. J. Cristol, W. L. Hause, and J. S. Meek, *J. Amer. Chem. Soc.*, **73**, 674 (1951).
78. C. H. DePuy, G. F. Morris, J. S. Smith, and R. J. Smat, *J. Amer. Chem. Soc.*, **87**, 2421 (1965).
79. H. Kwart, T. Takeshita, and J. L. Nyce, *J. Amer. Chem. Soc.*, **86**, 2606 (1964).
80. S. J. Cristol and N. L. Hause, *J. Amer. Chem. Soc.*, **74**, 2193 (1952).
81. *Organic Reaction Mechanisms 1967* (ed. B. Capon, M. J. Perkins, and C. W. Rees), Interscience, London, 1968, p. 143.
82. *Organic Reaction Mechanisms 1966* (ed. B. Capon, M. J. Perkins, and C. W. Rees), Interscience, London, 1967, p. 103.
83. D. Y. Curtin, R. D. Stolow, and W. Maya, *J. Amer. Chem. Soc.*, **81**, 3330 (1959).
84. W. H. Pauterbaugh and C. R. Hauser, *J. Org. Chem.*, **24**, 416 (1959).
85. G. M. Fraser and H. M. R. Hoffmann, *J. Chem. Soc. (B)*, **1967**, 425.
86. H. Oediger, H. J. Kabbe, F. Moller, and K. Eiter, *Chem. Ber.*, **99**, 2012 (1966).
87. K. A. Cooper, M. L. Dhar, E. D. Hughes, C. K. Ingold, B. J. MacNulty, and L. I. Woolf, *J. Chem. Soc.*, **1948**, 2043.
88. C. H. DePuy and R. W. King, *Chem. Rev.*, **60**, 431 (1960).
89. A. Maccoll, *The Chemistry of Alkenes* (ed. S. Patai), Interscience, London, 1964.
90. D. Y. Curtin and D. B. Kellom, *J. Amer. Chem. Soc.*, **75**, 6011 (1953).
91. R. Taylor, *J. Chem. Soc.*, **1962**, 4881.

92. A. Nickon and N. H. Werstiuk, *J. Amer. Chem. Soc.*, **89**, 3914 (1967).
93. C. L. Bumgardner, *Chem. and Ind.*, **1958**, 1555.
94. S. J. Cristol, J. K. Harrington, and M. S. Singer., *J. Amer. Chem. Soc.*, **88**, 1529 (1966).
95. D. J. Cram, *Fundamentals of Carbanion Chemistry*, Academic Press, New York, 1965, p. 243.
96. Ref. 95, p. 253.
97. F. G. Bordwell and J. M. Williams, Jr., *J. Amer. Chem. Soc.*, **90**, 435 (1968).

Chapter 4

Multicentre Reactions

As an introduction to the special characteristics of multicentre reactions (reviews in refs. 1—10) let us briefly compare three reactions already discussed in Chapters 2 and 3: an $E1$ elimination, an $E2$ elimination, and a pyrolytic syn-elimination. In an $E1$ reaction, the rate-determining step simply involves bond-breaking with creation of a carbonium ion and an anion. This bond-breaking requires the input of energy, and the ease with which it occurs depends on the stability of the ions produced (including their solvation) and on the strength of the bond being broken. In an $E2$ elimination:

$$B \overset{\curvearrowright}{\quad} H \overset{\curvearrowright}{-} C \overset{\curvearrowright}{-} C \overset{\curvearrowright}{-} X$$

the energy required to break the H—C and C—X bonds is to some extent compensated for by energy gained in forming the B—H and C=C(P) bonds. The bond-breaking and bond-making are concerted, and a system of linear conjugation or delocalization extends from B to X in the transition-state. However, ions are still being created with the attendant need for solvation. We should also note that the more favourable energy situation is to some extent compensated for by a more negative entropy of activation since a more strictly ordered transition-state is required. Finally, in the pyrolytic syn-elimination of an acetate (**1**) bond-breaking and bond-making are linked

$$(1)$$

in a cyclically conjugated transition-state, and there is no need for the creation of ions. This type of reaction is what we will call a multicentre reaction. From what has just been said we may select three likely characteristics of a multi-centre reaction mechanism. (a) It represents an efficient pathway in terms of

238

activation energy from reactants to products. (*b*) It usually demands a highly ordered transition-state (ΔS^+ is therefore often strongly negative). (*c*) There is usually little creation or loss of charged centres, so solvent effects are usually quite small. Experimentally, the main question about reactions which *appear* to be of this type is whether they really do go via a fully concerted mechanism or whether in practice a two-step non-concerted process occurs. Thus in the Diels–Alder reaction between cyclopentadiene and maleic anhydride to give *endo*-norborn-5-ene-2,3-dicarboxylic anhydride, does the reaction really go through transition-state **2**, with both **a—b** and **c—d** bonds partially formed, or is there an intermediate such as **3** or **4**? We can make use of a number of

        (2)   (3)   (4)

pieces of evidence to answer this question. Thus **2** will require ΔS^+ to be very negative, and in suitable cases strictly limits the number of possible stereoisomers which could be produced. The zwitterionic intermediate **3** would suggest a strong solvent effect on reaction rate. It is often possible to estimate from bond energies the heat of formation of a diradical like **4**, and thus decide if it is a reasonable intermediate. We will illustrate these points with a variety of multicentre reactions later, but first the expected properties of a cyclically conjugated transition-state must be considered in more detail.

We assume that the reader is familiar with the simple properties of linear and cyclic π-electron systems such as butadiene, hexatriene, cyclobutadiene, and cyclohexatriene (benzene). Molecular-orbital theory[11] satisfactorily explains why some cyclic systems are considerably stabilized compared to their acyclic counterparts (benzene vs. hexatriene) while others (cyclobutadiene vs. butadiene) are destabilized, and leads for simple monocyclic systems to the Hückel $4n+2$ rule. It had long been suspected that similar considerations should apply to cyclically conjugated transition-states, but it was Woodward and Hoffmann[12] in 1965 who first successfully worked out the implications of this and provided a set of rules (the orbital symmetry rules) which allow one to predict whether cyclic conjugation in a given transition-state will be stabilizing or destabilizing. Reaction via stabilized cyclically conjugated transition-states (and the transition-states for both **1** and **2** fall in this category) should therefore occur very readily; on the other hand, if cyclic conjugation is destabilizing ('antiaromatic'), it is likely that alternative non-concerted pathways for reaction will be chosen. The first part of this chapter explains how the orbital symmetry rules work, and mechanistic evidence about multicentre reactions is considered later. As a preliminary we need some definitions of the types of multicentre reactions which have been observed.

Perhaps the simplest type of multicentre reaction is the formation of a single σ-bond by joining the ends of a linearly conjugated π-electron system. This is known as an 'electrocyclic reaction'. Thus the cyclization of *trans,cis,-trans*-octa-2,4,6-triene to *cis*-5,6-dimethylcyclohexa-1,3-diene (**5**) is an electrocyclic reaction which like many others takes place in a highly stereospecific fashion with no detectable amount of the *trans*-isomer of the product being formed.[13] Reactions which take place in the reverse direction, a σ-bond breaking to give a conjugated π-system, are also known as electrocyclic reactions. Equation **6**, the conversion of a cyclobutene into a butadiene, is an example of this[14] and again is highly stereospecific. A general equation for electrocyclic reactions is shown in **7**. We shall see that the orbital symmetry rules provide a successful rationalization of the stereochemical results.

$$\text{(5)}$$

$$\text{(6)}$$

$$k\pi \rightleftharpoons (k-2)\pi \tag{7}$$

The second type of multicentre reaction is the 'cycloaddition' in which two (or occasionally more) new single bonds are formed by linking the ends of two (or more) linearly conjugated π-electron systems. In the familiar Diels–Alder reaction (**8**), a 4π-electron system (the diene) is joined to a 2π-electron system (the dienophile). The orbital symmetry rules permit this 2π+4π cycloaddition but forbid a simple 2π+2π addition to be concerted and, as we shall see later, the non-stereospecific addition in **9** is associated with the occurrence of a diradical intermediate. Another important kind of cycloaddition is shown in **10**; the azide is an example of a 1,3-dipole,[15] and the addition is known as a 1,3-dipolar cycloaddition. Many 1,3-dipolar cycloadditions have been discovered in the last 15 years, principally by Huisgen and his colleagues. 1,3-Dipoles are triatomic species a—b—c in which atom a, at least in one resonance form, is electron-deficient with only a sextet of electrons, while atom c has a full octet and is nucleophilic. The appropriate

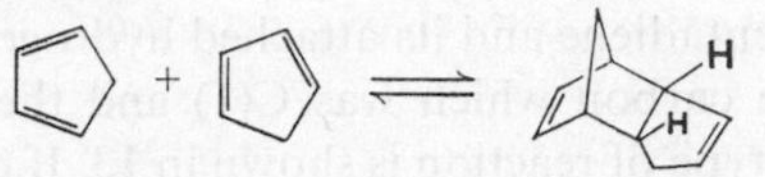

(8)

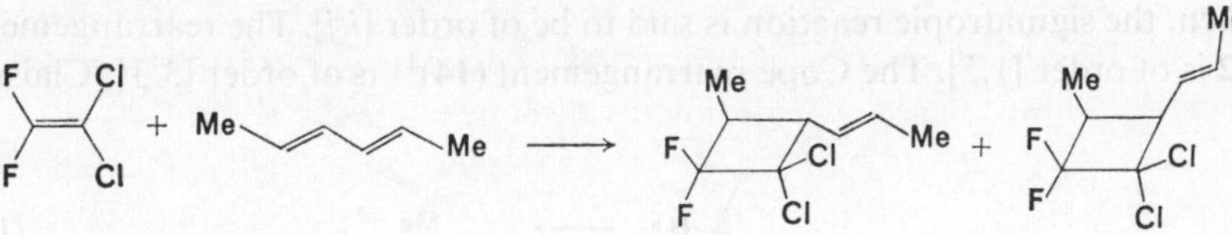

(9)

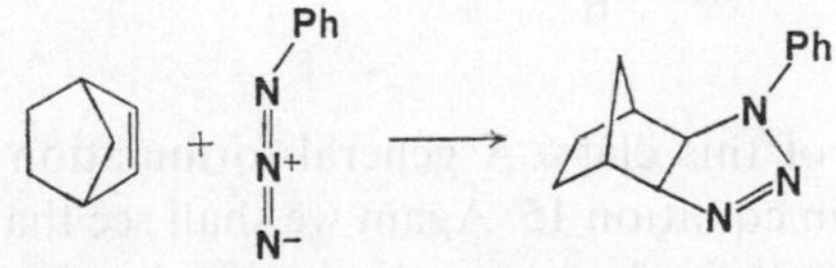

(10)

resonance form with phenyl azide is $PhN^+{-}N{=}N^-$. In most cases the central atom has a lone-pair of electrons which partially neutralizes the electron-deficiency at **a** as in $PhN{=}N^+{=}N^-$. Huisgen[3, 6] describes these cases as octet-stabilized, and it is really only in these cases that we can speak of cyclo-addition, since otherwise there is a break of conjugation at atom **b**. Other 1,3-dipoles are ozone, diazo-compounds, nitrile oxides ($RC{\equiv}N^+{-}O^-$), and nitrones $(\overset{|}{\underset{}{C}}{=}N^+{-}O^-)$. A general formulation of cycloaddition reactions is shown in **11**.

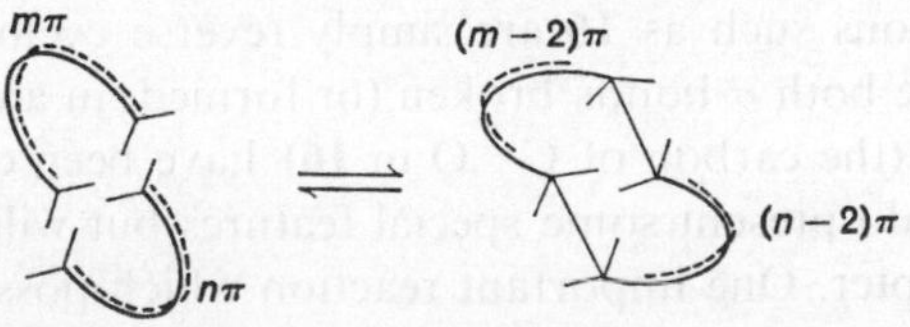

(11)

A third type of multicentre reaction is called a 'sigmatropic reaction'. This involves the switching of a single σ-bond to a new position when it is flanked by one or more π-systems. Thus in **12** the original σ-bond between

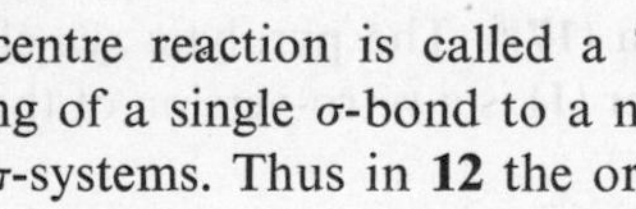

(12)

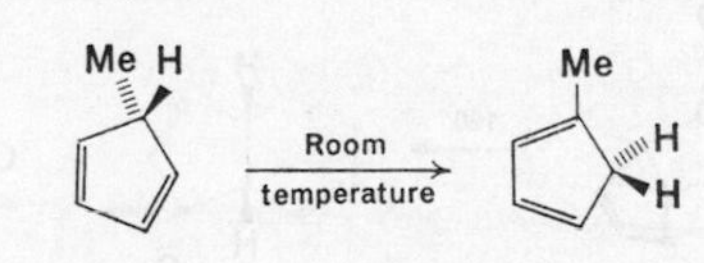

(13)

C(5) of the 5-methylcyclopentadiene and its attached hydrogen is replaced by a new σ-bond between the carbon which was C(1) and the hydrogen.[16] A general formulation of this type of reaction is shown in **13**. If the atoms joined by the new σ-bond are $(i-1)$ and $(j-1)$ atoms removed from the bond being broken, the sigmatropic reaction is said to be of order $[i,j]$. The rearrangement in **12** is of order [1,5]. The Cope rearrangement (**14**)[17] is of order [3,3]; Claisen

$$\text{(14)}$$

rearrangements are also of this class. A general formulation of sigmatropic reactions of order $[i,j]$ is in equation **15**. Again we shall see that, for concerted reactions, the orbital symmetry rules impose limitations on the stereochemistry of these reactions.

$$\text{(15)}$$

Most multicentre reactions fall into one of the three main types just described. Cyclic fragmentations such as **16** are simply reverse cycloadditions. Those, such as **16**,[18] where both σ-bonds broken (or formed, in a cycloaddition) are to the same atom (the carbon of C=O in **16**) have been called 'cheletropic reactions'.[1a] They do present some special features but will not be discussed further in this chapter. One important reaction which possesses some of the attributes of both cycloaddition and sigmatropic reactions is the 'ene' reaction or substitutive addition (**17**)[8]. The pyrolytic *syn*-elimination discussed at the beginning of the chapter (**1**) is a retro-version of this.

$$\text{(16)}$$

$$\text{(17)}$$

The Orbital Symmetry Rules: Electrocyclic Reactions

A convenient example to study is the ring-opening of cyclobutene to butadiene. There are two distinct geometrical processes by which this could occur. The two methylene groups of cyclobutene could rotate either (i) in the same direction or (ii) in opposite directions. It is clear that in **6** (page 240) the corresponding groups have rotated in the same direction; this is called the 'conrotatory mode' of ring-opening. The same *cis*-3,4-dimethylcyclobutene on 'disrotatory' ring-opening would give *cis,cis*- or (more likely) *trans,trans*-hexa-2,4-diene. The conrotatory mode maintains an axis of symmetry, **X** in **18**, as ring-opening

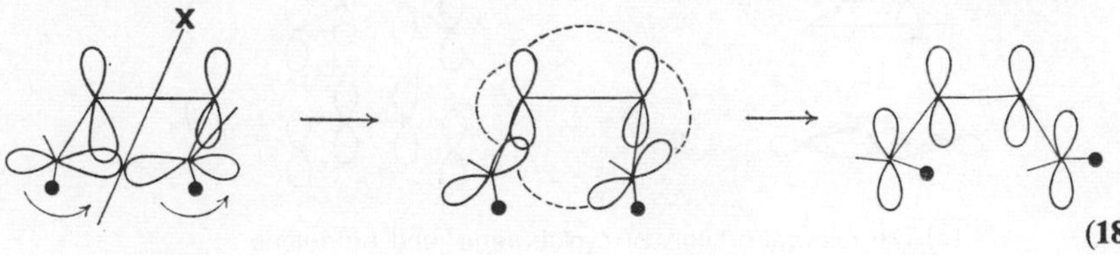

(18)

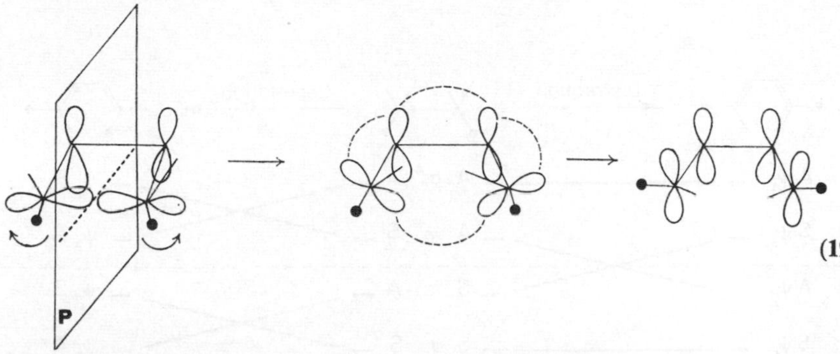

(19)

takes place, while in disrotatory ring-opening a plane of symmetry, **P** in **19**, is maintained. The equations **18** and **19** also make it clear that the relationships between the orbitals involved are different in the two cases. In the reaction the σ- and σ^*-orbitals of the 3,4 single bond of cyclobutene and the π- and π^*-orbitals of its double bond are transformed into the four π-orbitals of butadiene, ψ_1—ψ_4 (see Figure 1a). The 'principle of conservation of orbital symmetry' states that, in a cyclic concerted reaction, orbitals in the reactant can only transform into orbitals in the product which have the same symmetry properties with respect to the elements of symmetry preserved in the transformation. Thus in the disrotatory mode a plane of symmetry is preserved through the reaction; the σ- and π-orbitals of cyclobutene and ψ_1 and ψ_3 of butadiene are symmetric (S) with respect to this plane (reflection in the

plane causes no change of sign of the wave function of these orbitals), while
σ^*, π^*, ψ_2, and ψ_4 are antisymmetric (**A**; the wave function changes sign on
reflection in the plane). When an axis of symmetry is preserved (conrotation),

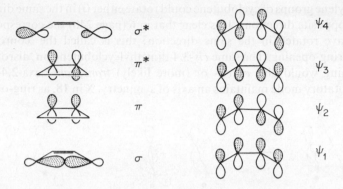

(a) The relevant orbitals of cyclobutene and butadiene

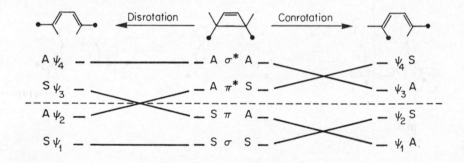

(b) Correlation scheme

Figure 1. Application of the principle of conservation of orbital symmetry to
the conversion of cyclobutene into butadiene.

the orbitals σ, π^*, ψ_2, and ψ_4 are **S**, σ^*, π, ψ_1, and ψ_3 are **A**. We can now correlate the orbitals in butadiene with those of cyclobutene according to the two
modes. It can be seen from Figure 1b that only by the conrotatory mode can
ground-state cyclobutene (in which σ and π are filled) be converted into ground-
state butadiene (ψ_1 and ψ_2 filled); disrotatory ring-opening would apparently
yield an electronically excited butadiene molecule and so would be energetically
most unfavourable. The preference in a thermal reaction for conrotatory

ring-opening as in **6** (page 240) is now readily understandable. It is apparent too that an electronically excited state of cyclobutene when formed might pass by disrotation into ground-state butadiene. This state $\sigma^2 \pi^{*2}$, is, however, not the first excited state which is probably $\sigma^2 \pi \pi^*$. To see what happens to this it is necessary to construct, not an *orbital* correlation diagram as we have done here, but an electronic *state* correlation diagram as was done by Longuet-Higgins and Abrahamson.[19] We will not do this but will merely state the result—that the first excited singlet state $\sigma^2(\uparrow\downarrow)\pi(\uparrow)\pi^*(\downarrow)$ does correlate by disrotation with ground-state butadiene. We also add a warning that it is more difficult to prove that a photochemical reaction is a truly concerted process and a reminder that many photochemical reactions involve triplet states, e.g. $\sigma^2(\uparrow\downarrow)\pi(\uparrow)\pi^*(\uparrow)$ which cannot possibly correlate directly with spin-paired ground-states.

There are several other equivalent methods[19-22a] of arriving at the results we have just derived, of which that of Longuet-Higgins and Abrahamson is possibly the most rigorous. Hoffmann and Woodward[1a] give a useful pictorial description of what happens to each orbital in the course of reaction. The same type of reasoning can be applied to other electrocyclic reactions, and is not limited to uncharged systems (for example cyclopropyl cation is converted into allyl cation by a disrotatory process). It might seem that the rules only apply to symmetrical cases but this is not so. Although 2-methylbutadiene lacks the necessary symmetry, the orbitals concerned in the reaction will in reality be only slightly perturbed by the substituent, and for example the slightly perturbed ψ_1 will still correlate with the perturbed π of the 1-methylcyclobutene in the conrotatory ring-closure.

When the hexatriene–cyclohexadiene process is examined in the same way, the result is precisely reversed and disrotatory closure is now thermally allowed (see **5**, page 240). In general, electrocyclic reactions should be thermally disrotatory when $k = 4q + 2$ (see **7**; q is an integer) and conrotatory when $k = 4q$.

Cycloaddition Reactions

Let us examine two simple examples, the dimerization of ethylene and addition of ethylene to butadiene. In the dimerization of ethylene two planes of symmetry (P_1, P_2) are preserved through the reaction (Figure 2a). As the two ethylene molecules approach one another the two filled π-levels interact forming a bonding (the SS level in Figure 2b) and an antibonding (**SA**) combination. As the reaction proceeds the SS level passes into the SS level of the cyclobutane (the combination of the σ bonding orbitals which is symmetrical with respect to P_1). The **SA** combination forms an antibonding

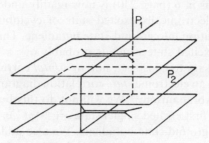

(a) Symmetry planes in ethylene dimerization

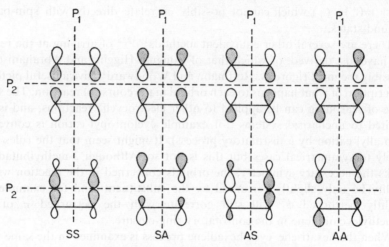

(b) Orbital combinations and their symmetries
(top row, two ethylenes; bottom row, cyclobutane)

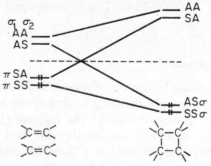

(c) Correlation diagram

Figure 2. Dimerization of ethylene to cyclobutane.

σ^*-level, however, so the reaction is thermally forbidden. The correlation diagram is shown in Figure 2c.

The situation is quite different for the cycloaddition of ethylene to butadiene. Figure 3a shows the mode of approach with symmetry plane **P** preserved in the reaction, and Figure 3b the derived correlation diagram. New bonding levels in the reactant correlate smoothly with the bonding levels of the product cyclohexene.

This type of analysis can be applied to other cycloadditions and to reactions in which one partner is in the first excited state. Although it was not explicitly stated above, a decision was, in effect, taken as to the geometrical mode of addition—*syn*-addition to both partners. If one of the partners undergoes *anti*-addition (i.e. bonding at either end to different faces of its π-system),

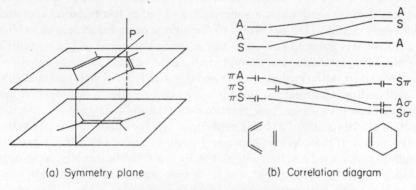

(a) Symmetry plane   (b) Correlation diagram

Figure 3. Cycloaddition of ethylene to butadiene.

the predictions of the rules are reversed; of course, for most systems *syn*-addition to one partner and *anti*-addition to the other is likely to be difficult or impossible because of the twisting involved and the non-bonded interactions generated. The selection rules for cycloadditions can be summarized as follows (see **11**, page 241):

$m+n$	Thermal	Photochemical
$4q$	*syn-anti*	*syn-syn*
	anti-syn	*anti-anti*
$4q+2$	*syn-syn*	*syn-anti*
	anti-anti	*anti-syn*

For butadiene + ethylene, $m = 4$, $n = 2$, so $m + n = 6 = 4q + 2$ $(q = 1)$ and *syn-syn* thermal cycloaddition can occur in a concerted and stereochemically

9

favourable manner. Again, as for electrocyclic reactions, nothing has been said to exclude charged species or their heteroatom analogues. For example allyl cation ($m = 2$) should undergo concerted *syn-syn* cycloaddition to a diene ($n = 4$) but allyl anion ($m = 4$) should add *syn-syn* to an olefin ($n = 2$). Although reactions of these types are known[22b], heteroatom analogues of the latter case are especially important since most of Huisgen's 1,3-dipolar cyclo-additions fall in this class. 1,3-Dipoles with octet stabilization like ozone, nitrones, and azides as in **10** (page 240) are 3-centre 4π-electron systems like allyl anion and readily undergo addition to 2π-electron systems.

Sigmatropic Reactions

In reactions in this class neither reactant nor product possesses usable symmetry and so we shall adopt a simplified, and rather less rigorous approach which only considers the behaviour of the highest occupied molecular orbital. If any orbital is going to become antibonding in the course of reaction this is the most likely candidate. For example, in the electrocyclic closure of butadiene, attempted disrotation pushes positive and negative lobes of ψ_2 together leading to antibonding. Conrotatory closure brings two positive lobes together and bonding occurs. In sigmatropic reactions the transition-states for a concerted reaction can be visualized as being made up of two radicals with orbital overlap at two points. If the overlaps between the half-occupied orbitals can both be bonding, a concerted reaction with relatively low activation energy is possible. Consider a [1,3]-hydrogen shift; the transition-state is the combination of an allyl radical with a hydrogen atom, and the symmetries of the half-occupied orbitals show that there can be no net bonding in the arrangement shown (Figure 4a). The symmetry of the half-occupied orbital of pentadienyl radical is such that bonding can be maintained throughout reaction and a concerted H-shift might occur (Figure 4b). We shall see that this is in excellent agreement with experiment. A [1,7]-hydrogen shift by this route is again prohibited but there is now another geometrically reasonable possibility—transfer from one face of the coiled π-system to the other (pictured in Figure 4c). In the [1,5]-shift the pentadienyl component is used 'suprafacially', bonding occurring at all times with the same face of the π-system. In the process shown in Figure 4c the heptatrienyl system is used 'antarafacially'. In the [3,3]-sigmatropic reaction (Figure 4d) the use of both allyl components suprafacially is allowed, giving a geometrically very satisfactory transition-state. We shall see later how the chair form shown is preferred to the alternative boat form.

In the [1,j]-shifts hydrogen was used as one component, and since this only has an s-orbital available for bonding, it could only conceivably be used suprafacially. This is not the only possibility for atoms with available p-

orbitals, such as carbon. For example, Figure 4e shows how a [1,3]-shift of a carbon group could occur, but with *inversion* at the carbon centre undergoing migration. This use of the σ-bond component in such a way that one carbon is inverted and the other has retained stereochemistry is really analogous to the use of the heptatrienyl component in Figure 4c, and the term antarafacial can be applied to both. This analysis of a multicentre reaction in terms of its components, characterized as being used suprafacially or antarafacially, can be extended to encompass the predictions of orbital symmetry for all 'pericyclic'[1a] (i.e. genuinely concerted multicentre) reactions. Each component,

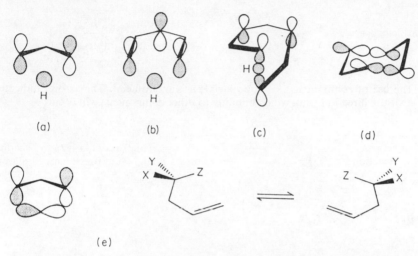

Figure 4. Highest occupied molecular orbitals in the transition-states for (a) [1,3]-hydrogen shift, (b) [1,5]-hydrogen shift, (c) antarafacial [1,7]-hydrogen shift, (d) [3,3]-sigmatropic shift, and (e) [1,3]-shift with inversion at the migrating centre.

which may be a system of π-orbitals (π-component) with $(4q + 2)$ or $(4r)$ electrons, a σ-bond (σ-component), or a filled or unfilled atomic orbital (ω-component), is examined to see if it is used suprafacially (s) or antarafacially (a) (see Figure 5a). *A pericyclic reaction is thermally allowed when the total number of $(4q + 2)_s$ and $(4r)_a$ components is odd.* Odd-electron systems obey the rule applying to the system with one more electron, and excited-state reactions require an even total. Apparently the number of $(4q + 2)_a$ and $(4r)_s$ components is not important. The operation of the rule is shown in Figure 5b.

The orbital symmetry rules allow us to decide, for any multicentre reaction, if a concerted one-step mechanism is permitted, but of course they do not

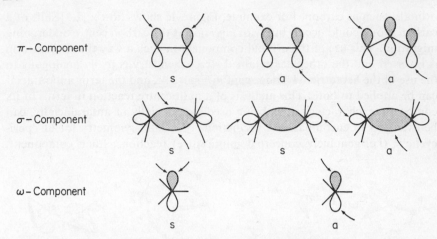

π– Component

σ– Component

ω– Component

(a) The use of components: s = suprafacial; a = antarafacial. The arrows indicate the direction from which bonding to other components will occur.

Reaction	Components	Number of $(4q+2)_s$ and $(4r)_a$ components	Thermally allowed
(diene with Me Me → cyclohexadiene with Me Me)	$\pi 6_s$	1	Yes
(Me Me H H → Me ... Me)	$\pi 2_s + \sigma 2_a$ or $\pi 2_a + \sigma 2_s$	1	Yes
(diene + ‖ $\xrightarrow{cis\text{-}cis}$ cyclohexene)	$\pi 4_s + \pi 2_s$	1	Yes
(H ... → H ...)	$\pi 2_s + \sigma 2_s$	0	No
(H ... → H ...)	$\pi 2_s + \pi 2_s + \sigma 2_s$	3	Yes

(b) The counting of components

Figure 5. The operation of the general rule for pericyclic reactions. If the *lowest* occupied molecular orbital of the component is considered, suprafacial use implies bonding across an even (0,2,...) number of nodes, and antarafacial use across an odd number.

say that in a permitted case this will be the mechanism followed. Sometimes a pathway via a bifunctional intermediate will have a lower free-energy of activation. When the bifunctional intermediate is a diradical it is often possible to make a fair guess, from bond-dissociation energies, at the activation energy required to form it. But in any case experimental evidence must now be sought, and the remainder of this chapter will present some of the evidence for a variety of multicentre mechanisms and, hopefully, show the reader how future problems in this field might be approached.

Mechanisms of Electrocyclic Reactions

Pyrolysis of cyclobutenes at temperatures between 100° and 200° leads to butadienes. With all the monocyclic cyclobutenes only the products of conrotatory ring-opening are observed. In the case of a *trans*-3,4-disubstituted cyclobutene there are two possible conrotatory products, a *cis,cis*-butadiene or a *trans,trans*-butadiene, and usually the latter is the major product owing to lower steric compression in the transition-state (see **20**).[23]

$$
\begin{array}{ccc}
\underset{\text{Cl Cl}}{\diagup} & \overset{\text{Cl}}{\underset{\text{Cl}}{\diagup}} & \text{Cl}\diagup\diagdown\text{Cl}
\end{array}
\qquad (20)
$$

In the bicyclic compounds in Figure 6*, ring-opening must occur by a disrotatory process (probably non-concerted), and it is interesting to compare the activation energy for these reactions and for *cis*-3,4-dimethylcyclobutene. When allowance is made for the extra strain energy of the bicyclic compounds,[24, 25] it seems that conrotatory ring-opening must be favoured in an unstrained case by $\geqslant$10 kcal mole^{-1}. This difference is more than enough to ensure complete stereoselectivity in the simple cyclobutenes because, even if careful product analysis shows that less than 0.1 % of 'forbidden' product is formed, this only requires a difference of 6 kcal mole^{-1} in the free-energies of activation for the 'allowed' and 'forbidden' paths at 150°.

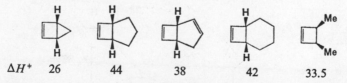

ΔH^* 26   44   38   42   33.5

Figure 6. Activation energies (kcal mole^{-1}) for ring-opening of some cyclobutenes.

* An alternative mode of ring-opening is likely for bicyclopentane; J. E. Baldwin and A. H. Andrist, *Chem. Comm.*, 1970, 1561.

Another impressive demonstration of the stereospecificity of these reactions is the example displayed in **21**. The interconversion of the labelled butadiene was followed by n.m.r.[26] and almost certainly takes place by the conrotatory ring-closure–ring-opening sequence shown; after 51 days at 124° less than 1% of the disrotatory product had formed. The 'allowed'/'forbidden' rate ratio is therefore at least 10^4 and the difference of free-energy of activation for the two paths at least 7.3 kcal mole^{-1}.

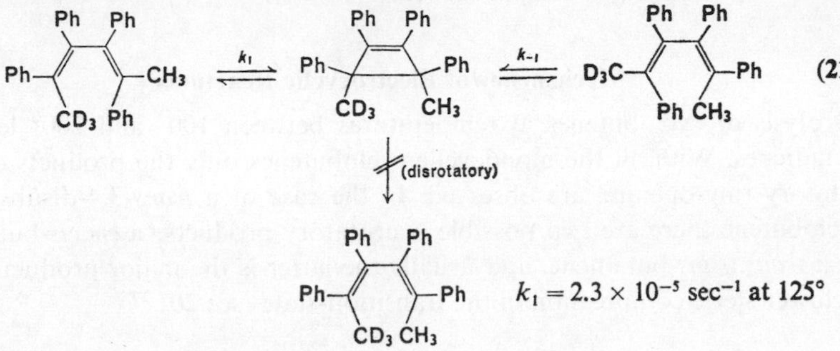

$$k_1 = 2.3 \times 10^{-5} \text{ sec}^{-1} \text{ at } 125°$$

(21)

All the evidence then points to the occurrence of a concerted thermal electrocyclic reaction for the cyclobutene–butadiene interconversion where this is sterically possible. Photocyclization of butadienes has often been observed; a good example which has been examined in detail is that of *cis,cis*-cycloocta-1,3-diene (**22**). In a sensitized reaction which serves to produce only the triplet excited state of the diene (with unpaired spins) and not the singlet state (with paired spins), the only reaction which occurs is conversion of the *cis,cis*- into the *cis,trans*-isomer. The latter is, however, quite strained and at 50—100° undergoes thermal conrotatory cyclization to bicyclo[4.2.0]-oct-7-ene (in this case the cyclobutene derivative is more stable than the butadiene derivative). When the photolysis of the *cis,cis*-isomer is conducted without a sensitizer with light of 254 nm wavelength so that the excited singlet

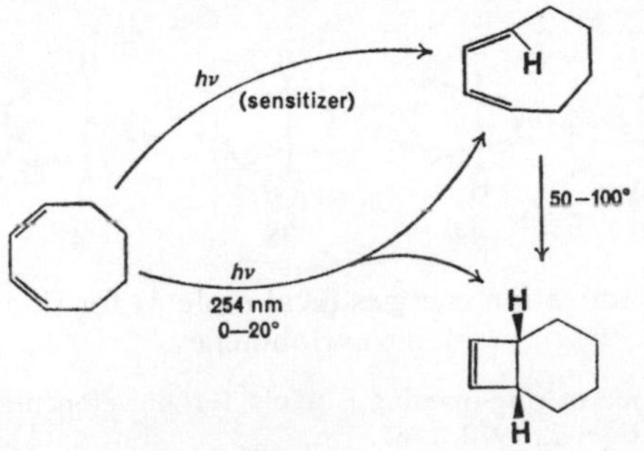

(22)

state produced, the bicyclo[4.2.0]oct-7-ene, and the *cis,trans*-diene are both formed even at low temperatures it seems that the singlet excited *cis,cis*-diene photocyclizes in a disrotatory manner directly to the bicyclic olefin.[27]

The cyclizations of *cis,cis*-octatetraenes to cyclooctatrienes and the subsequent cyclizations of the cyclooctatrienes to bicyclo[4.2.0]octadienes form an intricate set of electrocyclic reactions which have been examined by Huisgen

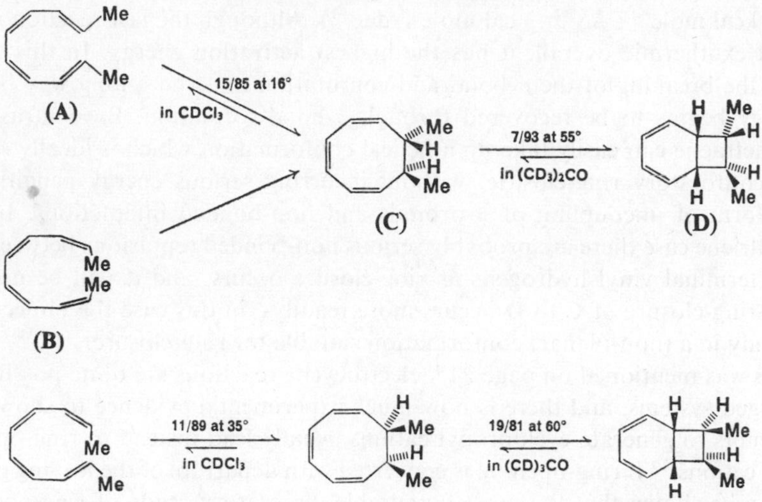

Activation parameters	$\Delta G^{\ddagger}$ (kcal mole^{-1})	$\Delta H^{\ddagger}$ (kcal mole^{-1})	$\Delta S^{\ddagger}$ (cal mole^{-1} deg^{-1})
A → C	20.1	15.1	−19
B → C	25.1	21.8	−12
E → F	22.5	17.8	−17

Figure 7. Equilibria and activation parameters for the interconversion of some $C_{10}H_{14}$ isomers.

and coworkers[28] and they make a most rewarding example for study. The reactions involved are set out in Figure 7; each reaction is stereospecific within the limits of detection and in the sense expected from the orbital symmetry rules. Rate and equilibrium constants have been found in most cases. At much higher temperatures (in fact on passage through a gas-chromatography column) the rearrangements between **A**, **B**, **C**, and **D** are so rapid that these compounds emerge as a single peak; in addition, however, there is a small peak which was shown to be due to **E**, **F**, and **G**. Leakage from one set (**A**,

B, C, and **D**) to the other (**E, F,** and **G**) must occur by some 'forbidden' process and it was possible to estimate that the difference between the free-energy of activation for the 'forbidden' process and that for the allowed conrotatory cyclization of **A** was about 11 kcal mole^{-1}. It is also worthwhile comparing the activation parameters for the cyclization of **A** to **C** with those for the cyclization of the triene in **5** (page 240) ($\Delta H^{\ddagger}$ 28.6 kcal mole^{-1}; $\Delta S^{\ddagger}$ −7 cal mole^{-1} deg^{-1}) and for the ring-opening of cis-3,4-dimethylcyclobutene ($\Delta H^{\ddagger}$ 33.5 kcal mole^{-1}; $\Delta S^{\ddagger}$ −3 cal mole^{-1} deg^{-1}). Although the last reaction is the most exothermic overall, it has the highest activation energy. In this small ring the breaking of the σ-bond and conrotation must go a long way before energy begins to be recovered through π-bond formation. By contrast the octatetraene can easily take up a helical conformation which is ideally suited for conrotatory ring-closure, without incurring serious energy penalties in the form of uncoupling of π-orbitals and non-bonded interactions. In the hexatriene case there are probably serious non-bonded repulsions between the two terminal vinyl hydrogens as ring-closure occurs, and it will be noticed that ring-closure of **C** to **D** occurs more readily—in this case the molecule is already in a (non-planar) conformation suitable for ring-closure.

As was mentioned on page 245, electrocyclic reactions are quite possible in charged systems, and there is now much experimental evidence to show that attempts to generate cyclopropyl cations usually lead instead to ring-opened allyl cations.[29] If ring-opening is concerted with departure of the leaving group it can be shown that the most favourable disrotatory mode of ring-opening is that in which the substituents cis to the leaving group rotate inwards (see **23**). By dissolving the cyclopropyl halides shown in **24** in SbF$_5$–SO$_2$ClF at −100° it was possible to obtain the n.m.r. spectra of the allyl cations shown, thus confirming the expected stereochemistry.[30] In solvolysis experiments the relative rates for the tosylates corresponding to chlorides **A, B,** and **C** are 1/65/4500 at 150° consistent with the steric repulsions introduced in the transition-state for this mode of ring-opening. On the other hand, in a bicyclic compound in which the leaving group is exo (**25**), ring-opening according to **23** will produce a highly strained trans-trans fused allyl cation, and this isomer therefore reacts very slowly (exo/endo rate ratio about $1/2.5 \times 10^6$ at 100°).

In the examples discussed so far quite a range of rate constants and thus

(23)

(A)   (B)   (C)   (24)

↓   ↓   ↓

(25)

'barriers' to reaction have been encountered. It seems possible that in a suitably designed system the 'transition state' for the electrocyclic process might turn out to be of lower energy than either the fully open or fully closed form. Just such a case where the ground-state of the system is one of partial ring-closure is found in the ion produced by protonation of cyclooctatetraene with strong acids.[31] This is neither the cyclooctra-2,4,6-trienyl cation (26) nor the bicyclo[5.1.0]octa-3,5-dienyl cation (27) but has an intermediate structure (28) representing some point along the disrotatory path connecting the two. Evidence for this comes mainly from the n.m.r. spectrum and in particular from the remarkable shielding of the 8-*endo*-proton (τ 10.67) caused by the ring-current. **28** is called the homotropylium ion (tropylium interrupted by an extra CH_2 group) and is one example of a homoaromatic species. Largely owing to the work of Winstein[32] a variety of homoaromatic and non-classical species are now known which may also be regarded as examples of arrested multicentre reactions. The bicyclo[3.2.1]octadienyl

(26)   (27)   (28)

anion (29) represents an arrested cycloaddition,[33] while the non-classical 2-norbornyl cation (see Chapter 2) is in a state of arrested allowed sigmatropic reaction.

Neither   nor   but   (29)

Mechanisms of Cycloaddition Reactions

In this section we will concentrate on thermal $2\pi+2\pi$ and $2\pi+4\pi$ reactions; in recent years many more exotic cycloadditions have been discovered, and a selection of examples is shown in Figure 8. Various aspects of cycloaddition are covered in several excellent reviews.[2-4, 6] Let us begin with a general question: "What experimental methods will distinguish between concerted and stepwise processes?". The work of Bartlett and his coworkers[40] on the reaction of some olefins which give both $2\pi+2\pi$ (vinylcyclobutane) and $2\pi+4\pi$ (cyclohexene) adducts with butadienes illustrates the problems nicely. Reaction of 1,1-dichloro-2,2-difluoroethylene (30) with butadiene at about 100° gives a product consisting of 98.7% 2,2-dichloro-3,3-difluoro-1-vinylcyclobutane (31), the 1,2-adduct, with 1.3% 4,4-dichloro-5,5-difluorocyclohexene (32), the 1,4-adduct. Reaction of 30 with the three geometrical isomers of hexa-2,4-diene at 80° gives only vinylcyclobutanes but is not stereospecific and the various isomers shown in Figure 9 are produced. The ratio of products **A** and **B**, from the *trans,trans*-diene is not the same, however, as the ratio **A**/**B** from the *cis,trans*-isomer. These results fit in well with a diradical mechanism, for if it is supposed that 33 is formed from *trans,trans*-hexa-2,4-diene and 30, there is no reason to expect it to be formed in a conformation suitable for ring-closure. If rotation about the vertical bond **a** in 33 occurs before rotation about **b**, product **A** will be formed, but rotation about **b** could occur first leading to product **B**. Notice that, because rotation about the partial double bonds in the allylic radical is much more difficult, **C** and **D** are not formed from *trans,trans*-hexa-2,4-diene at this temperature (80°). At a higher temperature (120°), **C** and **D** are formed but, in addition, recovered diene is found to be geometrically isomerized. This can occur if, in addition to going on to product, the diradical can revert to starting materials. If this occurs for 33 after rotation about bond **b**, *cis,trans*-hexa-2,4-diene will be formed.

The vinylcyclobutane products arise, as expected, from a two-step mechanism via an intermediate diradical. What about the 1,4-addition to butadiene? A concerted reaction is now allowed, but the evidence still points to a diradical process. If rotation within the allylic radical cannot occur, 1,4-addition can only occur in those diradicals formed from cisoid butadiene. The amount

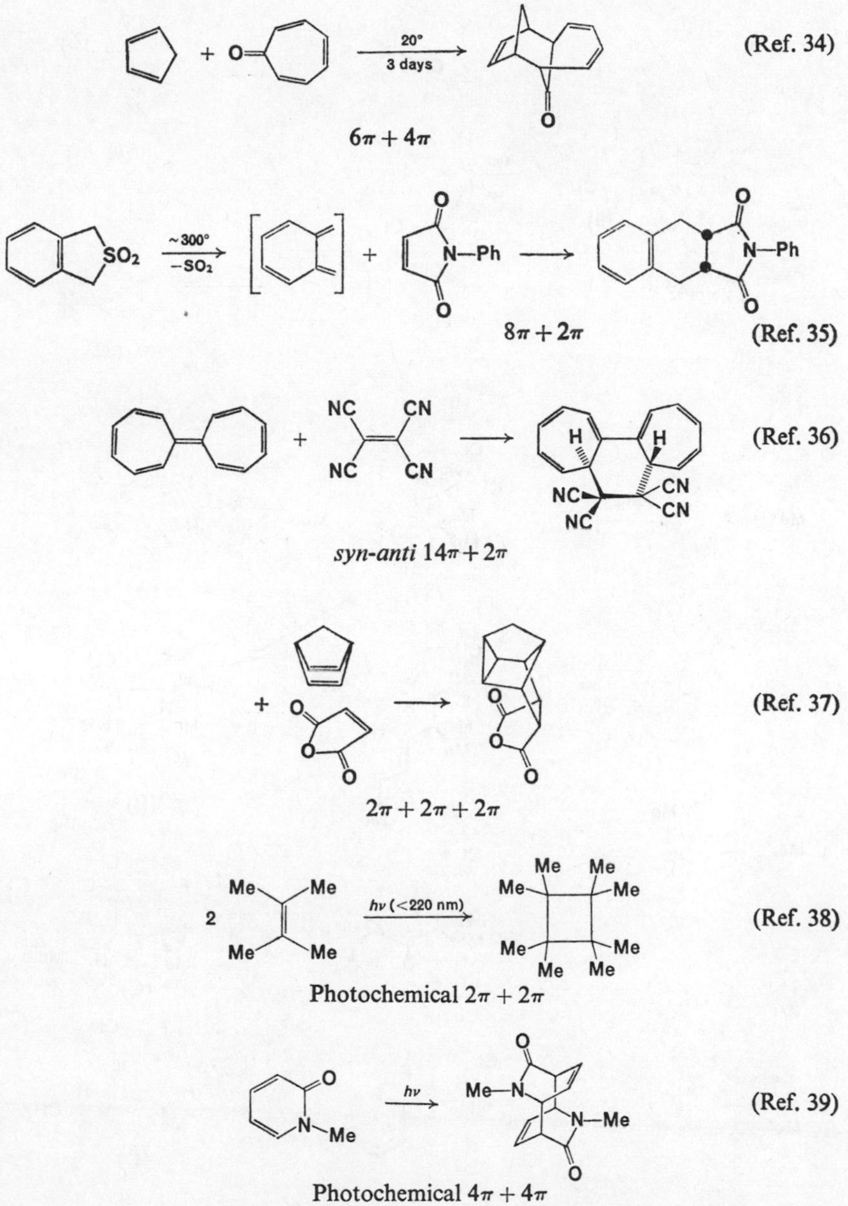

$6\pi + 4\pi$ (Ref. 34)

$8\pi + 2\pi$ (Ref. 35)

syn-anti $14\pi + 2\pi$ (Ref. 36)

$2\pi + 2\pi + 2\pi$ (Ref. 37)

Photochemical $2\pi + 2\pi$ (Ref. 38)

Photochemical $4\pi + 4\pi$ (Ref. 39)

Figure 8. Examples of cycloaddition reactions.

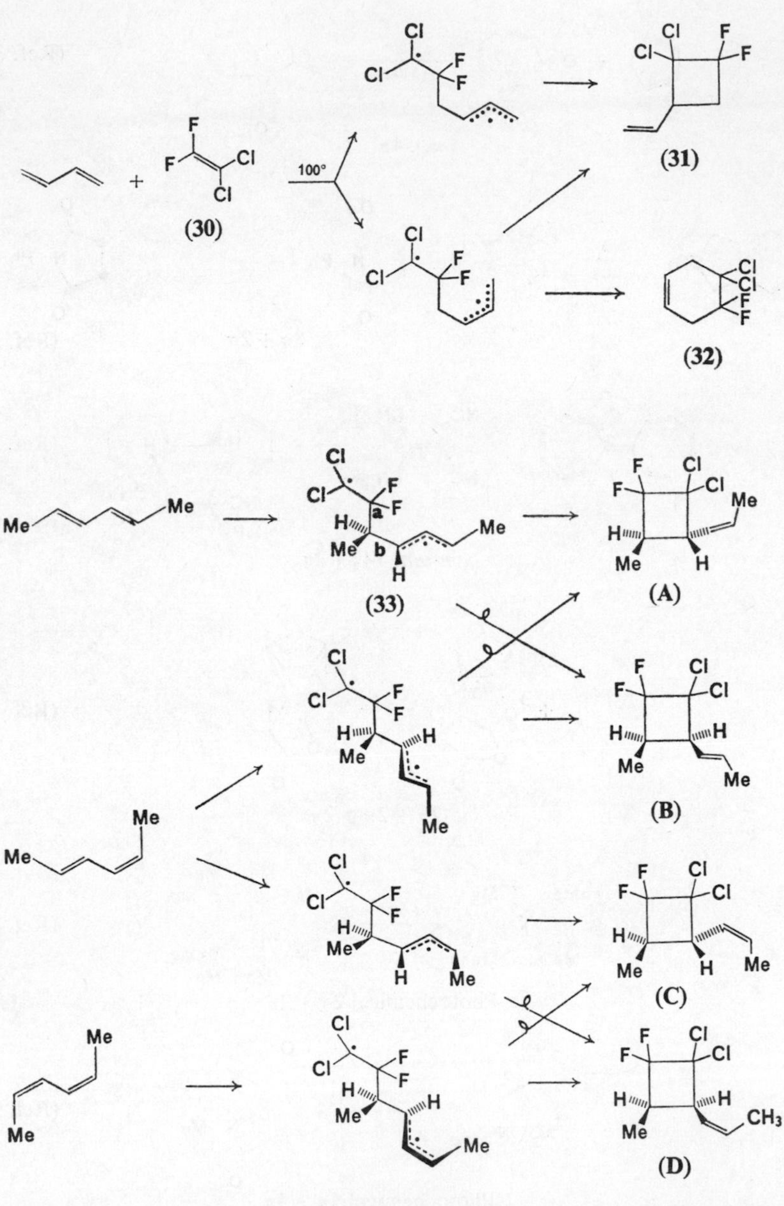

Figure 9. Diradical mechanisms for cycloadditions with 1,1-dichloro-2,2-difluoroethylene (30).

of cisoid butadiene present at equilibrium varies from 5.8% at 60° to 13.0% at 176°, and it is found that the per cent of 1,4-adduct formed varies in step with this from 0.86% at 60° to 2.32% at 176°. Also, a butadiene with a bulky 2-substituent such as 2-t-butylbutadiene gives a larger amount (45%) of 1,4-adduct, and is known to exist largely in a cisoid conformation. We cannot be sure, however, that only the diradical mechanism still operates because a high proportion of cisoid conformation should also favour the concerted addition process (see page 263).

1,1-Dichloro-2,2-difluoroethylene is a rather unusual olefin; even with a diene like cyclopentadiene which is usually very reactive in giving 1,4 (Diels–Alder) adducts, it gives 9% of 1,2-addition product and 44% of 1,4-addition product together with 47% of cyclopentadiene dimer. An olefin intermediate in reactivity between **30** and maleic anhydride (a typical Diels–Alder dienophile) is α-acetoxyacrylonitrile. With butadiene at 150° this olefin gives about 20% 1,2-adduct and 80% 1,4-adduct. The amount of 1,4-adduct formed exceeds the per cent cisoid butadiene present, so the cisoid diene which is present must be much more reactive than the transoid form, probably because a concerted cycloaddition to it can take place. 1,4-Addition to *trans,trans*-hexa-2,4-diene is stereospecific giving two cyclohexenes with *cis*-methyl groups (see **34**), again pointing to a concerted reaction or at least to a situation where ring-closure is now faster than the possible bond rotations in

the diradical intermediate. In fact the distinction between reaction through an intermediate with these properties and a concerted reaction is rather a hazy one!

We can see that drawing a sharp dividing line between a concerted mechanism and reaction via a diradical could become quite difficult in some cases. What of reaction via a zwitterion (e.g. **3** on page 239)? Distinguishing reaction through a dipolar intermediate should be relatively easy, for if the transition-state is much more polar than the starting materials, strong solvent and substituent effects should appear. An example is the reaction of tetracyanoethylene with 4-methoxystyrene (**35**). These give a blue charge-transfer complex, which disappears as the actual cycloaddition takes place. Disappearance of the blue colour takes about 1 minute in nitromethane, 1 week

(35)

in toluene, and is incomplete after 1 month in cyclohexane. The ρ-value for reaction of a series of *para*-substituted styrenes with tetracyanoethylene is about -7 which suggests a lot of positive charge on the benzylic carbon in the transition-state. These facts strongly suggest the zwitterionic intermediate shown. Identification of this mechanism might not be so easy in those cases, though, where the zwitterion and the diradical are of comparable stability. Electron-transfer converting one into the other *might* be a faster process than any other 'reaction' of the system, but we do not know.

A rather special class of $2\pi + 2\pi$ cycloadditions are those in which one partner is a cumulated system such as a ketene or an allene. The dimerization of dimethylketene to 2,2,4,4-tetramethylcyclobutane-1,3-dione is a reaction for which a very plausible zwitterionic intermediate can be written (36). In fact the reaction rate is surprisingly insensitive to solvent polarity, increasing only 30-fold on going from CCl_4 to MeCN. $\Delta S^{\ddagger}$ is also very negative (-42 cal mole^{-1} deg^{-1}) suggesting a highly ordered transition-state and therefore

(36)   (a)   (37)   (b)

possibly a concerted cycloaddition.[41] It is also interesting that reaction of 1,1-dimethylallene with dimethyl fumarate (*trans*-MeOCOCH=CHCO$_2$Me) is stereospecific giving only the two products (addition occurs to either double bond of the allene) with *trans* ester groups.[42] It has been suggested that the reaction of a ketene (or allene) with another 2π-system could be a $_{\pi}2_s + _{\pi}2_s + _{\pi}2_s$ (or $_{\pi}2_s + _{\pi}2_a + _{\pi}2_a$) process (see 37a) using *both* bonds of the ketene, or that it is a $_{\pi}2_s + _{\pi}2_a$ process with the π-bond of the ketene or allene used antarafacially (see **37b**). Some evidence in support of this comes from measurement of the isotope effects in reaction **38**. At the β-position of styrene k_H/k_D is 0.91—an inverse effect as is expected for a reaction in which the β-carbon is

$$
\begin{array}{c}
Ph \\
\diagdown \\
Ph
\end{array} C\!\!=\!\!C\!\!=\!\!O \;\; + \;\;
\begin{array}{c}
Ph \quad H(D) \\
\diagup \\
\diagdown H(D) \\
H(D)
\end{array}
\;\longrightarrow\;
\begin{array}{c}
Ph \quad O \\
Ph \\
Ph
\end{array}
\tag{38}
$$

changing from sp^2- to sp^3-hybridization (Chapter 1, page 19). At the α-position, however, k_H/k_D is 1.23, a most unexpected value. For a two-step reaction via a diradical or zwitterion, k_H/k_D should be about 1.0, and the large value observed is thought to be associated with the twisting of the sp^2-carbon out of conjugation with the neighbouring p-orbitals.[43] Such a twisting is certainly required by a $_\pi2_s + _\pi2_a + _\pi2_a$ or $_\pi2_s + _\pi2_a$ cycloaddition (**37**).

The $_\pi2_s + _\pi2_a$ viewpoint is especially attractive for ketenes since the transition-state may be further stabilized by interaction of the electrophilic (carbon) end of the orthogonal π-C$=$O orbital with the olefin (dotted lines in **37b**). This geometry of approach has been used to explain the stereochemistry of ketene addition to cyclopentadiene (see **39**).[44]

$$
\bigcirc \;\; + \;\;
\begin{array}{c}
H \\
\diagdown \\
Cl
\end{array} C\!\!=\!\!C\!\!=\!\!O
\;\longrightarrow\;
\tag{39}
$$

The familiar Diels–Alder reaction has received great attention in the past, with the result that a wide range of dienes and dienophiles have been investigated. Some of the facts[2b] which satisfactory mechanisms must account for are:

(*i*) In essentially every case examined, the addition to both diene and dienophile is *cis* or *syn*.

(*ii*) Dienes must take up a cisoid conformation for reaction to occur. Thus *cis*-1-methylbutadiene, which is sterically hindered in its cisoid conformation, gives only 4% of adduct with maleic anhydride under conditions where its *trans*-isomer reacts quantitatively and exothermically. Fixed cisoid dienes such as cyclopentadiene are especially reactive.

(*iii*) The reactivity in a series of dienophiles with a 'typical' diene such as cyclopentadiene is increased by electron-withdrawing substituents (see the examples of cyanoethylenes in Figure 10). With a given dienophile, such as maleic anhydride, reaction rates rise with increasing electron-supply in the

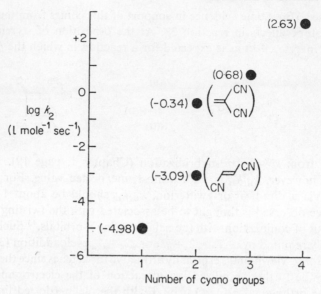

Figure 10. Rate of reaction of the cyanoethylenes with cyclopentadiene at 20°. Compare this with Figures 22 and 23 in Chapter 5.

diene. In a limited set of examples with dienes containing strongly electron-withdrawing substituents (such as hexachlorocyclopentadiene) these reactivity trends may be reversed. With this diene cyclopentene reacts faster than maleic anhydride.

(*iv*) As mentioned in Chapter 1 (page 46), solvent effects on reaction rates are quite small.

(*v*) Entropies of activation ($\Delta S^{\ddagger}$) are highly negative (-35 ± 6 cal mole^{-1} deg^{-1} for 17 additions to cyclopentadiene).

(*vi*) In unsymmetrical reactions where two position-isomers are possible there is usually a moderate degree of selectivity. Thus isoprene with methyl acrylate at 20° (conditions of kinetic control) gives **40** and **41** in a 5.4/1 ratio.

Me   Me
   CO$_2$Me   CO$_2$Me

(**40**)   (**41**)

(*vii*) There is also a moderate preference for *endo*-addition in the appropriate cases. Cyclopentadiene with maleic anhydride gives the products shown in **42** under conditions of kinetic control.

$$endo \ (98.5\%) \qquad exo \ (1.5\%)$$ (42)

These facts fit with a picture of most Diels–Alder reactions as concerted $_\pi 2_s + _\pi 4_s$ cycloadditions (see Figure 5, page 250). By a concerted reaction we mean one in which all the new bonds are in the process of forming at the transition-state; there is no need to assume that they have reached the same degree of completion. Thus in 2 (page 239) a one-step reaction via a transition-state with bond a—b 90% formed and bond c—d only 10% formed would still be a concerted reaction (in this example it is likely that the transition-state is more symmetrical than this). The concerted process accounts well for $\Delta S^\ddagger$ being strongly negative and for the cisoid requirement for the diene (though this would probably fit a non-concerted process; see page 259). The solvent effects show a non-polar transition-state being formed from non-polar reactants, but would also fit a diradical hypothesis. We have already seen on pages 256–259 that 2+4 cycloaddition via a diradical intermediate *can* occur, and the question therefore arises again, can we really eliminate this? The position selectivity in (*vi*) above is readily accounted for by considering the stabilities of the possible diradical intermediates. On the other hand, substituent effects on an unsymmetrical but concerted process might be expected to be very similar.[45] The problem is essentially similar in considering the reactivities in (*iii*) above. The greater reactivity of 1,1-dicyanoethylene than the 1,2-isomers suggests an unsymmetrical (possibly zwitterionic) transition-state for the former—but how unsymmetrical we do not know. We can only answer these questions for certain specific examples at present (we examine one case below); in general it may turn out that the borderlines between the mechanisms are far from sharp. For a moment, however, let us look at another aspect of the reactivity patterns in (*iii*). They suggest that charge-transfer (usually from diene to dienophile) stabilizes the transition-state, though it is not obvious why this should be so. It is possible that the real cause is slightly different—energy matching between the highest occupied and lowest unoccupied orbitals on the diene and dienophile, which mix to generate the interaction energy, may be more efficient when the dienophile (or occasionally the diene) has electron-withdrawing substituents. Calculations seem to bear this idea out.[46]

One way to try to discover how concerted and symmetrical the transition-state of a Diels–Alder reaction is, is to measure secondary deuterium isotope effects. This has been done by Seltzer[2a, 47] for the retro-Diels–Alder decomposition of the 2-methylfuran–maleic anhydride adduct (the reverse reaction was studied for convenience; the conclusions should apply equally well to the formation reaction by the principle of microscopic reversibility). The secondary deuterium isotope effects k_H/k_D reflect the degree of hydridization change (from sp^3 to sp^2 in this case) at the carbon atom to which the H (or D) is attached (see Chapter 1). The various isotope effects measured by Seltzer are summarized in Figure 11. Thus $k_I/k_{II} = 1.16$ means that the rate of decomposition of the all-protium adduct was 1.16 times that of the adduct with $X = Y = D$. The ratios k_I/k_{II}, k_I/k_{III}, and k_I/k_{VI} were found by normal kinetic methods. The ratio k_{IV}/k_V was found by taking a 1:1 mixture of **IV** and **V** (prepared from monodeuteriomaleic anhydride), allowing its decomposition to go to 90% completion, recovering the remaining (**IV** + **V**) and determining

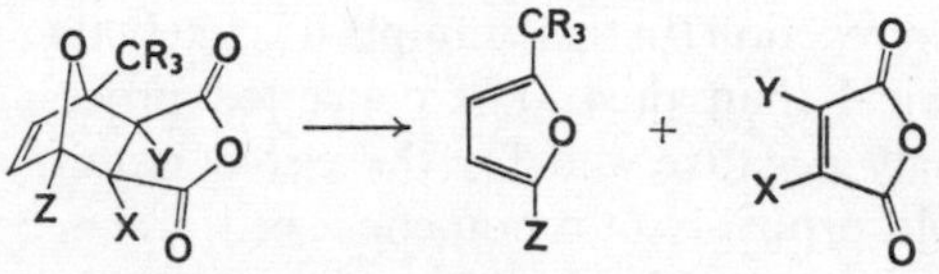

	X	Y	Z	R	
(I)	H	H	H	H	
(II)	D	D	H	H	$k_I/k_{II} = 1.16$
(III)	H	H	D	H	$k_I/k_{III} = 1.08$
(IV)	D	H	H	H	$k_I/k_{VI} = 1.03$
(V)	H	D	H	H	$k_{IV}/k_V = 1.00$
(VI)	H	H	H	D	

Figure 11. Isotope effects in the decomposition of the adduct from 2-methylfuran and maleic anhydride.

the ratio **IV**/**V** in this recovered material by n.m.r. integration of the signals due to the X and Y protons. A 1/1 ratio was found. It can be seen that the observed effects fit in well with there being an isotope effect k_H/k_D of 1.08 at each carbon C—X, C—Y, and C—Z. This means that both the carbon bonds have broken to nearly equal extents at the transition-state. The β-deuterium isotope effect k_I/k_{VI} is very small and agrees with there being little charge or radical density at the carbon bearing the methyl group in the transition-state. All in all, a concerted, symmetrical process is suggested for this example.

The one remaining feature of the Diels–Alder reaction which has to be discussed is the common preference for *endo*-stereochemistry. This is interesting because it often runs counter to expected steric effects; it is worth keeping in mind, however, that it is unlikely that the difference in free-energy of activation for *endo*- and *exo*-addition ever exceeds 4—5 kcal mole⁻¹. Possible reasons for the *endo*-preference have been discussed by Woodward and Hoffmann, by Salem, and by Herndon and Hall.[48] The last authors maintain that better overlap of the orbitals of the olefin with those on the 1- and 4-positions of cyclopentadiene can be achieved by an *endo*-approach (**43** vs. **44**). Their

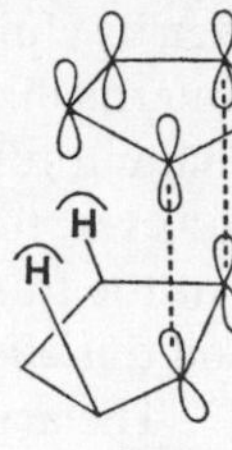

(43) 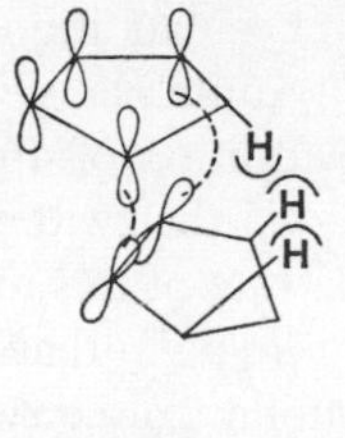   (44) 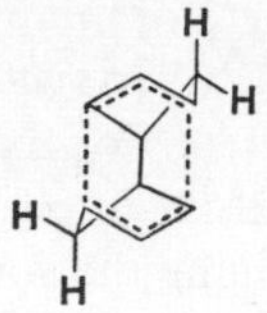   (45)

explanation applies to mono-olefins such as cyclopentene (which does give mainly *endo*-adduct). The other authors consider the possibilities for secondary attractive forces between the orbitals at positions 2 and 3 of the diene and π-orbitals on the substituents of the dienophile (e.g. π-orbitals on the −COOCO− portion of maleic anhydride). They differ, however, as to which attractive forces are important but Salem points out the possibility of a symmetrical transition-state for cyclopentadiene dimerization with one σ-bond well formed and the other interactions resembling those in a Cope rearrangement (**45**). Diels–Alder adducts do sometimes undergo Cope rearrangements. Quite possibly all the effects suggested may play their part in suitable cases.

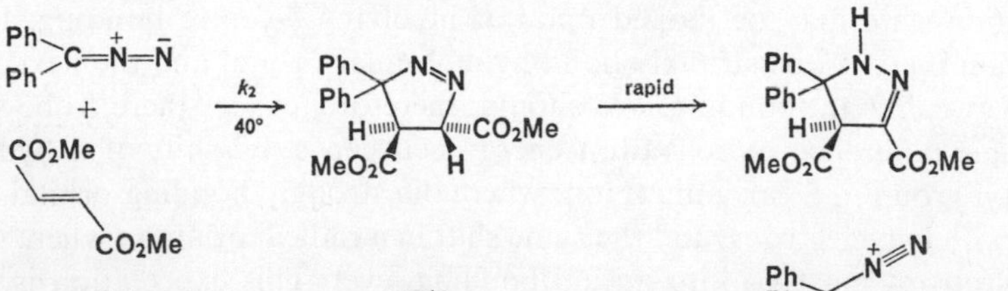

Solvent	$10^4 k_2$
Benzene	1.44
Dioxan	1.15
Ethyl acetate	1.25
Ethanol	5.24
Acetone	1.10
Acetonitrile	2.63
Dimethylformamide	2.45

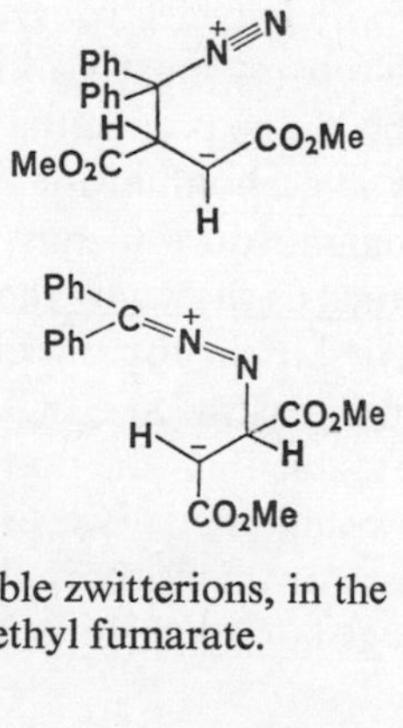

Figure 12. Solvent effects on the rate, and possible zwitterions, in the addition of diphenyldiazomethane to dimethyl fumarate.

Another type of cycloaddition which has become synthetically quite important is the 1,3-dipolar cycloaddition. Octet-stabilized 1,3-dipoles, as explained on page 248 are hetero-analogues of allyl anion, a 4π-electron system, and concerted supra-supra cycloaddition to an olefin is allowed. However, for many 1,3-dipolar cycloadditions (see the lists given by Huisgen[3, 6] and by Roberts and Caserio[49]) quite plausible zwitterionic intermediates can be written. For example the addition of diphenyldiazomethane to dimethyl fumarate (Figure 12) might go via the zwitterions shown. However, the effect of changing the solvent on the rate of this reaction shows that the transition-state can be only marginally more polar than the reactants, and this effectively eliminates a transition-state similar to the zwitterion shown.[50] The activation parameters for this reaction (in dimethylformamide) ($\Delta H^{\neq}$ 8.4 kcal mole^{-1}; $\Delta S^{\neq}$ −39 cal mole^{-1} deg^{-1}) also strongly support a concerted reaction through a tightly constrained transition-state. Similar evidence has been produced for some other 1,3-dipolar cycloadditions, but it would be unwise to assume that all such reactions follow the same mechanism.

Mechanisms of Sigmatropic Reactions

The simplest sigmatropic reaction is a [1,2]-shift within a cation, radical, or anion (see Figure 13). Such reactions may proceed by a two-step dissociation–recombination sequence, a two-step association–dissociation sequence, or by a synchronous process in which bond-formation and bond-breaking are concerted. The orbital symmetry rules apply to the last mechanism. The transition-state may be looked upon as involving 3-centre bonding. In this situation there is one strongly bonding molecular orbital and the next higher level is weakly antibonding. We should, therefore, expect there to be a con-siderable difference in activation energy between a 1,2-shift of a hydrogen or alkyl group in a carbonium ion, where the strongly bonding orbital is just filled with two electrons, and the same shift in a radical or anion where one or two electrons must go into an antibonding level. This expectation is borne out by the experimental facts. [1,2]-Hydrogen and alkyl shifts occur very easily in carbonium ions. The degenerate hydrogen shift in the methylethyl-carbonium ion (46) has an activation energy below 6 kcal mole^{-1} if $\log A$ is normal as shown by the time-averaged n.m.r. at $-110°$,[51] and the degenerate methyl shift in the dimethyl-t-butylcarbonium ion (47) is still rapid on the n.m.r. time-scale at $-180°$.[52] In these cases the classical cations are still slightly more stable than the non-classical bridged ion which is the transition-state in these reactions. However, the non-classical ion becomes the ground-state in the case of the norbornyl ion (see Chapter 2, page 119). By contrast no [1,2] hydrogen or alkyl shifts in radicals or carbanions are really firmly established

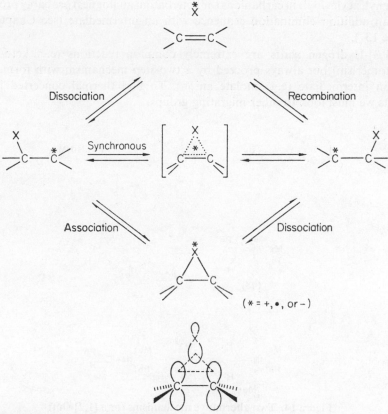

Figure 13. Possible mechanisms for a [1,2]-shift. The transition-state for the synchronous mechanism showing the 3-centre bonding.

as being concerted.[53] Thus the Stevens rearrangement of ylid **48**, itself generated by the addition of benzyne to *N,N*-dimethylbenzylamine, apparently goes by a radical dissociation–recombination mechanism (Figure 14) since CIDNP (see Chapter 2, page 168) has been observed for the product (which must arise therefore from a radical precursor).[54] [1,2]-Shifts of unsaturated groups

$$ \underset{Me}{\overset{H}{>}}\!\overset{+}{C}\!-\!\underset{Me}{\overset{H}{C}}\!\!<^{H} \quad \rightleftharpoons \quad \underset{Me}{\overset{H}{>}}\!C\!-\!\overset{+}{\underset{Me}{C}}\!\!<^{H} \tag{46} $$

$$ \underset{Me}{\overset{Me}{>}}\!\overset{+}{C}\!-\!\underset{Me}{\overset{Me}{C}}\!\!<^{Me} \quad \rightleftharpoons \quad \underset{Me}{\overset{Me}{>}}\!C\!-\!\overset{+}{\underset{Me}{C}}\!\!<^{Me} \tag{47} $$

(phenyl, acyl, vinyl) in carbanions (or heteroatom analogues) probably proceed by an addition–elimination sequence with an intermediate (see Chapter 2, page 137).

[1,3]-Hydrogen shifts are extremely common reactions (e.g. keto–enol tautomerism) but always proceed by a two-step mechanism with formation of an intermediate (e.g. enolate anion). To find thermal concerted [1,3]-shifts we must look to other migrating groups.

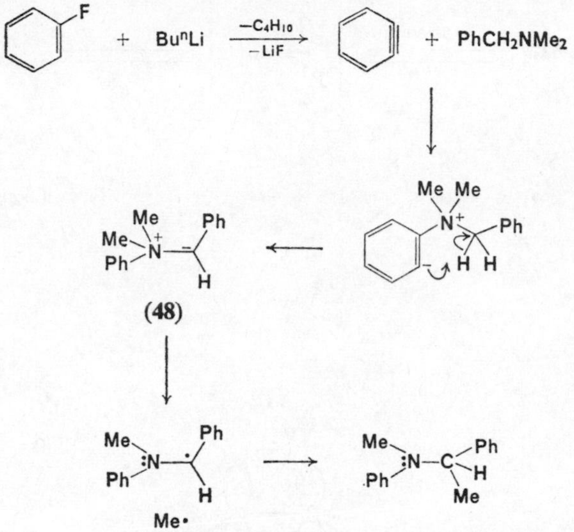

Figure 14. Two alternative mechanisms for a [1,2]-shift in a carbanion; dissociation into a radical pair, and an associative reaction in which bond-making is complete before bond-breaking begins.

On page 249 it was predicted that in a concerted [1,3]-shift migration should occur with inversion at the migrating carbon. This has now been proved in several cases[55, 56] (Figure 15) and must be regarded as one of the triumphs of the orbital symmetry rules, because the transition-states, particularly in the first example,[55] have severe non-bonded interactions and migration with retention would seem to be easier on all other counts. A [1,4]-sigmatropic shift within a carbonium ion should also occur with inversion. In the hepta-methylbicyclo[3.1.0]hexenyl cation a series of shifts of this type causes time-averageing of the C(1) to C(5) methyl groups above −100° in the n.m.r. spectrum. In this process 6-*exo*- and 6-*endo*-methyl groups do not exchange positions, as would occur if the migration occurred with stereochemical retention at C(6). In fact the allowed process, with inversion, is at least 50,000 times faster than the forbidden, retention, process.[57]

[1,5]-Sigmatropic shifts of hydrogen occur widely. They are particularly rapid in cyclopentadienes (see 12, page 241). Many of the best examples of

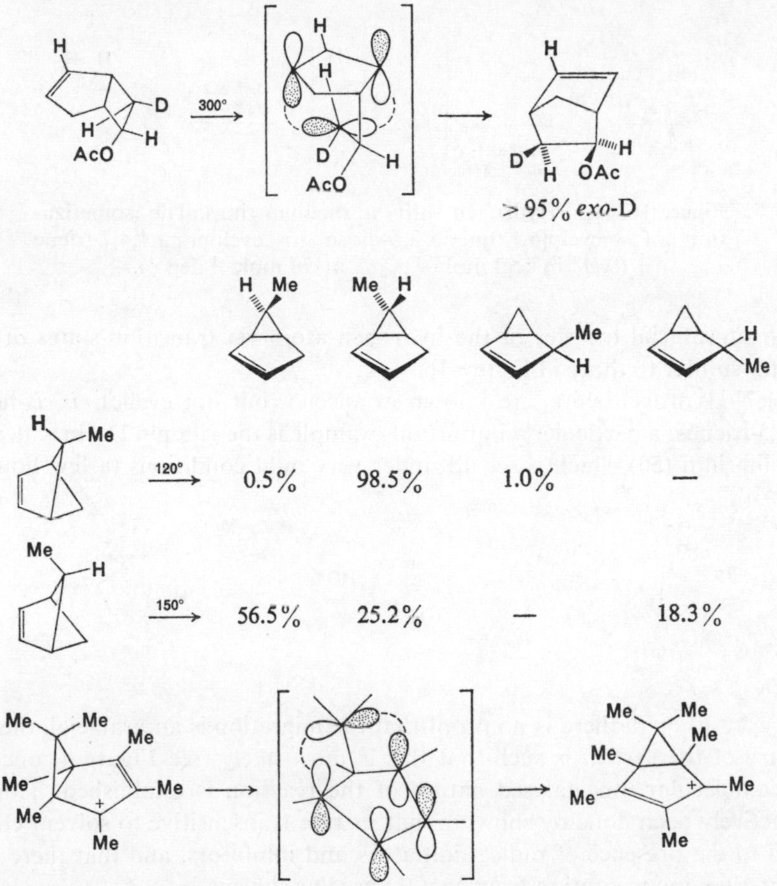

Figure 15. [1,3]- and [1,4]-Shifts with inversion at the migrating carbon.

[1,5]-hydrogen shifts come from medium-ring chemistry, where they occur in both dienyl and homodienyl systems (Figure 16). In the latter a cyclopropane ring replaces one of the double bonds, and the reaction could equally well be considered as the reverse of an intramolecular 'ene' reaction (see 17, page 242). Glass, Boikess, and Winstein[7] have reviewed these [1,5]-hydrogen shifts and shown that the variation of activation energy with ring size is consistent

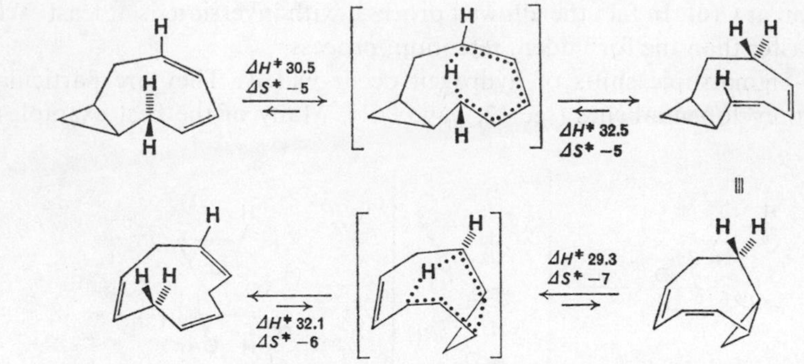

Figure 16. [1,5]-Hydrogen shifts in medium rings. The isomerization of bicyclo[6.1.0]nona-3,5-diene to cyclonona-1,4,7-triene ($\Delta H^{\neq}$ in kcal mole^{-1}; $\Delta S^{\neq}$ in cal mole^{-1} deg^{-1}).

with suprafacial transfer of the hydrogen atom via transition-states of geometry similar to those in Figure 16.

[1,7]-Hydrogen shifts are known in acyclic (but not cyclic) *cis,cis*-hepta-1,3,5-trienes; a particularly important example is the vitamin D_2–precalciferol equilibrium (**50**) which is set up under very mild conditions (a few hours at

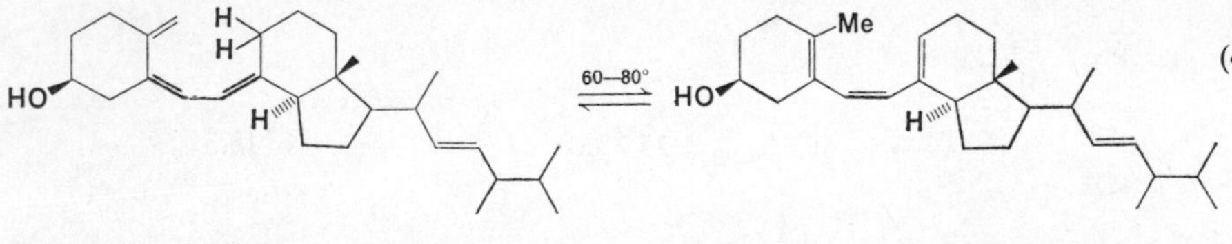

(49)

$60°$).[58] Although there is no proof that the migration is antarafacial, the geometry of the system is such that this is most likely (see Figure 4) once the intramolecular, uncatalysed nature of the reaction is established. This has effectively been done by showing that its rate is insensitive to solvent change and to the presence of radical initiators and inhibitors, and that there is no deuterium incorporation from MeOD used as solvent.

If we turn from [1,x]-sigmatropic rearrangements to [x,y]-rearrangements, the main interest lies in [3,3]-rearrangements of which the Cope and Claisen rearrangements are the major examples. Rearrangement should be suprafacial in both components, and geometrically this is easy to achieve. Rearrangement of 1,1-dideuteriohexa-1,5-diene to 3,3-dideuteriohexa-1,5-diene occurs with $\Delta H^{\neq}$ 35 kcal mole^{-1} and $\Delta S^{\neq}$ −11 cal mole^{-1} deg^{-1},[59] whereas the fragmentation of hexa-1,5-diene to two allyl radicals requires 62 kcal mole^{-1} of energy.[60] It is obvious from these figures that the two partial bonds holding the two allyl fragments together at the transition-state must be worth about 13 kcal mole^{-1} each. The $\Delta S^{\ddagger}$ value for the rearrangement is also in keeping

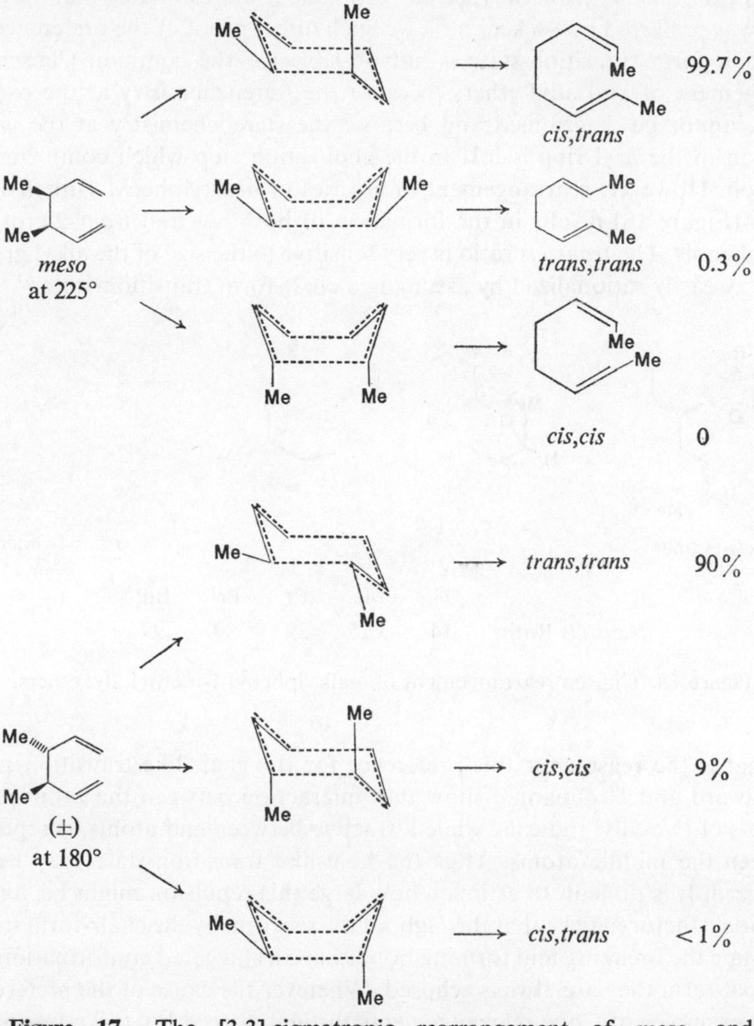

Figure 17. The [3,3]-sigmatropic rearrangement of *meso*- and (±)-3,4-dimethylhexa-1,5-diene.

with a 'tight' transition-state. As was mentioned on page 248, there are two conceivable geometrical possibilities for this transition-state, a boat or chair form. Doering and Roth[17] showed that the chair form was preferred by re-arranging both *meso* and racemic isomers of 3,4-dimethylhexa-1,5-diene

(Figure 17). The amounts of products formed through a boat-form transition-state are very small (0.3% *trans,trans*-octa-2,6-diene from the *meso*-isomer; 1% *cis,trans*-isomer from the racemic form) and it was estimated that the chair form was preferred by ~6 kcal mole^{-1}. Such direct proof of the preference for the chair-form transition-state is not possible for the common Claisen re-arrangement of aryl allyl ethers, because the stereochemistry at the oxygen atom cannot be determined and because the stereochemistry at the *ortho*-position in the aryl ring is lost in the enolization step which completes the reaction. However, rearrangement of a series of *o*-alkylphenyl 1-methylallyl ethers (Figure 18) results in the formation of both *cis*- and *trans*-2-crotyl-6-alkylphenols. The *trans/cis* ratio is very sensitive to the size of the alkyl groups in a way easily rationalized by assuming a chair-form transition-state.[61]

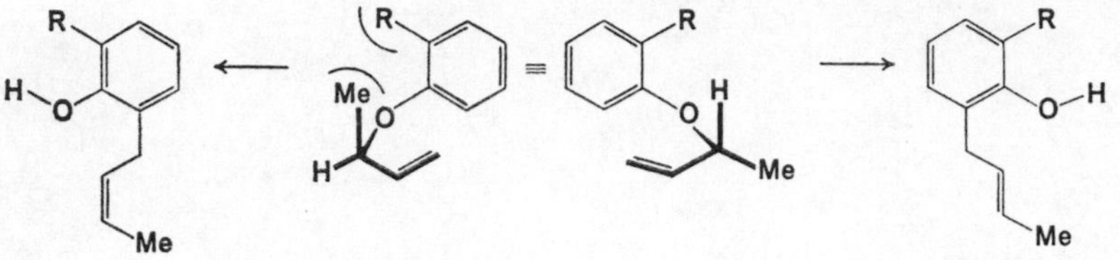

cis-Isomer   trans-Isomer

R	H	Me	Et	Pr^i	Bu^t
trans/cis Ratio	14	37.5	39	39	99

Figure 18. Claisen rearrangement of *o*-alkylphenyl 1-methylallyl ethers.

What is the reason for this preference for the chair-like transition-state? Woodward and Hoffmann[62] show that interaction between the π-molecular orbitals of two allyl radicals, while attractive between end atoms, is repulsive between the middle atoms. Thus the boat-like transition-state may be de-stabilized. It is difficult to estimate how large this repulsion might be, and an additional factor may be that throughout the reaction by the chair-form transition-state the breaking and forming bonds have a staggered conformation—in the boat-form they are always eclipsed. Whatever the cause of the preference, many examples of Cope rearrangements through forced boat-like transition-states are known, of which *cis*-divinylcyclopropane rearrangements are excellent examples. Two spectacular cases (Figure 19) are the degenerate rearrangements of tricyclo[3,3,0,0^{2,8}]octa-3,6-diene (semibullvalene),[63] which is fast on the n.m.r. time-scale even at −100°, and that of tricyclo[3,3,2,0^{2,8}]-deca-3,6,9-triene (bullvalene) which is slower but ultimately results in the interconversion of all of the possible 1.2×10^6 isomers.[64] Any C—H group

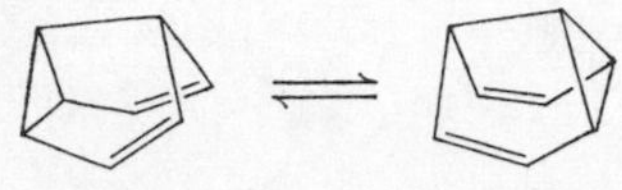

(A)

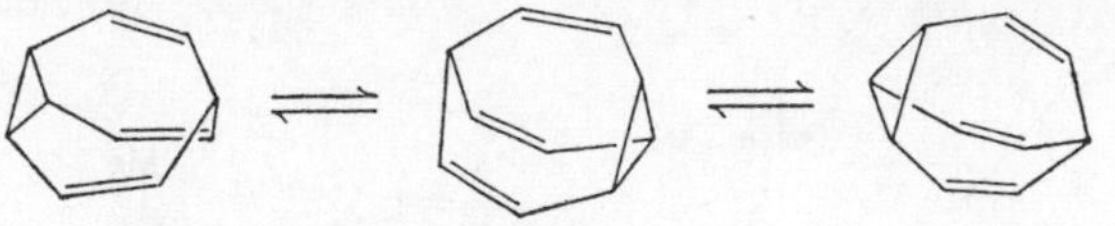

(B)

Figure 19. Degenerate Cope rearrangement of semibullvalene (A) and bullvalene (B).

may have as its neighbours any selection of three from the nine remaining C—H groups.

Besides these [3,3]-sigmatropic rearrangements, several other $[x,y]$-shifts have been observed; three examples are shown in **50**—**52**. Rearrangement of the phenyl pentadienyl ether shown in **50** leads to both [3,3]- and [5,5]-shifts, probably with both components used suprafacially.[65] The benzidine rearrangement[66] is apparently a [5,5]-shift (**51**) though, when the *para*-positions of the hydrazobenzene are blocked, other products (e.g. the *ortho*-semidine shown) are formed by rearrangements which are formally forbidden to be concerted. Probably an intermediate is formed the two halves of which can rotate before collapsing to give the various products and which probably lives long enough to forget its origins in orbital symmetry terms. The bonding in the intermediate is still uncertain and will probably vary according to

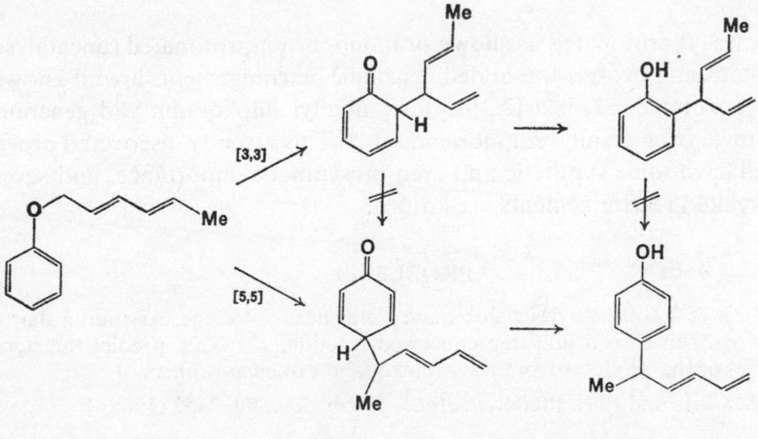

(50)

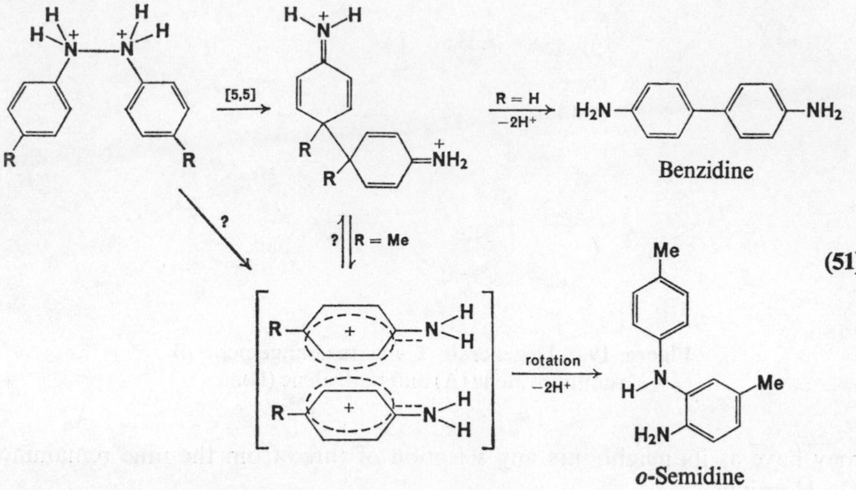

(51)

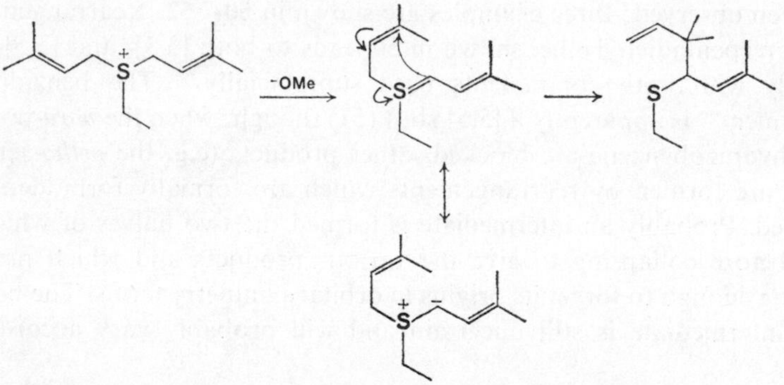

(52)

whether it is diprotonated as shown or mono- or un-protonated (uncatalysed, one-proton- and two-proton-added benzidine rearrangements are all known). The final example (52) is a [2,3]-shift in an allyl sulphonium ylid generated, for example, from an allyl sulphonium salt.[67] This recently discovered process may well have some synthetic and even biosynthetic importance, and several closely related rearrangements are known.

PROBLEMS

1. Pyrolysis of 2,4-dimethylbicyclobutane yields hexa-2,4-diene. Assuming that the double ring-opening is a one-step concerted reaction, $_\sigma 2_s + _\sigma 2_a$, predict the stereo-chemistries of the products from the *exo,exo-* and *exo,endo-*isomers.

[See G. L. Closs and P. E. Pfeffer, *J. Amer. Chem. Soc.*, **90**, 2452 (1968).]

2. Use the data of Figure 7 (ignoring solvent and temperature effects) to construct an approximate free-energy diagram interrelating isomers **A** and **G** and the various transition-states connecting them.

3. How many possible isomeric cyclohexenes can arise from combination of the two compounds shown? Make drawings of those that can be formed by *cis*-addition

to each partner. Which of these is likely to be the major product in a kinetically controlled Diels–Alder reaction?

4. The reaction shown occurs on heating at 100° for 14 hr.

$$(Ar = C_6H_4OMe\text{-}p)$$

Suggest a likely pathway and account for the observed stereochemistry. (*Hint*: an electrocyclic reaction followed by a cycloaddition.) Heating the aziridine in CCl_4 in the absence of dimethyl acetylenedicarboxylate leads to equilibration with the *cis*-isomer; can your reaction scheme accommodate this?

[See R. Huisgen, W. Scheer, and H. Huber, *J. Amer. Chem. Soc.*, **89**, 1753 (1967).]

5. (a) Predict the two products of rearrangement of:

[See p. 831 of ref. 1a.]

(b) The diene shown racemizes with a half-life of 24 hr at 50°.

Show the mechanism of this racemization.

[See P. S. Wharton and R. A. Kretchmer, *J. Org. Chem.*, **33**, 4258 (1968).]

6. The n.m.r. spectrum of benzene $+ HF–SbF_5$ in $SO_2ClF–SO_2F_2$ at $-134°$ shows peaks at τ 4.16 (two protons), 1.60 (two protons), 0.80 (one proton), and 0.58 (two protons). At $-80°$ however there is a single line at τ 1.91, due to rapid rearrangement by $[1,x]$-hydrogen shifts.

Explain the spectra and suggest plausible values for x.

[See G. A. Olah, R. H. Schlosberg, D. P. Kelly, and G. D. Mateescu, *J. Amer. Chem. Soc.*, **92**, 2546 (1970).]

REFERENCES

1. (a) R. B. Woodward and R. Hoffmann, *Angew. Chem. Internat. Edn. Engl.*, **8**, 781 (1969). The definitive statement of the orbital symmetry rules. (b) Other interesting discussions on this topic: S. I. Miller, *Adv. Phys. Org. Chem.*, **6**, 185 (1968); G. B. Gill, *Quart. Rev.*, **22**, 338 (1968); L. Salem, *Chem. in Britain*, **5**, 449 (1969).
2. (a) S. Seltzer, *Adv. Alicyclic Chem.*, **2**, 1 (1968). (b) J. Sauer, *Angew. Chem. Internat. Edn. Engl.*, **6**, 16 (1967). Discussions of the mechanism of the Diels–Alder reaction.
3. R. Huisgen, R. Grashey, and J. Sauer, *The Chemistry of the Alkenes* (ed. S. Patai), Wiley-Interscience, London, 1964, pp. 739–953. Cycloaddition reactions of olefins.
4. P. D. Bartlett, *Science*, **159**, 833 (1968). 2π and 2π cycloadditions.
5. H. M. Frey and R. Walsh, *Chem. Rev.*, **69**, 103 (1969). Gas-phase, thermal, unimolecular reactions of hydrocarbons; includes many electrocyclic and sigmatropic reactions.
6. L. I. Smith, *Chem. Rev.*, **23**, 193 (1938); R. Huisgen, *Angew. Chem. Internat. Edn. Engl.*, **2**, 565 (1963); *Helv. Chim. Acta*, **50**, 2421 (1967). 1,3-Dipolar cycloadditions.
7. D. S. Glass, R. S. Boikess, and S. Winstein, *Tetrahedron Letters*, **1966**, 999. 1,5-Hydrogen shifts.
8. H. M. R. Hoffmann, *Angew. Chem. Internat. Edn. Engl.*, **8**, 556 (1969). The ene reaction.
9. H.-J. Hansen and H. Schmid, *Chem. in Britain*, **5**, 111 (1969). Aromatic sigmatropic rearrangements, especially Claisen rearrangements.
10. G. Schroeder, J. F. M. Oth, and R. Merényi, *Angew. Chem. Internat. Edn. Engl.*, **4**, 752 (1965). Fast, reversible, valence tautomerism.
11. C. A. Coulson, *Valence*, 2nd Edn., Oxford University Press, Oxford, 1961.
12. R. B. Woodward and R. Hoffmann, *J. Amer. Chem. Soc.*, **87**, 395, 2046, 2511 (1965).
13. E. N. Marvell, G. Caple, and B. Schatz, *Tetrahedron Letters*, **1965**, 385.
14. R. E. K. Winter, *Tetrahedron Letters*, **1965**, 1207.
15. K. Alder and G. Stein, *Ann. Chem.*, **485**, 223 (1931); R. Huisgen, L. Möbius, G. Müller, H. Stangl, G. Szeimies, and J. M. Vernon, *Chem. Ber.*, **98**, 3992 (1965).
16. S. McLean and P. Haynes, *Tetrahedron*, **21**, 2329 (1965).
17. W. von E. Doering and W. R. Roth, *Tetrahedron*, **18**, 67 (1962).
18. J. E. Baldwin, *Can. J. Chem.*, **44**, 2051 (1966).
19. H. C. Longuet-Higgins and E. W. Abrahamson, *J. Amer. Chem. Soc.*, **87**, 2045 (1965).

20. K. Fukui, *Tetrahedron Letters*, **1965**, 2009; K. Fukui and H. Fujimoto in *Mechanisms of Molecular Migrations* (Ed. B. S. Thyagarajan), Wiley-Interscience, New York, 1969, Vol. 2, p. 117.
21. H. E. Zimmerman, *J. Amer. Chem. Soc.*, **88**, 1564 (1966).
22a. M. J. S. Dewar, *Tetrahedron*, Suppl. 8, 75 (1966).
22b. H. M. R. Hoffmann and D. R. Joy, *J. Chem. Soc.* (B), **1968**, 1182; C. F. Huebner and E. M. Donoghue, *J. Org. Chem.*, **33**, 1678 (1968).
23. R. Criegee, W. Hörauf, and W. D. Schellenberg, *Chem. Ber.*, **86**, 126 (1953).
24. J. I. Brauman and D. M. Golden, *J. Amer. Chem. Soc.*, **90**, 1920 (1968).
25. E. C. Lupton, *Tetrahedron Letters*, **1968**, 4209.
26. G. A. Doorakian and H. H. Freedman, *J. Amer. Chem. Soc.*, **90**, 5310, 6896 (1968).
27. W. J. Nebe and G. J. Fonken, *J. Amer. Chem. Soc.*, **91**, 1249 (1969): R. S. H. Liu, *ibid.*, **89**, 112 (1967).
28. R. Huisgen, with A. Dahmen and H. Huber, *J. Amer. Chem. Soc.*, **89**, 7130 (1967); *Tetrahedron Letters*, **1969**, 1461, 1465.
29. C. H. DePuy, *Accounts Chem. Res.*, **1**, 33 (1968).
30. P. von R. Schleyer, T. M. Su, M. Saunders, and J. C. Rosenfeld, *J. Amer. Chem. Soc.*, **91**, 5174 (1969).
31. C. E. Keller and R. Pettit, *J. Amer. Chem. Soc.*, **88**, 604, 606 (1966); S. Winstein, C. G. Kreiter, and J. I. Brauman, *ibid.*, p. 2047.
32. S. Winstein, *Quart. Rev.*, **23**, 141 (1969).
33. J. M. Brown, *Chem. Comm.*, **1967**, 638; J. M. Brown and J. L. Occolowitz, *J. Chem. Soc.* (B), **1968**, 411.
34. R. C. Cookson, B. V. Drake, J. Hudec, and A. Morrison, *Chem. Comm.*, **1966**, 15.
35. M. P. Cava and A. A. Deana, *J. Amer. Chem. Soc.*, **81**, 4266 (1959).
36. W. von E. Doering, quoted in reference 1(a), p. 816.
37. R. C. Cookson, J. Dance, and J. Hudec, *J. Chem. Soc.*, **1964**, 5416.
38. D. R. Arnold and V. Y. Abraitys, *Chem. Comm.*, **1967**, 1053.
39. L. A. Paquette and G. Slomp, *J. Amer. Chem. Soc.*, **85**, 765 (1963).
40. P. D. Bartlett *et al.*, *J. Amer. Chem. Soc.*, **86**, 616, 622, 628 (1964); *ibid.*, **90**, 2049, 2056, 6067, 6071, 6077 (1968); *ibid.*, **91**, 405, 409 (1969); *ibid.*, **92**, 3822 (1970), *J. Org. Chem.*, **32**, 1290 (1967); *Quart. Rev.*, **24**, 473 (1970).
41. R. Huisgen and P. Otto, *J. Amer. Chem. Soc.*, **90**, 5342 (1968).
42. E. F. Kiefer and M. Y. Okamura, *J. Amer. Chem. Soc.*, **90**, 4187 (1968).
43. J. E. Baldwin and J. A. Kapecki, *J. Amer. Chem. Soc.*, **91**, 3106 (1969).
44. P. R. Brook, J. M. Harrison, and A. J. Duke, *Chem. Comm.*, **1970**, 589.
45. J. Feuer, W. C. Herndon, and L. H. Hall, *Tetrahedron*, **24**, 2575 (1968).
46. W. C. Herndon and L. H. Hall, *Theor. Chim. Acta*, **7**, 4 (1967).
47. S. Seltzer, *J. Amer. Chem. Soc.*, **87**, 1534 (1965).
48. L. Salem, *J. Amer. Chem. Soc.*, **90**, 543, 553 (1968); R. B. Woodward and R. Hoffmann, *ibid.*, **87**, 4388 (1965); W. C. Herndon and L. H. Hall, *Tetrahedron Letters*, **1967**, 3095; see also K. N. Houk, *Tetrahedron Letters*, **1970**, 2621.
49. J. D. Roberts and M. C. Caserio, *Basic Principles of Organic Chemistry*, Benjamin, New York, 1964, p. 1013.
50. R. Huisgen, H. Stangl, H. J. Sturm, and H. Wagenhofer, *Angew. Chem.*, **73**, 170 (1961).
51. M. Saunders, E. L. Hagan, and J. C. Rosenfeld, *J. Amer. Chem. Soc.*, **90**, 6882 (1968).
52. G. A. Olah and J. Lukas, *J. Amer. Chem. Soc.*, **89**, 4739 (1967).

53. U. Schöllkopf, *Angew. Chem. Internat. Edn. Engl.*, **9**, 763 (1970); G. F. Hennion and M. J. Shoemaker, *J. Amer. Chem. Soc.*, **92**, 769 (1970).
54. A. R. Lepley, *J. Amer. Chem. Soc.*, **91**, 1237 (1969).
55. J. A. Berson and G. L. Nelson, *J. Amer. Chem. Soc.*, **89**, 5503 (1967); *ibid.*, **92**, 109 (1970); J. A. Berson, *Accounts Chem. Res.*, **1**, 152 (1968).
56. W. R. Roth and A. Friedrich, *Tetrahedron Letters*, **1969**, 2607.
57. R. F. Childs and S. Winstein, *J. Amer. Chem. Soc.*, **90**, 7146 (1968).
58. J. L. M. A. Schlatmann, J. Pot, and E. Havinga, *Rec. Trav. Chim.*, **83**, 1173 (1964).
59. W. von E. Doering and V. Toscano, quoted in W. von E. Doering and J. C. Gilbert, *Tetrahedron*, Suppl. 7, 397 (1966).
60. D. M. Golden, N. A. Gac, and S. W. Benson, *J. Amer. Chem. Soc.*, **91**, 2136 (1969).
61. G. Fráter, A. Habich, H.-J. Hansen, and H. Schmid, *Helv. Chim. Acta*, **52**, 335 (1969).
62. R. B. Woodward and R. Hoffmann, *J. Amer. Chem. Soc.*, **87**, 4389 (1965).
63. H. E. Zimmerman and G. L. Grunewald, *J. Amer. Chem. Soc.*, **88**, 183 (1966).
64. G. Schroeder, *Chem. Ber.*, **97**, 3140, 3150 (1964).
65. G. Fráter and H. Schmid, *Helv. Chim. Acta*, **51**, 190 (1968).
66. H. J. Shine, *Aromatic Rearrangements*, Elsevier, Amsterdam, 1967; D. V. Banthorpe in *The Chemistry of the Amino Group* (ed. S. Patai), Wiley, London, 1968, p. 585.
67. W. Kirmse and M. Kapps, *Chem. Ber.*, **101**, 994, 1004 (1968); J. E. Baldwin, R. E. Hackler, and D. P. Kelly, *Chem. Comm.*, **1968**, 537, 538; G. M. Blackburn, W. D. Ollis, C. Smith, and I. O. Sutherland, *Chem. Comm.*, **1969**, 99.

Chapter 5

Associative Reactions

We define associative reactions as those where bond-making to carbon is occurring at the rate-limiting transition-state without concurrent bond-breaking. Reactions in this class are most frequently encountered in unsaturated systems, and interaction with the reagent results in the fission of π-bonds. Two sub-classifications may be considered: the π-bond being broken can be polar with significant difference in electronegativity of the bonded atoms as in a carbonyl group, or it may be the relatively non-polar unsaturation of an olefin or arene. In the former systems the reagent is most frequently nucleophilic but in the latter electrophilic and free-radical agents are dominant, and we shall discuss these first.

ASSOCIATIVE REACTIONS OF NON-POLAR BONDS

First consider the consequences of attack of an electrophile on a π-system. At some stage in the reaction full disruption of the π-system may occur, when a carbonium-ion intermediate is produced (Figure 1). Our experience of carbonium ions suggests that one of two paths may then be followed: either loss of a proton with the overall effect of substitution, or attack of a nucleophile resulting in formation of an addition product. It is to be expected that the tendency to follow one or other of these paths will, to some extent, relate to the thermodynamic stabilities of the alternative products. If the unsaturated

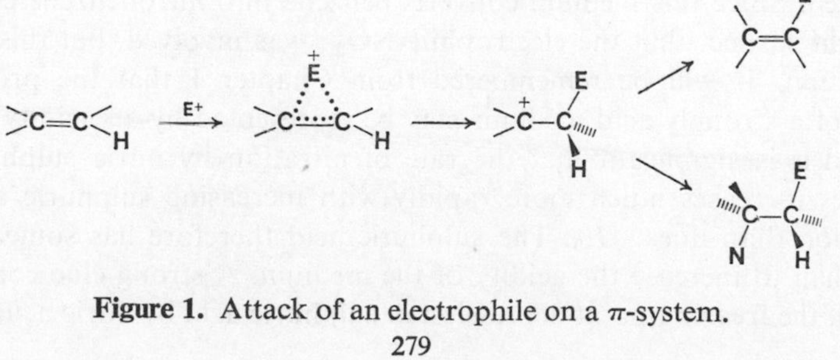

Figure 1. Attack of an electrophile on a π-system.

system is a stable one, there will be a strong tendency for it to be retained in the product and thus substitution will be the result. Indeed, the production of substitution rather than addition products has been the classical criterion for aromatic π-electron delocalization in a cyclic unsaturated system. This tendency is not unique to aromatic systems however, since electrophiles occasionally cause substitution in reaction with olefins, or addition in reaction with aromatic systems. Despite this, it remains a useful general guideline to whether a given molecule is aromatic (as, for example, in the annulene 1).[12] However, the modern chemist may be inclined to use instrumental techniques for this diagnosis, and in particular, the presence of a diamagnetic ring-current, indicating aromaticity may be discerned by study of proton chemical shifts in the nuclear magnetic resonance spectrum.

(1)   (2)

Electrophilic Aromatic Substitution[1-11]

Electrophilic aromatic substitution had been very widely studied in the older chemical literature and by 1950 the broad principles were well understood. For reasons of synthetic utility and practical convenience, certain substitutions had been in focus, notably Friedel–Crafts reactions, halogenation, nitration, and sulphonation. A primary concern was to identify the precise nature of the electrophile, and although it is not our intention to offer a systematic survey of the field here, a brief reference to nitration will be illustrative.

The nitration of aromatic systems may be carried out under a variety of conditions, depending on the reactivity of the aromatic substrate. Commonly, mixtures of strong aqueous solutions of nitric and sulphuric acid may be employed. Since this medium converts benzene into nitrobenzene efficiently, we might suspect that the electrophile NO_2^+ was involved, but this requires verification. It will be remembered from Chapter 1 that the protonating power of a strongly acid medium may be represented by an acidity function H_0, and it is significant that the rate of nitration by nitric–sulphuric acid mixtures increases much more rapidly with increasing sulphuric acid concentration than does $-H_0$. The sulphuric acid therefore has some function other than to increase the acidity of the medium. A strong clue comes from study of the freezing-point depression of sulphuric acid by nitric acid which is consistent with production of four particles from each molecule of nitric acid, very reasonably explained by the equilibrium:

$$HNO_3 + 2H_2SO_4 \rightleftharpoons NO_2^+ + H_3O^+ + 2HSO_4^- \tag{3}$$

and the function of the sulphuric acid is to protonate nitric acid to promote dissociation, and also to protonate the water formed on dissociation and displace the equilibrium.

More recently, the existence of NO_2^+ in these media has been substantiated by Raman spectroscopic observations. Nitration is as normally expected, governed by a second-order rate law, first-order in the concentration of both nitric acid and the aromatic substrate. In some organic solvents such as acetic acid and nitromethane, nitration rates may depend only on the nitric acid concentration. Here the rate-limiting step appears to be *production* of nitronium ions which are then rapidly consumed by substitution. There are subtle mechanistic variations in nitration which we shall return to later.

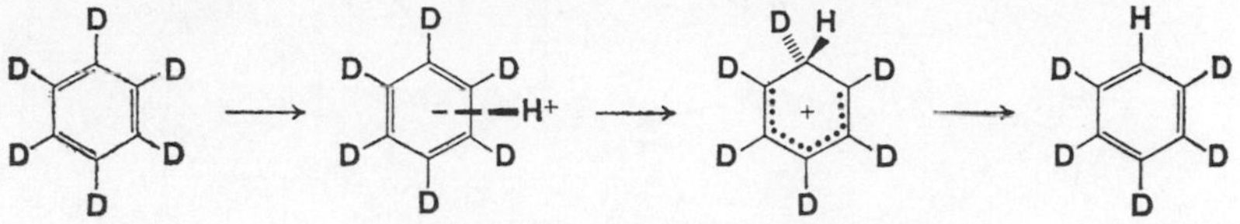

Figure 2. Electrophilic substitution of H for D.

In general, the course of electrophilic substitution may be described by a mechanistic sequence with three possible stages, and our first concern should be to attempt to define the shape of the energy profile along the reaction coordinate. It is instructive to consider the simplest electrophilic substitution, which is the replacement of one hydrogen isotope by another. We could imagine two possibilities, one in which the rate-determining step involved a non-specific association of the acid with the ring, and the other in which the transition-state resembled a cyclohexadienyl cation, or σ-complex (Figure 2). Fortunately the relative equilibrium constants are known for two processes resembling these two possible transition-states in a variety of alkylated aromatic systems. Hydrogen chloride shows enhanced solubility in n-heptane at low temperatures when such compounds are added, and since no ions would be expected in this medium, the effect is attributed to formation of a π-complex (4).[13] In very strongly acidic media such as HF–BF$_3$, protonation of arenes occurs, and their equilibrium basicity (that is, their tendency to produce the conjugate acid 5) may be measured.[6, 14] Comparison of relative rates of deuteriation of benzene, toluene, and the xylenes with π- and σ-complexing ability suggests that the transition-state for isotope exchange closely resembles the σ-complex, and is not related to the π-complex. The data for xylene isomers is particularly instructive (Figure 3).

In this particular example we might expect that the intermediate σ-complex would more readily expel a proton than a deuteron because of the attendant primary isotope effect, making it more difficult to break the carbon–deuterium bond. In the general case it is not clear without further experimental evidence whether the addition of electrophile is reversible, i.e. whether formation of σ-complex or its breakdown is rate-determining. Consider the addition of

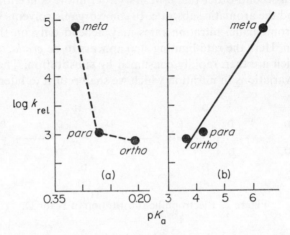

Figure 3. Rates of electrophilic deuteriation of xylene; log$k$ (relative to benzene) plotted against (a) pK_a for π-basicity, and (b) pK_a for σ-basicity, relative to the same standard.

bromonium ion to benzene, where the final product is bromobenzene. If breakdown of the σ-complex (**6**) was rate-determining, then we might expect that the reaction with hexadeuteriobenzene under identical conditions would be considerably slower because a carbon–deuterium bond would be broken in the rate-determining stage. In fact, the rates for C_6H_6 and C_6D_6 are virtually identical, proving that *formation* of **6** is rate-determining. This behaviour represents the most usual mechanism of aromatic substitution although by no means the only one, as we shall see.

The proposal of cyclohexadienyl cations as intermediates in electrophilic substitution reactions has received considerable support by characterization of such intermediates in the past few years. Carbonium ions formed by the protonation of aromatic rings in very strongly acidic media are well known and have been characterized by n.m.r. and electronic spectra (Figure 4). They are now so adequately defined that heat of formation data have been described by Arnett, and these follow expectations based on thermodynamic basicities of the parent arenes.[15] Not only proton σ-complexes are isolable, for Doering showed that Friedel–Crafts alkylation of hexamethylbenzene with methyl chloride produced the heptamethylbenzenonium ion (7) stable in the solid state.[16]

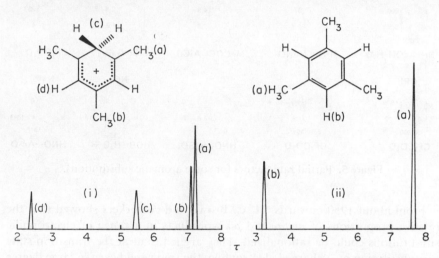

Figure 4. Schematic n.m.r. spectra of (i) mesitylene in HF–BF$_3$ and (ii) mesitylene in CDCl$_3$.

Substituent Effects on Reactivity

It is well known that an electron-releasing substituent on an aromatic ring will tend to enhance the rate of electrophilic aromatic substitution, and to direct reaction to *ortho-* and *para*-positions; an electron-withdrawing substituent depresses the rate, and is *meta*-directing. With the advent of modern analytical techniques and the possibility of precise determination of product ratios, there have been attempts to place reactivities on a more quantitative basis. For proper assessment, some quantitative index of reactivity is required and it is customary to express the reactivity of a specific site relative to the reactivity of one position in benzene as a 'partial rate factor', symbolized for the *meta*-position in chlorobenzene as m_f^{Cl} and for *ortho*- and *para*-positions in toluene as o_f^{Me} and p_f^{Me} respectively. Some examples of partial rate factors for aromatic substitutions (Figure 5) suggest that there are many problems.

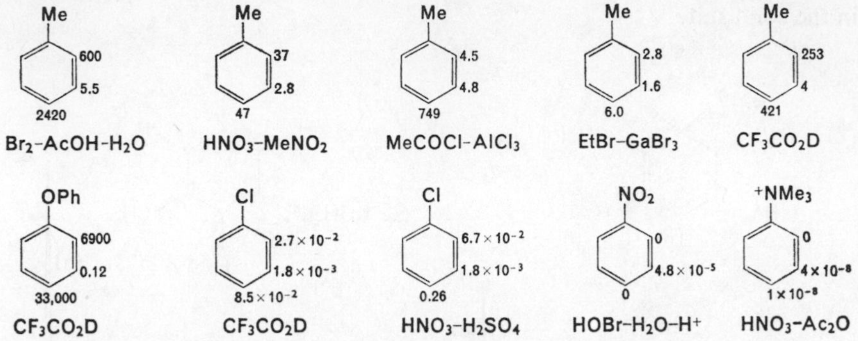

Figure 5. Partial rate factors for some aromatic substitutions.

From about 1950 onwards, H. C. Brown and coworkers showed that the relative proportions of *meta*- and *para*-products in a range of electrophilic substitutions could be rationalized. They argued that, at the transition-state for substitution by a charged electrophile, the ring would carry a large degree of the charge, and effects of substituents at the positions *meta* and *para* to the point of attack would be similar to the effects of the same substituents on the rate of solvolysis of cumyl chloride. We have already seen in Chapter 1 that *para*-substituents capable of stabilizing positive charge do not correlate with a standard Hammett $\sigma\rho$ treatment of the reaction, and a new set of substituent constants σ^+ is necessary to allow for positive charge delocalization to the *para*-position. *meta*-Substituents cannot stabilize a charge conjugatively and σ_m constants may still be used.

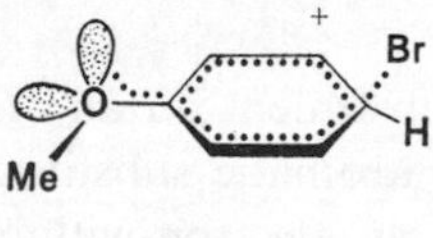

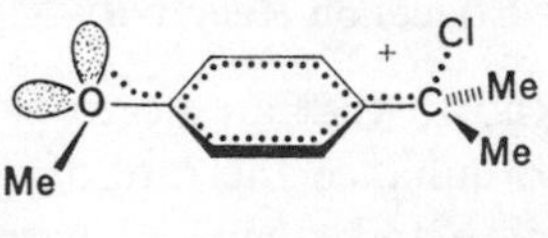

A given electrophilic substitution reaction in toluene may therefore be treated in the same way as Hammett treated side-chain reactions, using a reaction constant ρ^+ to describe the substitution. Then:

$$\log k_m/k_H = \rho^+ \sigma_m; \quad \text{and} \quad \log k_p/k_H = \rho^+ \sigma_p^+ \quad (10)$$

where k_m and k_p are the rates of reaction at *meta*- and *para*-positions of toluene, and k_H the rate of reaction at one site in benzene. This equation may be rewritten:

$$\log k_p/k_m = \rho^+(\sigma_p^+ - \sigma_m) \quad (11)$$

and in terms of partial rate factors:

$$\log p_f^{Me} - \log m_f^{Me} = \rho^+(\sigma_p^+ - \sigma_m) \quad (12)$$

This equation is termed the 'selectivity relationship', and is applicable to substitution reactions of monosubstituted benzenes (with slight modification

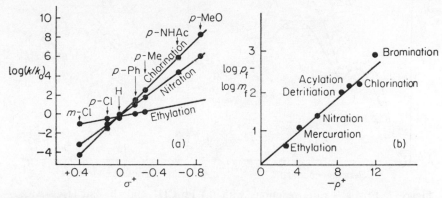

Figure 6. The selectivity relationship. (a) Plot of σ^+ against log(partial rate factor) for ethylation, nitration, and chlorination. (b) Verification of equation **12** for reactions of toluene.

it is applicable to more heavily substituted systems). What it implies is that there is a linear free-energy relationship between *meta/para* ratios in, say, substitution of toluene and the relative reactivity of toluene and benzene in the same reaction. Recourse to the data of Figure 5 makes this clear; in bromination, where toluene is much more reactive than benzene, little *meta*-substitution occurs. In Friedel–Crafts alkylation, where the reactivity spread is small, *meta* and *para* proportions are similar.

Values of $-\rho^+$ obtained in electrophilic substitution reactions may be obtained for a variety of reactions, and it is obvious that selectivity varies widely (Figure 6). If $-\rho^+$ is large (implying a low *meta/para* ratio), then conjugative interactions must be important at the transition-state. Indeed, the larger the value, the *later* the transition-state on the reaction profile and the closer its resemblance to a cyclohexadienyl cation.

We may reach this conclusion by application of Hammond's postulate (Chapter 1). With a typical selective reagent like molecular bromine, the difference in free-energy between ground-state and σ-complex is high, and the transition-state structure is close to the latter. With nitration or Friedel–Crafts alkylation, the ground-state energy is higher owing to the high energy of the electrophilic entity, and the transition-state is correspondingly closer to it. The point may be further illustrated by variation of the electrophile in Friedel–Crafts reactions. With the system benzyl chloride–aluminium chloride in nitromethane, where benzyl cation is the electrophile, mesitylene is only

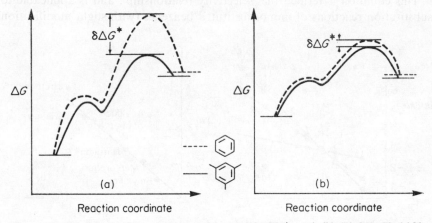

Figure 7. Friedel–Crafts reactions with (a) Ph_2CH^+ and (b) $PhCH_2Cl$–$AlCl_3$ as electrophile.

five times more reactive than benzene. In contrast, if the electrophile is the much more stable benzhydryl cation, then mesitylene is about 100,000 times more reactive than benzene (Figure 7).[17]

Isotope Effects

The quantitative success of the selectivity relationship depends on the assumption that the rate-determining stage is the same for all electrophilic substitutions treated. Our previous discussion suggests that this is largely true, but exceptions have been noted, particularly in the last few years. It was suggested earlier that reversibility in the electrophilic addition might lead to breakdown of the σ-complex becoming rate-determining, and that this would result in a kinetic isotope effect. A number of electrophilic substitutions with attendant isotope effects are now known. One situation in which this might occur is where the incoming electrophile forms a bond of low strength to carbon so that addition is readily reversed. This is the case in mercuration (Figure 8) where benzene reacts six times faster than perdeuteriobenzene under the same conditions.[18]

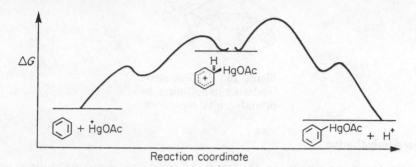

Figure 8. Mercuration of benzene.

Occasionally, an isotope effect may be observed for steric reasons. Brominations of benzene and perdeuteriobenzene occur at similar rates, as we have seen, but a different picture emerges in the bromination of 1,3,5-tri-t-butylbenzene where deuteriation of the ring reduces the reaction rate by a factor of 3.6.[19] This isotope effect suggests that the σ-complex (**13**), once formed, experiences a reluctance to proceed to the bromo-product, since this process will inevitably increase the non-bonded interactions between the bromo group and adjacent t-butyl substituents (Figure 9).

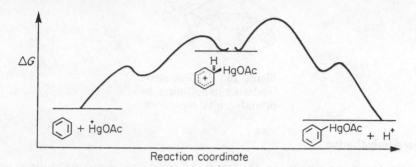

(**13**)

$$k_{-1} > k_2$$

Figure 9. Bromination of 1,3,5-tri-t-butylbenzene.

The steric requirements of the reacting electrophile cannot be ignored, and in Friedel–Crafts acylation there is the combination of a stable reagent (acylium ion) and considerable steric interactions between π-electrons in the arene and the carbonyl group. Consequently many reactions in this class show kinetically reversible formation of σ-complex resulting in an observable isotope effect.[20] Such steric requirements are directly relevant to the following section.

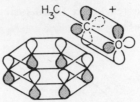

Figure 10. Possible steric hindrance in acylation, by orbital–orbital repulsions.

ortho-Substitution

In the foregoing discussion the ratio of *meta*- to *para*-substitution was readily rationalized, but the proportion of *ortho*-substitution achieved in a given reaction is less predictable, rather wide differences being encountered. We have previously classified reactions in terms of the selectivity relationship, but it is clear from the data of Figure 11 that further factors have to be taken into consideration. If we compare alkylation and acylation where the bulk of the electrophile is relatively similar, the more selective reagents give far less *ortho*-product. Nitration similarly gives a predominance of *ortho*-products, whereas bromination, where the electrophile should not suffer more steric hindrance, is heavily biased towards *para*-substitution. Part of this explanation

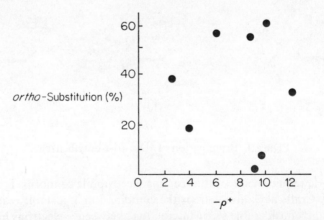

Figure 11. An apparent random relationship between *ortho*-substitution and selectivity.

undoubtedly lies in the argument that the more selective reagents operate through transition-states which are further advanced, and therefore close encounter between *ortho*-substituent and electrophile is forced, inhibiting *ortho*-attack.

Even where the reagent might be expected to have small steric demands, as in hydrogen-isotope exchange, there is a predominance of *para*-substitution. This has been rationalized by the results of calculations of charge-distribution in cyclohexadienyl cations. Simple Hückel calculations predict similar charge-density at *ortho*- and *para*-positions, but more sophisticated calculations[21] suggest that the *para*-carbon has accumulated more positive charge (**14**). In

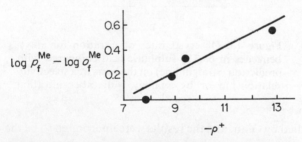

(14)

consequence, an electron-releasing charge-stabilizing substituent will prefer to be oriented *para* to an incoming electrophile; the data of Figure 12 show that, for hydrogen isotope exchange reactions of varying selectivity, there is a trend relating increasing *para*-substitution to increasing selectivity, $(-\rho^+)$.

$$\log p_f^{Me} - \log o_f^{Me}$$

Figure 12. A selectivity relationship for *ortho* hydrogen isotope exchange under a variety of conditions.

'Anomalous' Substitution

A number of electrophilic substitution reactions have been studied by Olah and coworkers which apparently violate the selectivity relationship. Thus, metal-catalysed halogenations, Friedel–Crafts alkylations, and nitrations carried out in dipolar aprotic solvents show high positional selectivity but low substrate selectivity. Most of the work was done on nitrations by nitronium ion. This reactive electrophile is capable of forming isolable salts, and Olah[22] studied reactions of $NO_2^+BF_4^-$ with aromatic substrates in nitromethane or sulpholan (tetrahydrothiophen 1,1-dioxide). Reactions were too fast to measure directly and a competitive approach was adopted wherein pairs of aromatic compounds were added to the nitronium fluoroborate solution and reactivities determined from the comparative yields of products. Under these conditions in sulpholan, the relative reactivities of benzene, toluene, and mesitylene were 1/1.67/2.71, suggesting a very low value of $-\rho^+$. However, the observed product distribution in toluene was 65.4% *ortho*, 2.8% *meta*, and 31.5% *para*. Given the toluene/benzene rate ratio, there is clearly far less *meta*-substitution than would be predicted by the selectivity relationship. A similar blend of relatively high *para/meta* selectivity was found in the other

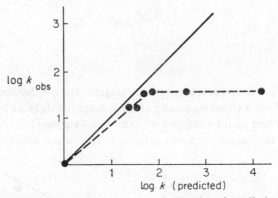

Figure 13. Observed rates of nitration for alkyl-benzenes in 68.3% sulphuric acid compared with prediction (straight line) on the basis of a selectivity relationship or by analogy with other nitration reactions at 25°.

systems studied by Olah, and the results were interpreted to be the consequence of rate-determining π-complex formation. π-Complexation would not show very much discrimination between substrates, and *meta/para* ratios would be determined after the rate-determining stage.

Competitive experiments of this type can be a rather dangerous way of obtaining relative rate data, and subsequent work by Schofield and coworkers[23] has questioned the validity of Olah's conclusions. The first hint came in a careful study of nitrations with nitric acid in 68.3% aqueous sulphuric acid, under conditions where NO_2^+ was known to be the effective electrophile, albeit at a concentration of ca. 10^{-8}M, and the reaction accurately first-order in both aromatic compound and nitric acid concentrations. A plot of relative rates observed versus those expected on the basis of a selectivity relationship is shown in Figure 13. It will be seen that there is a limiting rate for reactive aromatic substrates, and this was interpreted as being due to the onset of diffusion control. At the upper limit, rate is not being controlled by substrate reactivity but by the rate of encounter between reactants.

Furthermore, a similar situation was shown to exist in sulpholan. In classic work by Ingold and coworkers,[24] nitronium ion had been shown to be the nitrating agent in solutions of nitric acid in organic solvents, produced by the equilibria:

$$2HNO_3 \rightleftharpoons H_2\overset{+}{N}O_3 + NO_3^- \tag{15}$$

$$H_2\overset{+}{N}O_3 \rightleftharpoons NO_2^+ + H_2O \tag{16}$$

Schofield showed that with pure nitric acid in sulpholan, mesitylene reacted by a process zero-order in arene (**15** or **16** rate-determining), but in 7.5% aqueous sulpholan, where the nitronium ion concentration is considerably reduced by the operation of **16**, first-order kinetics could be established for a range of substrates. Again, a diffusion-controlled limit could be attained (Figure 13) and the mesitylene/benzene rate ratio was shown to be the same as that observed in a competitive experiment with nitric acid in pure sulpholan. Since both Olah and Schofield were studying reactions between arene and NO_2^+ in the same solvent, and Schofield has clearly established the importance of diffusion control, Olah's low rate ratios must be due to diffusion control, rather than an abnormal mechanism involving rate-determining π-complex formation.

Substitution in Fused and Heteroatom Systems

On the basis of our previous discussion we might generally expect that the rate of substitution in fused aromatic systems would depend on the relative stability of the intermediate cationic σ-complexes. Since polycyclic systems offer a greater opportunity for delocalization of charge than benzene, it might be expected that all electrophilic substitutions in naphthalene and higher fused systems will occur faster than in benzene, and this is indeed found to be the case. This is an area where considerable efforts have been made to predict relative reactivities in terms of molecular-orbital theory,[25] the simplest approach being in terms of a parameter termed cation localization energy, L_r^+, which is a measure of the difference between the Hückel resonance integral, for the hydrocarbon and for the σ-complex formed by the attack of an electrophile at a given position in that hydrocarbon. A reasonable correlation between theory and experiment is found (Figure 14).

In substituted fused systems, it is important to realize that the substituent will affect reactivity at all sites in the molecule and not just in its own ring.

Recent quantitative work on hydrogen isotope exchange in naphthalene derivatives[26] serves to illustrate the nature and direction of substituent effects (Figure 15).

It is beyond the scope of the present book to offer a general quantitative picture of substitution in heteroaromatic systems. The system of classification

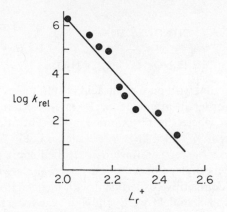

Figure 14. Relationship between rates of deuteriation of polycyclic aromatic compounds in CF_3CO_2H and cation localization energies.

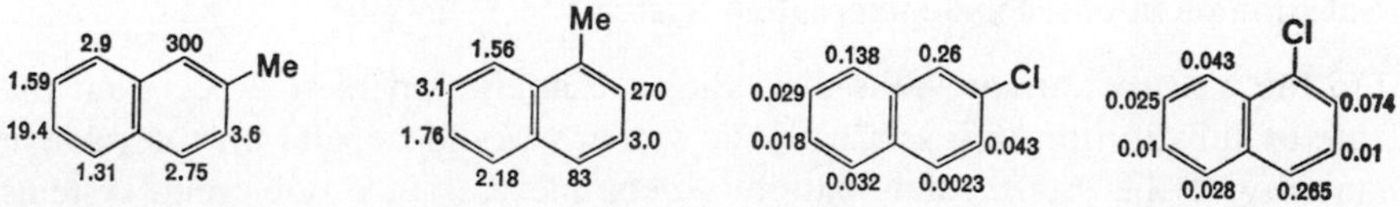

Figure 15. Directive effects in naphthalene. Partial rate factors for detritiation of methyl- and chloro-naphthalenes compared with corresponding positions in naphthalene.

into π-deficient heterocyclic systems such as pyridine and π-excessive systems such as furan is well known. What is less well appreciated is the enormous spectrum of reactivity obtained in electrophilic substitution reactions. Little is available by way of direct comparison, but extrapolations make possible a rough estimate of relative reactivities towards a selective electrophile such as molecular bromine. The span of $>10^{25}$ is equivalent to an extreme difference in half-life between 1 microsecond and 10,000 million years!

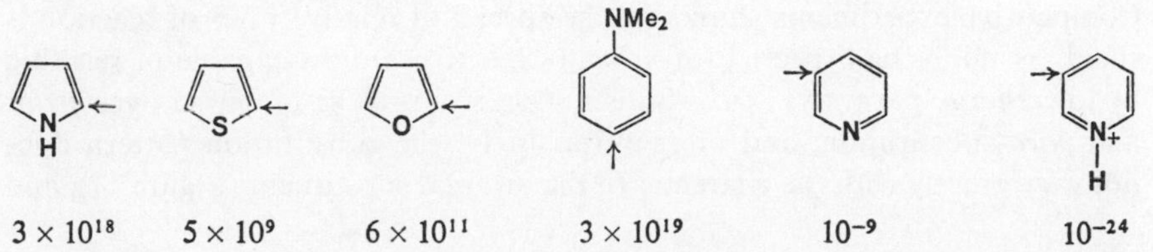

$$3 \times 10^{18} \qquad 5 \times 10^9 \qquad 6 \times 10^{11} \qquad 3 \times 10^{19} \qquad 10^{-9} \qquad 10^{-24}$$

Figure 16. Partial rate factors for bromination relative to benzene. For reactive systems, values are extrapolated from data on derivatives with deactivating substituents, and for unreactive systems from known σ^+ constants assuming $\rho^+ = -12$ (for bromination).

Homolytic Aromatic Substitution

Free-radicals attack aromatic rings readily with the overall consequence of substitution, but the reaction may be mechanistically very different from electrophilic substitution. Thus, phenyl radicals may be prepared in a variety of reactions (17—20) and the selectivity of their addition to arenes assessed.[27]

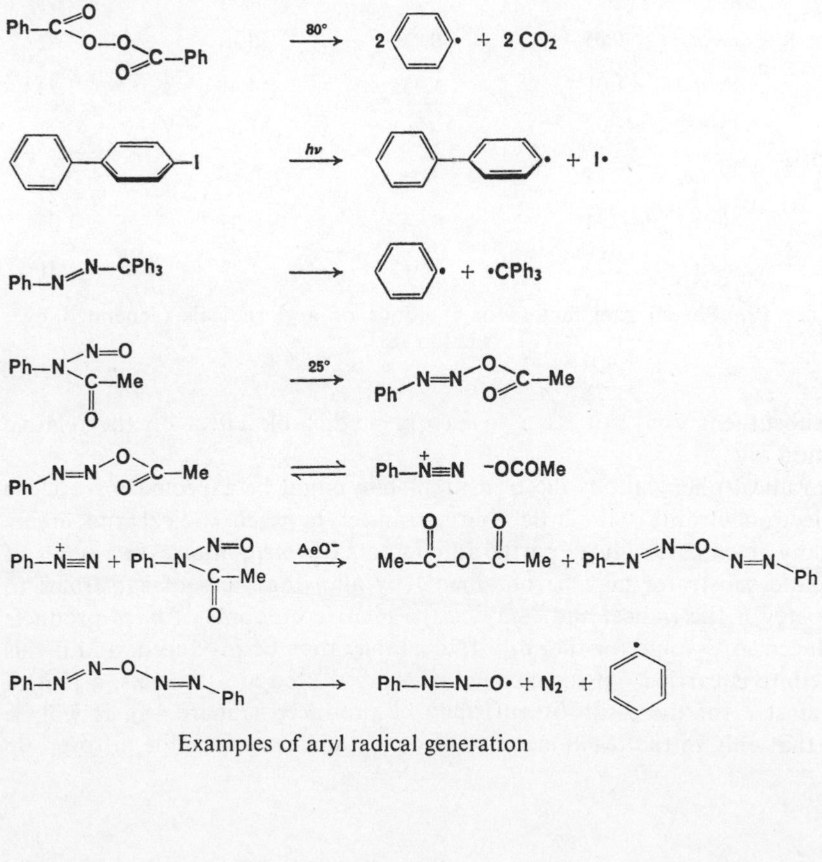

Examples of aryl radical generation

Competitive experiments show that the spread of relative rates of reaction is small, as might be expected for a highly reactive radical capable of reacting by low-energy pathways. *ortho*-Substitution seems to be favoured over *meta*- and *para*-substitution, and more surprisingly, the substitution pattern does not vary greatly with the character of the original substituent (Figure 17), and

		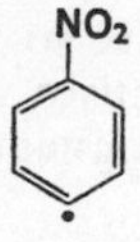 NO₂	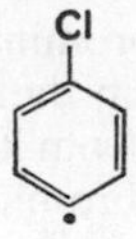 Cl	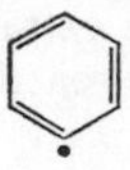 (H)	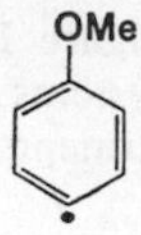 OMe
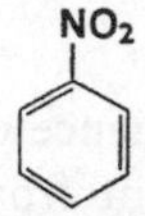 **NO₂**	*ortho*	0.93	4.35	9.38	6.45
	meta	0.35	0.61	1.16	1.19
	para	1.53	6.18	9.05	8.36
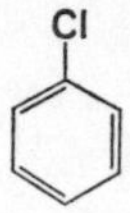 **Cl**	*ortho*	1.53	2.70	3.09	3.08
	meta	0.65	0.87	1.01	1.82
	para	1.01	1.33	1.48	1.74
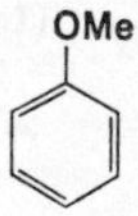 **OMe**	*ortho*	5.17	3.93	3.56	3.68
	meta	0.84	0.94	0.93	1.03
	para	2.30	1.54	1.29	1.31

Figure 17. Partial rate factors in reactions of aryl radicals (generated by reaction **18**).

the substituent does not seem to have a predictable effect on the relative reaction rate.

In radical phenylations the transition-state might be expected to be close to electroneutrality with little charge-transfer between radical and arene. For any given aryl radical, partial rate factors for *meta*-attack on a series of aromatic substrates may be obtained[28] by allowing pairs of substrates to compete for the radical and assaying the relative amounts of *meta*-products produced. A ρ-value for that aryl radical may thus be produced, and if this procedure is carried out for a range of *p*-substituted aryl radicals, a plot of ρ against σ (of the *p*-substituent) may be produced (Figure 18). It will be seen that only in the reactions of *p*-nitroaryl radicals, where the nitro group exerts a powerful electron-withdrawing effect and makes the radical weakly electrophilic, is there any appreciable polar character to the transition-state.

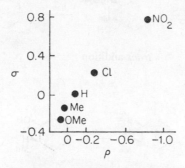

Figure 18. Plot of ρ for *meta*-attack on aromatic substrates by *para*-substituted phenyl radicals against σ for the *para*-substituent of the radical.

Electrophilic Addition to Unsaturated Systems[29-38]

The typical reagents we encountered in electrophilic aromatic substitution reacted via a transition-state where bonding to one of the ring atoms was important, and considerable charge was transferred to the ring. Many of the reagents used in olefinic additions are similar, but detailed mechanisms of the two types of reaction often bear only superficial resemblance to one another. Two features are outstandingly different; one is the very frequent importance of bonding of electrophile to *both* olefinic carbon atoms at the transition-state, and the other the occurrence of a number of electrophilic additions where the electrophile is uncharged, and little charge-separation exists at any point on the reaction profile. Examples of the wide variety of known addition reactions are illustrated in Figure 19.

A quantitative study of olefin additions will require that we give attention to three aspects: the relative rates of reaction of different olefins and its interplay with the nature of the reagent; the direction of addition of a reagent XY to an unsymmetrical olefin (one lacking a mirror plane orthogonal to the π-bond and bisecting the C=C bond) (Figure 20); and the overall stereochemical consequence of the addition process. Thus, in the general case, four possible products may arise in the addition (Figure 20).

There are therefore more possible variables than in aromatic substitution, and certainly the amount of detailed quantitative work available is rather less. Aside from the fundamental importance to the physical organic chemist, a

Oxymercuration–Reduction

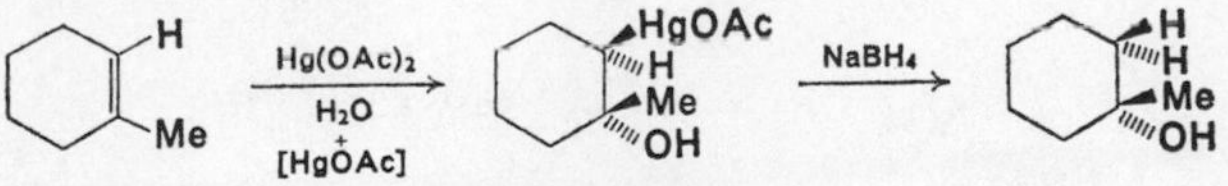

trans-addition

Hydroboration–Oxidation

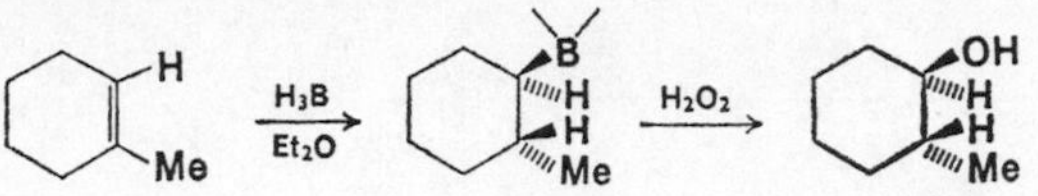

cis-addition

$^1\Delta_g$ *Singlet Oxygen*

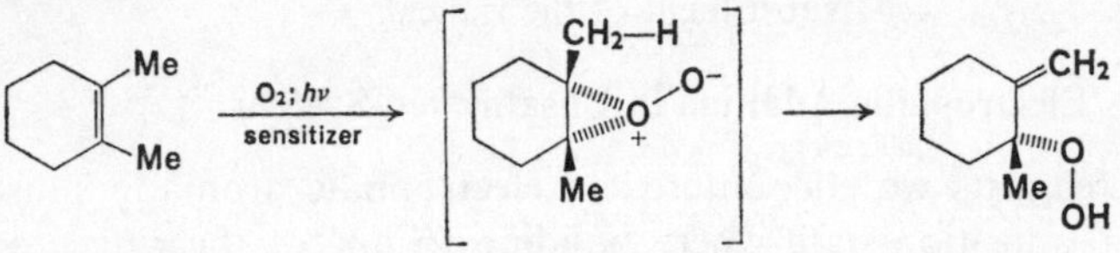

Ozonolysis

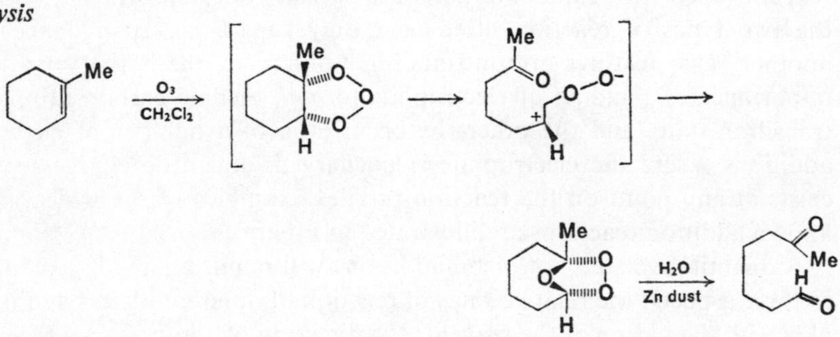

Figure 19. Some synthetic procedures based on electrophilic additions to olefins.

knowledge of the factors governing olefin additions will be valuable to the synthetic chemist who is frequently confronted with the need to carry out reaction selectively and with stereochemical control at a specific unsaturated centre in a polyfunctional molecule.

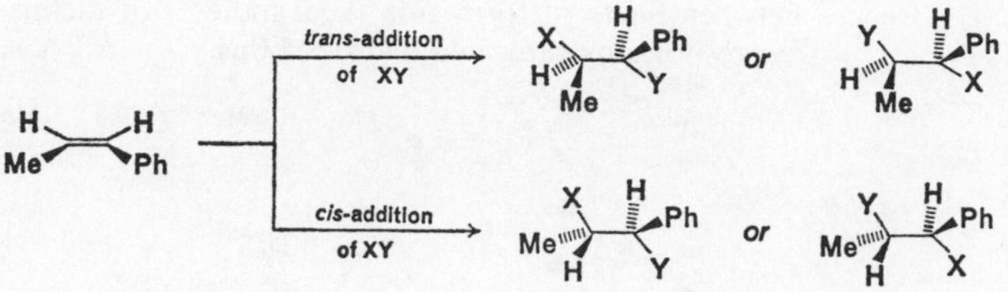

Figure 20. Addition to an unsymmetrical olefin.

Reagent Types

The reactivity of olefins towards electrophilic addition is affected by the degree of alkyl substitution but the nature of the effect depends very largely on the reagent. If we take as typical reactions the bromination of alkylated olefins, and their hydroboration with diisopentylboron, then markedly different trends appear to be operative (Table 1).

Table 1. Bromination of olefins, and their hydroboration, compared with that of *trans*-but-2-ene.

				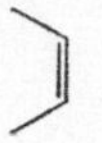		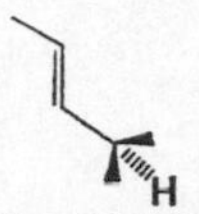	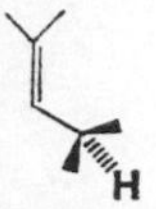
Br_2 MeOH	0.04	0.03	0.016	1.4	1.0	0.91	19
Diisopentyl-boron hydride	275	150	12.3	5.9	1.0	1.2	0.25

We might expect that increasing alkylation would enhance the reactivity of olefins towards an electrophile by stabilizing incipient positive charge, and conclude that charge-stabilization therefore plays little part in hydroboration reactions. Inhibition of reaction rate by alkylation could be due to either or both of two factors: an increasing steric hindrance towards the incoming reagent, and an increased stability of the π-bond counteracting the tendency to undergo addition by lowering the energy of the ground-state. The origin of this latter effect is variously argued as hyperconjugation (resonance between a π-bond and neighbouring C—H) or a preference of an alkyl group for forming a bond to an sp^2- rather than an sp^3-centre; its reality is made evident by quantitative study of the heat of hydrogenation of olefins which decreased with increasing alkylation (Table 2).[39]

Table 2. Differences between heats of formation (kcal mole^{-1}) of olefins and the corresponding saturated hydrocarbons.

$\parallel$	$\parallel$ Me	Me $\diagdown$ Me	Me $\diagup$ Me	Me $\diagdown$ Me	Me $\diagdown$ Me
32.7	29.7	28.5	27.5	28.1	26.7

Me $\diagdown$ Me Me $\diagup$ Me 25.8 $\triangleleft$ 53.9 $\square$ 31.1   (octene) 23.5   (norbornene) 33.0   Me–C≡C–Me 65.1

'Charge-controlled' Addition

We have seen in Chapter 2 (page 107) how Bartlett and coworkers measured the solvolysis rates of **21**, **22**, and **23** and showed these to be in the ratio 1/7/39. They considered this to be evidence for symmetrical participation by the double-bond in carbonium-ion formation, and compared the rates with those of a

(21)   **(22)**   **(23)**

number of electrophilic addition reactions as a function of the degree of methyl substitution.[40] A plot of this kind may give useful information on the selectivity of reactions, and the importance of transfer of charge to olefinic carbon atoms at the transition-state (Figure 21).

This may be taken as evidence that the approach of Br$^+$ to an olefin is substantially symmetrical, and that an intermediate 'bromonium ion' is the logical precursor of product. The existence of an intermediate of this kind was considered by Roberts and Kimball in 1937 in order to explain the *trans*-orientation of the two carbon–bromine bonds formed in the overall reaction.[41] Intermediates of this kind have had more direct supporting evidence recently. In the bromination of the very hindered olefin adamantylideneadamantane, an intermediate is produced which has the structure **24**; a bromonium ion is stable in this instance because of steric hindrance to attack by bromide ion.[42] Low-temperature n.m.r. studies by Olah and coworkers indicated that treat-

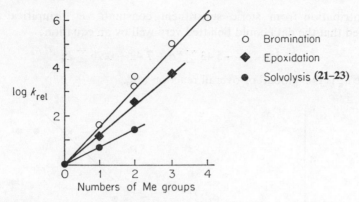

Figure 21. Electrophilic reaction rates plotted against the degree of alkylation for olefins.

ment of 1-bromo-2-fluoropropane with antimony pentafluoride in sulphur dioxide gave a solution whose n.m.r. spectrum was consistent with **25**. The methine hydrogen was shifted to much lower field than the methylene hydrogens, suggesting stabilization of positive charge by the methyl group.[43]

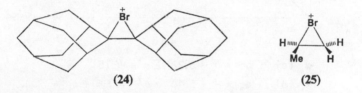

The kinetics of bromination have been studied in detail, particularly by Dubois and coworkers.[44] In methanol solution, a mixture of dibromide and α-bromo-ether may be formed depending on whether the bromonium ion is intercepted by bromide ion or by methanol. The kinetics of reaction suggest that the major reactant is molecular bromine, and that a small amount of addition is occurring with Br_3^- as electrophile. Their extensive data make it clear that the major factor governing relative reactivity is the degree of alkylation of the olefin, but this alone cannot be the determining factor. Alkyl groups may operate to stabilize a partially positively charged centre inductively, and this is allowed for in the assessment of inductive substituent constants, σ^*. Examination of a selection of Dubois' data (Figure 22) suggests that increasing steric bulk in an alkyl substituent inhibits the reaction, and therefore the construction of a linear free-energy relationship for the reaction will require a contribution from steric substituent constants, E_s. Empirical analysis showed that the data could be fitted very well by an equation:

$$\log k_g = -5.43 \sum {}^*\sigma + 7.42 + 0.96 \sum E_s$$

where k_g is the 'global' or overall reaction rate.

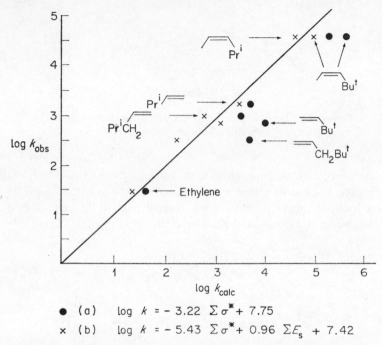

● (a) $\log k = -3.22 \sum \sigma^* + 7.75$

× (b) $\log k = -5.43 \sum \sigma^* + 0.96 \sum E_s + 7.42$

Figure 22. Linear free-energy correlation of olefin bromination rates in methanol, both in purely inductive terms (●) and including steric parameters (×) at 25°.

The excellent reactivity correlations observed here are not of general application. A possible clue to deviations lies in the observation that iso-butene brominates rather faster than *cis*- or *trans*-butene, tending to suggest that the bromonium ion is rather unsymmetrical in the former case with the tertiary carbon atom having enhanced carbonium-ion character. It may well be imagined that in a more extreme case the reaction intermediate could actually be a carbonium ion, and that the reagent could approach the olefin unsymmetrically so as to produce the most stable carbonium-ion. In this instance, the 'Bartlett plot' would take on a very different appearance since the crucial factor would be the comparative stabilities of 'open' carbonium ions (Figure 23). Many addition reactions fall into this category, and it is particularly pertinent to consider those where the proton is the electrophile, as in acid-catalysed hydration. In much older literature proton π-complexes are written as addition intermediates, but it is clear from Figure 23 that these are less stable than the isomeric carbonium ions; reinforcement of this idea is provided by the n.m.r. spectra of simple carbonium ions (Chapter 2) which

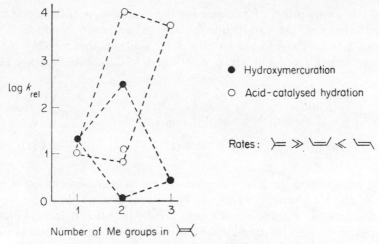

Figure 23. Electrophilic additions proceeding via unbridged ions.

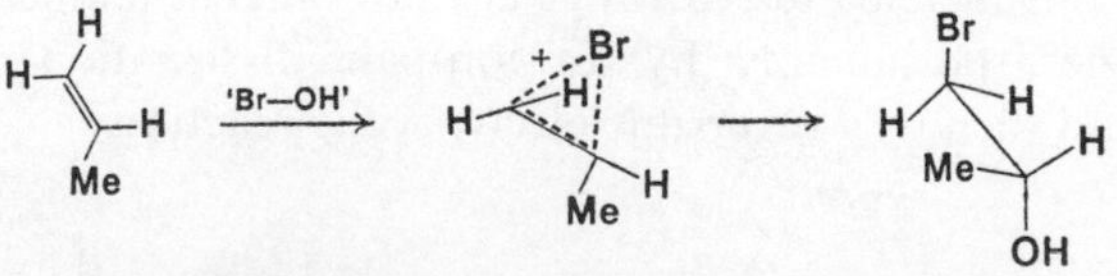

Figure 24. Addition of an unsymmetrical reagent to an unsymmetrical olefin.

do not exist as protonated olefins. One point which is evident from Figure 23 is that alkylation at the site of attack may decrease the rate; thus, accumulation of positive charge at the site must be minimal, and its effect on the rate over-ridden by steric interaction with the reagent.[45]

It will be apparent that the direction of addition of an unsymmetrical electrophile such as HBr to an unsymmetrical olefin such as propene will be determined by the attack of nucleophile at the centre of highest carbonium ion stability, and that this is the basis of the classic 'Markownikov rule'. What is not so obvious is that an unsymmetrical reagent such as BrOH, which reacts via bromonium ion intermediate, also gives product by selective attack of nucleophile at the centre of higher carbonium-ion stability. The bromonium ion may be considered to be a delocalized species, and attack occurs at the carbon centre possessing more positive charge (Figure 24).

Stereochemistry of 'Charge-controlled' Addition

One of the classic pieces of evidence for the intervention of bromonium-ion intermediates was the production of *trans*-substitution products;[41] thus *trans*-but-2-ene is brominated to give RS-2,3-dibromobutene (26) and *cis*-but-2-ene gives a racemic mixture of RR- and SS-2,3-dibromobutane (27).[46]

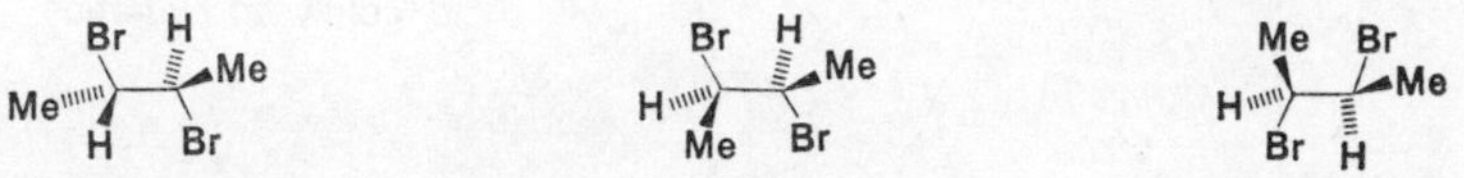

| (26) | R/R-(27) | S/S-(27) |

Even in bromination there are exceptions to the general observation of *trans*-addition, and for a variety of reasons. Thus, the addition of bromine to cyclooctatetraene is *cis*, and occurs much more rapidly than the addition of bromine to cyclooctene, which is *trans*. Addition of bromine at −55° occurs not to give a bromonium ion, but the stable, non-classical homotropylium ion, which is then attacked stereospecifically by bromide ion. The initial product (28) (Figure 25) is inclined to rearrange at rather higher temperatures, first to the *trans*-isomer, presumably by an ion-pair dissociation–recombination mechanism, and then by a reversible electrocyclic reaction.[47]

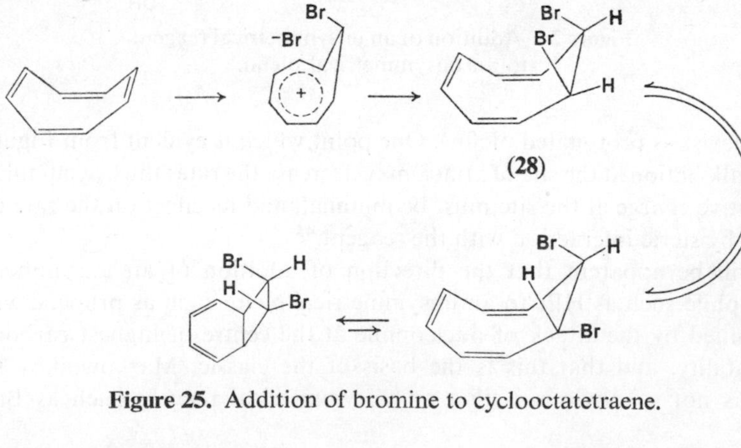

Figure 25. Addition of bromine to cyclooctatetraene.

In the addition of bromine to norbornene the major course of reaction is to produce a non-vicinal dibromide. This may readily be rationalized if a non-classical intermediate is involved, which preferentially reacts with bromide at the carbon remote from the bromine atom (Figure 26).[48]

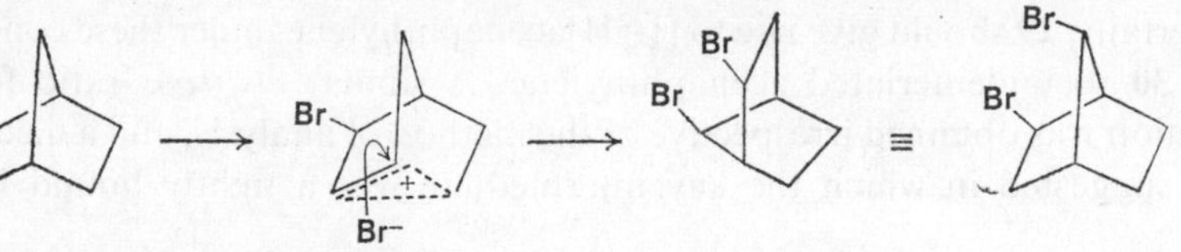

Figure 26. Addition of bromine to norbornene.

What of the stereochemistry of other addition reactions? If *trans*-1-phenyl-propene is subjected to a variety of additions, it transpires that with PhSCl (where a phenyl-episulphonium ion is the bridged intermediate) 100% *trans*-addition occurs, but with Br_2 in CCl_4 the proportion drops to 88%. This may well reflect a greater stability for the bridged episulphonium ion relative to a bridged bromonium ion. In addition of deuterium bromide in non-polar media, the proportion of *trans*-addition dropped to 12%.[49] Since this is a reaction expected to occur via an open carbonium ion, it is of interest to enquire why the reaction is *cis*-stereoselective, and perhaps simpler to consider the entirely related reaction with acenaphthylene.

Addition of DBr to acenaphthylene could give either the *cis*- (29) or *trans*-isomer (30), and the problem arises of how to distinguish these.[50] This was done by Dewar and Fahey in two ways. The Karplus relationship between n.m.r. coupling constants of vicinal protons and the torsional angle between

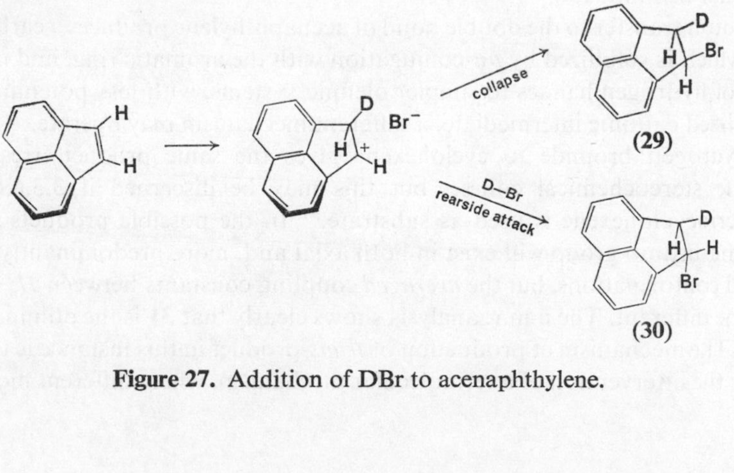

Figure 27. Addition of DBr to acenaphthylene.

their C—H bonds suggests that the *cis*-isomer should have a larger coupling constant than the *trans*-isomer (deuteron coupling is much smaller than proton coupling and may be ignored in this case). Making this assumption, the appearance of the adduct-spectrum certainly suggests a predominant *cis*-addition. To reinforce this evidence, the bromo-product was subjected to elimination under *E*2 conditions where a *trans*-stereochemistry was known to pertain; **29** should give rise to [1-^{2}H]acenaphthylene under these conditions, and **30** to undeuteriated acenaphthylene. A similar *cis/trans* ratio for DBr addition was obtained irrespective of the method of analysis, and a mechanism was suggested in which the key intermediate was a tightly bound ion-pair

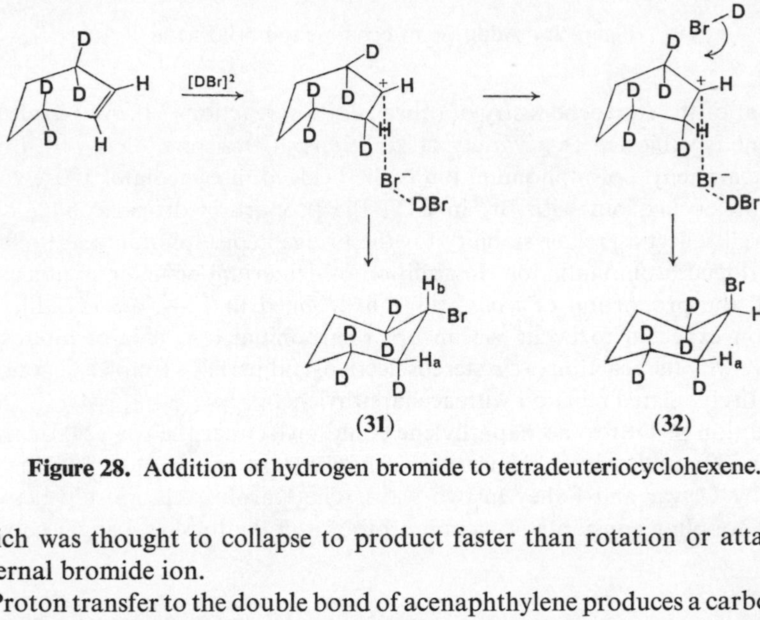

Figure 28. Addition of hydrogen bromide to tetradeuteriocyclohexene.

which was thought to collapse to product faster than rotation or attack by external bromide ion.

Proton transfer to the double bond of acenaphthylene produces a carbonium ion which is stabilized by $p\pi$-conjugation with the aromatic ring, and in addition of hydrogen halides to simpler olefinic systems, with less potential for a stabilized cationic intermediate, a different mechanism may operate. Addition of hydrogen bromide to cyclohexene gives the same product irrespective of the stereochemical course, but this may be discerned if 3,3,6,6-tetra-deuteriocyclohexene is used as substrate.[51] In the possible products **31** and **32**, the bromo group will exist in both axial and, more predominantly, equatorial conformations, but the *averaged* coupling constants between H_a and H_b will be different. The n.m.r. analysis shows clearly that **31** is the dominant isomer. The mechanism of production of *trans*-product in this instance is thought to be the intervention of a termolecular mechanism, where different molecules of HBr react as electrophile and nucleophile. On this basis, it might be expected that the kinetic form of the reaction would be

$$\text{rate} = k_3[\text{Olefin}][\text{HBr}]^2 \tag{33}$$

and this is observed in acetic acid and other polar media. In non-polar media the reaction may be third-order in HBr; presumably the function of the third molecule is to produce the electrophile $H^+HBr_2^-$.

'Overlap-controlled' Addition

The rate of addition of diborane to olefins decreases with increasing alkylation at the olefinic site. Rates of reduction of olefins with diimide follow a similar

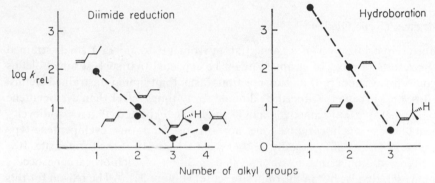

Figure 29. Effects of alkylation on the rates of 'overlap-controlled' addition reactions: the diimide reduction at 80° and hydroboration at 0°.

trend, and reactions of this type are most readily explained if it is assumed that little charge-transfer to the olefin has occurred at the transition-state. Factors controlling reaction rate will now be associated with relative ground-state energies of olefins and with steric and strain effects in ground and transition-states.[52]

Diimide is a transient, highly reactive intermediate which may be produced in a variety of ways, including the thermolysis of arenesulphonyl hydrazides. Garbisch carried out an extensive study of competitive rates of diimide additions to pairs of olefins, and was able to make generalizations about the nature of the transition-state in the reaction.[53] He very reasonably assumed a symmetrical cyclic transition-state with partial rehybridization at the olefinic carbon atoms (Figure 30) and applied a classical mechanical approach (Chapter 1) to assess relative torsional, angle bending, steric strain, and alkyl substituent stabilizing effects, at the ground- and transition-state. His conclusions were that steric effects could be ignored, and empirically using a model for transition-state structure which had progressed 34% along the reaction profile, he was able to correlate reactivities varying over a range of more than 10^4 within a factor of 2.

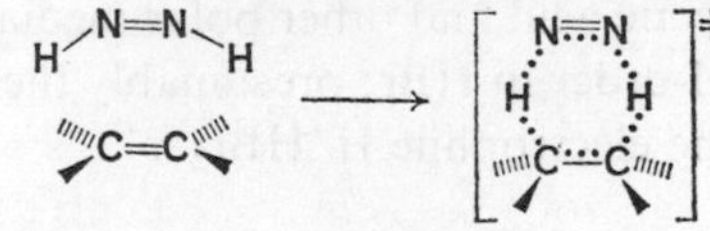

Figure 30. Transition-state model used by Garbisch in energy minimization calculations on diimide–olefin reactions.

Strained Cyclic Olefins

Since rehybridization of a carbon atom from sp^2 to sp^3 in a small strained ring results in relief of strain, it might be expected that such strained olefins would always undergo addition reactions faster than similar unstrained olefins. Thus, norbornene is reduced by diimide thirty times faster than cyclopentene in competitive experiments, and it might be expected that *endo*-dicyclo-pentadiene, which contains one norbornene- and one cyclopentene-type double bond, would generally tend to react selectively at the former site. It is therefore disconcerting to observe that addition of dichlorocarbene occurs highly selectively, but in the opposite sense! (Figure 31).[54] The reason for this

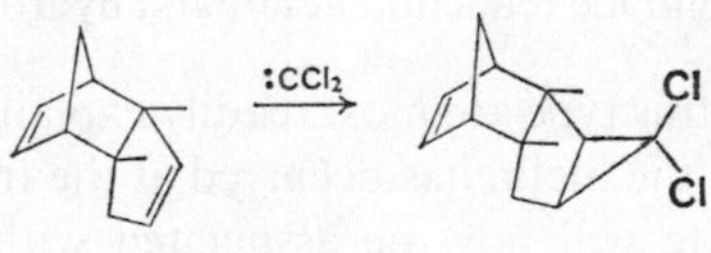

(33)

Figure 31. Selective addition of dichlorocarbene to *endo*-dicyclo-pentadiene.

apparent anomaly is that the transition-state in :CCl$_2$ additions offers no relief of angle strain, since a strained cyclopropane is being formed. This fact offers a mechanistic technique of some value in addition reactions of olefins which occur via a cyclic transition-state, since the relative reaction rates of a series of olefins of differing degrees of strain will give some indication of the ring-size in that transition-state. The point may be illustrated by consideration of relative rates of reaction of cyclohexene, cyclopentene, and norbornene in various addition reactions.[55]

Table 3. Relative reaction rates for cyclic olefins in various addition reactions.

Reaction	 ⬡	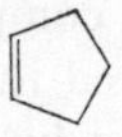 ⬠	 🔺	Transition-state model
Diimide reduction	1	15.5	450	Figure 30
Phenyl azide cycloaddition	1	64	6500	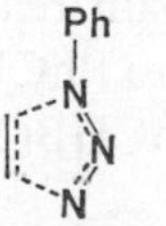 (Ph–N=N–N structure)
Epoxidation	1	1.5	1.2	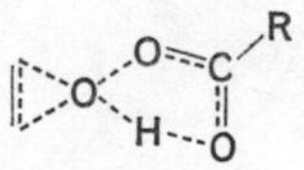 (O=C–R structure)
Osmium tetroxide addition	1	21.3	72.3	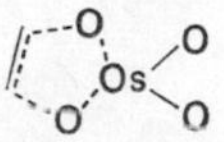 (OsO$_4$ structure)

Electrophilic Addition to Acetylenes

Relative rates of addition to substituted acetylenes are less well defined than is the case for olefins, but several generalizations may be made. Since the enthalpy of hydrogenation of acetylenes to olefins is somewhat higher than that of olefins to paraffins (Table 2), rates of addition reactions of acetylenes might be expected to be at least comparable to those of corresponding reactions of olefins. This is certainly true in many instances; for example, rates of electrophilic addition of trifluoroacetic acid to disubstituted acetylenes are faster than of the corresponding disubstituted olefins, and rates of hydroboration and diimide reduction are somewhat faster. An important proviso exists, however. With reagents which form a small-ring transition-state, as in epoxidation or bromination, additions to acetylenes can be up to 10^5 slower than to comparable olefins. This fact can be of synthetic utility, e.g. as in the specific production of 35 in addition of dichlorocarbene to 34.[56]

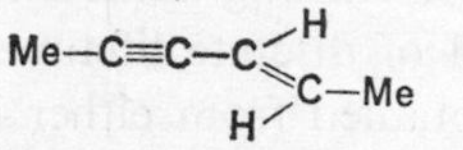

Me—C≡C—C(H)(C—Me)(H)

(34)

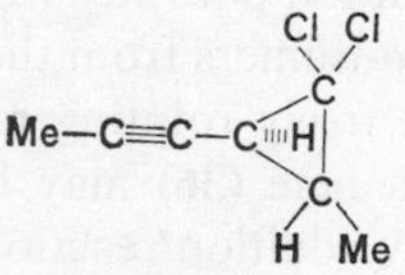

Me—C≡C—C(H)...(Cl Cl)(H Me)

(35)

Free-radical Addition to Olefins

The chemistry of free-radical and electrophilic additions to olefins may, in certain circumstances, be interlinked. In much of the older literature on the addition of HBr to olefins, variable orientation was observed, and reactions were noted which occurred in a completely anti-Markownikov manner. Kharasch and Mayo in 1933 showed that this 'abnormal' addition was, in fact, a free-radical reaction induced by traces of peroxides in the olefin.[57] Initiation occurs by oxidation of HBr to bromine atoms which then set up a chain reaction. When this mode of addition operates, styrene is converted into $PhCH_2CH_2Br$, whereas in the absence of peroxides the product is entirely PhCHBrMe (Figure 32).

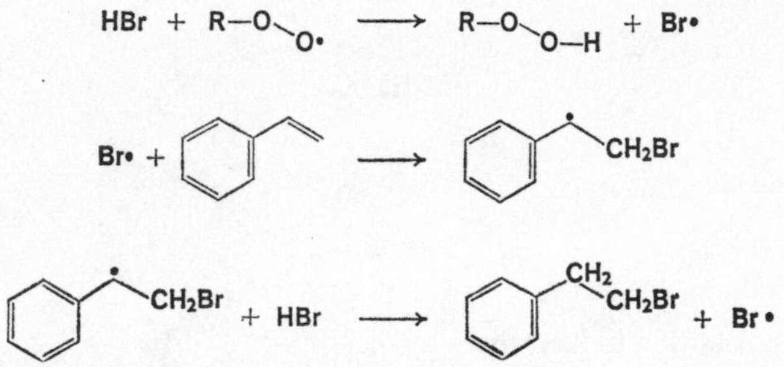

Figure 32. Peroxide-induced radical-chain addition of HBr to styrene.

A wide variety of radical reagents add to olefins and examples are shown in Figure 33. In many cases the addition stage is reversible, which may have the stereochemical consequence of isomerizing the olefin since an intermediate is produced which has free rotation about the relevant carbon–carbon bond. If olefins are irradiated with long-wavelength ultraviolet light in the presence of diphenyl disulphide, rapid *cis–trans* isomerization to a thermodynamic equilibrium mixture occurs (Figure 34).[59]

The stereochemistry of radical additions to olefins has been studied and in a typical reaction, the homolytic addition of $BrCCl_3$, it has been shown that the same product is obtained starting with either *cis*- or *trans*-but-2-ene. A stereospecific addition pathway would have given chemically distinguishable *erythro*- and *threo*-isomers from these two starting materials. This result is to be expected of a freely rotating radical of discrete lifetime in the reaction, since this intermediate (36) may be obtained from either isomer of olefin. In contrast, many addition reactions of hydrogen bromide are highly stereoselective and it may well be that these reactions involve a bridged radical

Additions of ·CCl₃

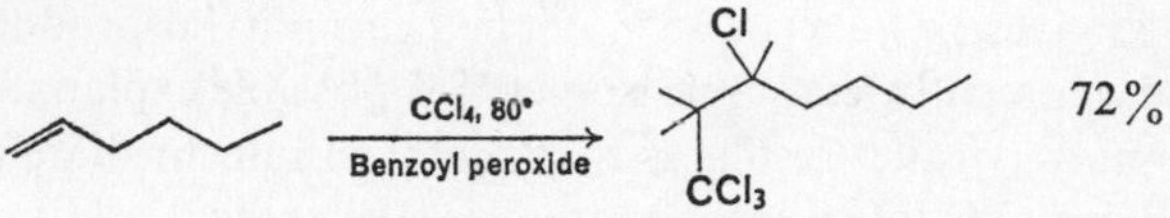

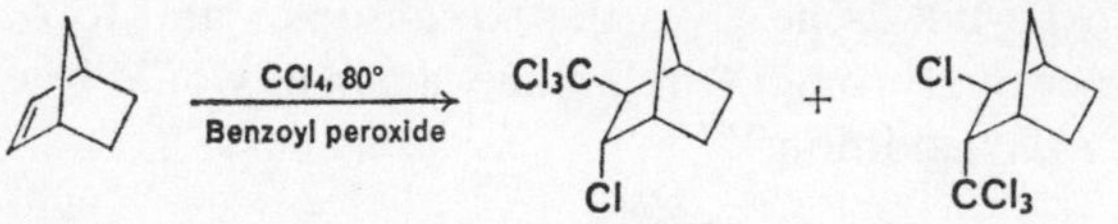

(major) (minor)

Addition of RĊO

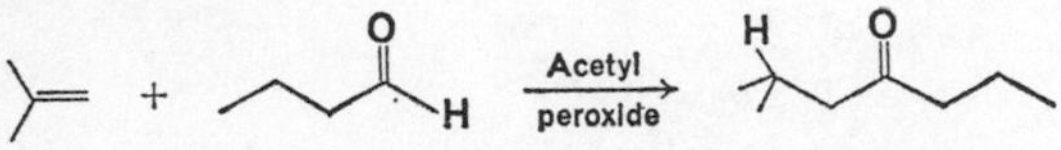

Addition of PhS·

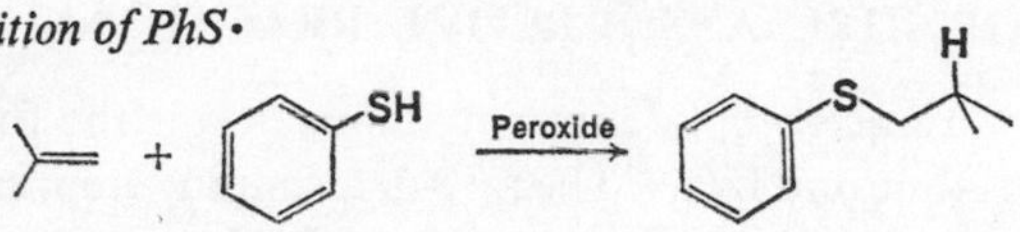

Intramolecular Addition

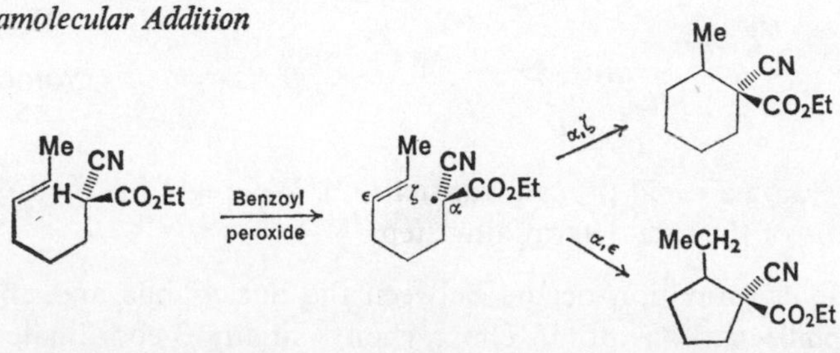

Figure 33. Examples of radical addition to olefins.[58]

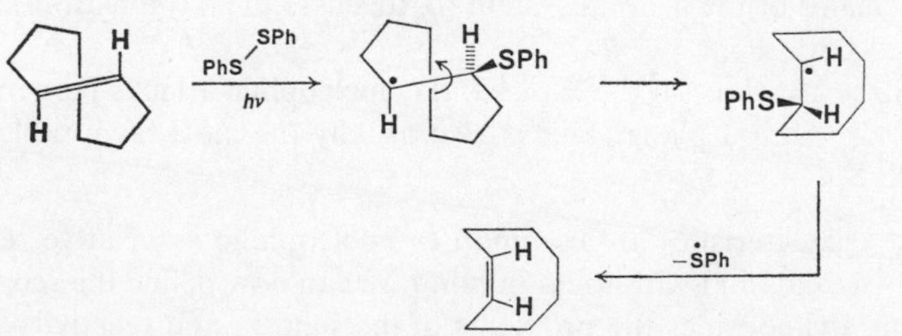

Figure 34. Geometrical isomerization of *trans*-cyclooctene by ·SPh.

since, for example, the radical addition of HBr to 1-chloro-4-t-butylcyclo-hexene at −78° gives more than 95% of the *trans*-diaxial product (37) which is the thermodynamically less stable isomer.[60] A possible explanation suggested for this and many parallel results is that a radical intermediate analogous to a bromonium ion (38) intervenes. Moreover, the radical addition of deuterium bromide to *cis*-but-2-ene gives the *threo*-isomer and to *trans*-but-2-ene the *erythro*-isomer of 2-bromo[3-²H]butane, again as would be expected from a stereospecific *trans*-addition.[61]

(36)   (37)   (38)

NUCLEOPHILIC ASSOCIATIVE REACTIONS[62-77]

A simple example of a nucleophilic associative reaction is the formation of a cyanohydrin from acetone and HCN. The rate-determining step, as Lapworth[78] showed in 1903, is nucleophilic attack by CN^-, with the reaction subsequently

completed by a rapid proton-transfer (39). We might list the following characteristics of the rate-determining step.

(*a*) Bond formation occurs between the nucleophile and an unsaturated electrophilic carbon atom. The carbon, initially 3-coordinate and planar, becomes 4-coordinate and tetrahedral,

(*b*) The only bond broken is the π-bond; σ-bonding is not much affected. What remains of the π-bond system (in this case the oxygen atom) acquires a pair of electrons,

(*c*) Since attack is on the π-bond, the nucleophile attacks perpendicular to the C—C(=O)—C plane. This is also usually the most favourable direction sterically.

These characteristics are common to nucleophilic associative reactions in general (see 40), and with them in mind we can now define the scope of these reactions and consider the problems of mechanism and reactivity that arise.

The electrophilic substrate can be quite varied in nature, the essential requirement being a carbon atom multiply bonded to an atom, Z in 40, which

$$
\underset{\substack{\| \\ Z}}{\overset{R \quad X}{\underset{}{C}}} + Y^- \longrightarrow \left[\overset{R \quad X}{\underset{C \quad Z}{C \cdots Y}} \right]^- \longrightarrow \underset{\substack{\| \\ Z^-}}{\overset{X}{\underset{}{R \diagdown C - Y}}} \qquad (40)
$$

can act as an electron sink. By far the most important substrates are carbonyl compounds with $Z = O$. Thus, rate-determining nucleophilic association occurs in addition and substitution (41) reactions of aldehydes and ketones, in ester hydrolysis, and in most other reactions of acyl (RCOX) derivatives, e.g. 42. Substrates with $Z = NR$ include imines and a variety of heterocycles

$$
\underset{Me}{\overset{Me}{\diagdown}} C=O + H_2NNHCONH_2 \;\rightleftharpoons\; \underset{Me}{\overset{Me}{\underset{}{\overset{\text{\tiny OH}}{\underset{\text{NHNHCONH}_2}{C}}}}} \longrightarrow \underset{Me}{\overset{Me}{\diagdown}} C=N \underset{H}{\diagdown} N - C \underset{NH_2}{\overset{O}{\diagup}} \quad (41)
$$

$$
PhNH_2 + (MeCO)_2O \rightarrow PhNHCOMe + MeCO_2H \qquad (42)
$$

$$
Et_2NH + H_2C=CHCN \rightarrow Et_2NCH_2CH_2CN \qquad (43)
$$

related to them (e.g. consider attack on the 2-position in pyridine). Substrates with $Z = CR_2$ generally require one or more electron-withdrawing and anion-stabilizing groups ($R = COR, CN, NO_2, SO_2R, CF_3$, etc.), but when these are present nucleophilic vinyl addition (43) and substitution (44) reactions generally occur quite readily. Nucleophilic aromatic substitution (Z and R part of an aromatic system) also occurs when a similar type of substituent is present (45).

$$
EtO^- + \underset{Cl \quad CF_3}{\overset{F_3C \quad Cl}{C=C}} \;\rightleftharpoons\; \underset{Cl \quad CF_3}{\overset{F_3C}{\underset{Cl}{\overset{\text{\tiny OEt}}{\underset{\text{\tiny Cl}}{C - C}}}}} \longrightarrow \underset{Cl \quad CF_3}{\overset{F_3C \quad OEt}{C=C}} + \underset{Cl \quad OEt}{\overset{F_3C \quad CF_3}{C=C}} + Cl^- \quad (44)
$$

$$
O_2N-\!\!\!\left\langle\!\!\bigcirc\!\!\right\rangle\!\!-Cl + \left\langle\!\!\bigcirc\!\!\right\rangle\!\!NH \longrightarrow O_2N-\!\!\!\left\langle\!\!\bigcirc\!\!\right\rangle\!\!-N\!\!\left\langle\!\!\bigcirc\!\!\right\rangle \qquad (45)
$$

Several of the reactions just mentioned result in overall substitution, and in these cases it is necessary to consider several alternative mechanisms. Thus, for the hypothetical vinyl substitution shown in Figure 35, we could have (a) the associative–dissociative mechanism via a carbanion, (b) a synchronous (S_N2) reaction in which X^- departs as Y^- attacks, (c) an S_N1 reaction in which X^- departs to give a vinyl cation which subsequently captures Y^-, or (d) a bimolecular elimination caused by Y^- or some other base, yielding an allene or acetylene intermediate which subsequently adds HY. With a carbonyl

11

or aromatic substrate there are analogous possibilities, the corresponding intermediates in (c) and (d) being an acylium (RCO^+) or phenyl cation and a keten or benzyne. The possibilities in (c) and (d) are fairly simply distinguished from those in (a) or (b) and we shall see examples later where they occur. The problem of deciding between (a) and (b) is a more subtle one; in the limit the transition-state for (b) is very similar to the anionic intermediate in (a). We will discuss several ways of making this distinction later, but, to anticipate

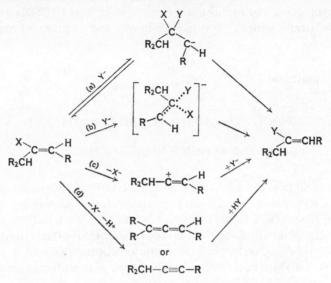

Figure 35. Possible mechanisms for nucleophilic vinyl substitution. (a) Associative–dissociative route via a carbanionic intermediate. (b) One-step synchronous mechanism. (c) Route via a vinyl cation. (d) Elimination–addition via an allene or an acetylene.

the results, there are at present no well tested examples of mechanism (b), and the associative mechanism described in detail on page 310 does seem to prevail in most cases.

Reactivity in the associative mechanism obviously depends on the electrophilicity of the substrate and on the nucleophilicity of the reagent. As we have already hinted, the former will largely depend on the stability of $-Z^-$ in **40** as an anion. It is interesting to enquire how far nucleophilicity towards carbonyl carbon runs parallel to nucleophilicity in S_N2 reactions (see Chapter 3, page 202), or to basicity towards a proton. The reactivity of five assorted nucleophiles, relative to H_2O, towards methyl bromide and p-nitrophenyl acetate in water[79] is plotted against their basicity in Figure 36. The correlation between nucleophilicity towards the carbonyl carbon of p-nitrophenyl acetate and basicity, while not perfect, is obviously far better than that between basicity and nucleophilicity in the S_N2 reaction. In terms of the Principle of Hard and Soft Acids and Bases, carbonyl carbon is nearer to the proton in hardness than is saturated carbon. Having said this, it is important to remember the warning

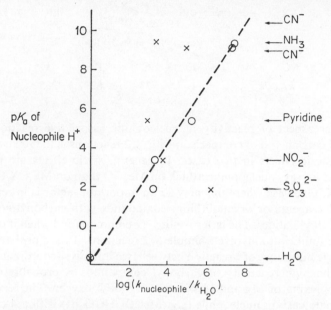

Figure 36.   Rates of nucleophilic attack (relative to H_2O) on methyl bromide ($\times$) and on p-nitrophenyl acetate ($\circ$) plotted against pK_a of nucleophile H^+.

expressed in Chapter 3 about the effect of solvents etc. on hardness and softness, and also to remember that we are comparing a kinetic parameter (nucleophilicity) with an equilibrium parameter (basicity). One factor which might tend to make carbonyl carbon a relatively hard acid is that nucleophilic attack goes to create a hard alkoxide anion centre in the tetrahedral intermediate. In nucleophilic vinyl and aromatic substitution, the corresponding anionic centre is usually more diffuse and polarizable, and it does seem to be true that soft nucleophiles are more effective in these cases. The polarizability and softness of the substrate can be affected too by the softness of what is to become the incipient leaving group. We see this in **46**: the soft nucleophilic PhS⁻ is relatively most effective with the substrate with the softest leaving group.[80]

(It is possible, however, that with 2,4-dinitrofluorobenzene the rate-determining step is not nucleophilic attack but the subsequent loss of F$^-$. This would invalidate the comparison in this case, and we will discuss this more fully on page 344 but this possible complication must be borne in mind in other cases.)

$$(Nu^- = PhS^- \text{ or } MeO^-)$$

X	F		Cl		Br		I
k_{PhS^-}/k_{MeO^-}	1	/	33	/	82	/	285

Another aspect of reactivity in nucleophilic associative reactions at unsaturated carbon and of comparison with S_N2 reactions is the relative importance of steric effects in two cases. In general, steric effects are much less important in, say, nucleophilic attack on Me_2CO than on Me_3CX or even on Me_2CHX. One case where this may be an important factor is in reactions of Grignard reagents or organolithium compounds with carbonyl compounds and with alkyl halides. The latter is not very common (and when it does occur it is quite unlikely to involve a simple S_N2 reaction). This is probably related to the steric hindrance of a reaction between two initially 4-coordinate carbons, though the high reactivity of carbonyl compounds is probably connected with complexing of Mg and Li with the carbonyl oxygen. The reaction of a 3-coordinate carbon nucleophile (e.g. $MeCO\bar{C}HCO_2Et$) with a 4-coordinate carbon electrophile (e.g. MeI) occurs quite readily as does combination of a 3-coordinate carbon electrophile with a 4-coordinate carbon nucleophile ($Me_2CO + MeMgBr$).

One final point needs to be made about nucleophilic associative reactions, and this concerns the opportunities for catalysis. It is clear that attachment of a proton or Lewis acid to the oxygen of a carbonyl group will considerably increase the electrophilic character of the carbon and also that detachment of a proton from a nucleophile such as ROH or RNH_2 will yield a much more potent nucleophile (RO^- or RNH^-). We can thus expect carbonyl reactions, and in fact nucleophilic associative reactions in general, to be frequently susceptible to acid and base catalysis. This catalysis may be either specific (pre-equilibrium formation of e.g. $>C{=}OH^+$ or RO^-) or general (proton-transfer concerted with nucleophilic attack), and much of the complexity (and fascination) of these reactions lies in unravelling these various forms of catalysis.

Nucleophilic Addition

Nucleophilic additions occur in a wide variety of molecular situations, for example with vinyl compounds (Michael addition) (see **43**, page 311) in covalent hydration[81] of some heterocyclic compounds, e.g. **47**, and in formation of Meisenheimer complexes (see **91**, page 343), but the most important

(47)

substrates are aldehydes and ketones. Some of these additions are irreversible (for example most additions of Grignard and organolithium reagents and reductions with $LiAlH_4$) but with many weaker nucleophiles reversible addition occurs. In general, addition is likely to be more or less irreversible if the attacking nucleophile, Y^-, is a more powerful base than $R_2C(Y)O^-$. Frequently, however, Y^- is not present in stoichiometric amounts but is formed in low equilibrium concentrations from HY. In this case one is usually concerned with the relative stabilities of $R_2C(Y)OH$ and $R_2CO + HY$.

A simple example of reversible addition is hydration of aldehydes and ketones.[63] The stability of the *gem*-diol, $R_2C(OH)_2$, is strongly dependent on structure. Thus, the equilibrium constants for dissociation of the *gem*-diol at 20° are 5×10^{-4} for formaldehyde, 0.7 for acetaldehyde, and 5×10^2 for acetone. The unstable hydrates of most ketones (and aromatic aldehydes) can best be detected as the intermediate responsible for ^{18}O exchange between $H_2^{18}O$ and the carbonyl compound.[82] Ortho-acid derivatives $RC(OH)_2X$ are even less stable with respect to RCOX (X = halogen, OR, NR_2) because of the loss of resonance energy involved in their formation. Electron-withdrawal, however, destabilizes $>C=O \leftrightarrow >C^+-O^-$, so that chloral forms a very stable hydrate (dissociation constant 3.6×10^{-5}) as do hexafluoroacetone and indanetrione. Another ketone which forms a stable hydrate is cyclopropanone; this is because angle strain is relieved in its formation [optimum C—C(O)—C angle $\approx 120°$; C—C(OH)$_2$—C angle $\approx 109°$]. These stability trends for hydration are a useful guide to the expected stabilities of other tetrahedral intermediates we shall encounter.

The kinetics of hydration of acetaldehyde have been carefully studied and have been mentioned in Chapter 1 (page 67). The reaction is both general-acid and general-base catalysed[63] and the transition-states in the two cases are probably **48** and **49**. In general we shall find that catalysis occurs where it is

(48)   (49)

needed; water is a relatively weak nucleophile and it can only attack the carbonyl group if it is itself made more nucleophilic by partial removal of a proton by a base or if the carbonyl group is made more electrophilic by partial proton-transfer to its oxygen atom.

The formation of acetals and ketals from carbonyl compounds and alcohols is closely related to hydration.[64] As with hydration, formation of hemiacetals RCH(OH)OR is both acid- and base-catalysed; an important and much investigated reaction related to this is the mutarotation of sugars (see Chapter 1, page 68). Most mechanistic studies have been concerned with the hydrolysis rather than formation of full acetals and ketals, R$_2$C(OR)$_2$. This is catalysed by acids but not bases (what role could they perform?) and general acid catalysis has not been detected, so that proton-transfer is not occurring in the rate-determining step (see Chapter 1, page 65). The mechanism shown in Figure 37 makes this a simple unimolecular decomposition of the conjugate acid of the acetal; this is just the reverse of the associative process (a) in Figure 35, but translated into other terms the reaction is an S_N1 solvolysis. Hydrolysis of *para*-substituted benzaldehyde acetals (Figure 37) is accelerated if the substituent is electron-donating (ρ negative). The electron demand in the transition-state, ArCH=OR$^+$ is sufficient to require a combination of σ^+ and σ for successful correlation of the rates (see Chapter 1, page 33).[83]

Formation of cyanohydrins has already been mentioned (see page 310). The rate is proportional to [CN$^-$][R$_2$CO] and there is no acid catalysis. Apparently the relatively strong nucleophile, CN$^-$, is quite capable of attacking the carbonyl group without any acidic assistance. The equilibrium R$_2$C(CN)O$^-$ $\rightleftarrows$ R$_2$CO + CN$^-$ generally strongly favours dissociation so that it is normally necessary to use HCN to get reasonable conversion into the cyanohydrin; the equilibrium $R_2CO + HCN \rightleftarrows R_2C(OH)CN$ is much more favourable. Cyanohydrin formation is an especially simple case of reaction of a carbon nucleophile with a carbonyl group. Many other reactions in this class are important in synthesis, e.g. aldol condensations,[71] reactions with Grignard and other organometallic reagents, and Wittig reactions.[72] The mechanism and overall kinetic equation for these reactions depend on how

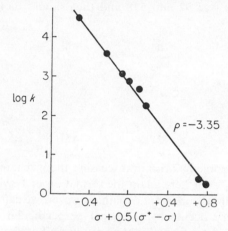

Figure 37. Plot of the logarithms of the second-order rate constants for hydrolysis of substituted benzaldehyde diethyl acetals against $\sigma + \frac{1}{2}(\sigma^+ - \sigma)$, in 50% dioxan–water at 30°. The suggested mechanism is shown.

$$ArCH(OEt)_2 + H_3O^+ \rightleftharpoons \underset{\substack{|\\H}}{\overset{OEt}{Ar\!-\!\overset{|}{C}\!-\!\overset{+}{O}\!-\!Et}}\,\underset{H}{} \xrightarrow{\text{slow}} \underset{\substack{|\\H}}{Ar\!-\!\overset{+}{C}\!=\!\overset{+}{O}\!-\!Et} \xrightarrow[\text{H}_2\text{O}]{\text{fast}} ArCHO + EtOH$$

the carbon nucleophile is generated, but the actual carbon–carbon bond formation step always occurs by the π-attack process. The aldol condensation and related reactions are often complicated by subsequent dehydration steps, and we will therefore postpone discussion of these until after we have discussed reactions with nitrogen nucleophiles which also usually involve dehydration and thus overall substitution (see **41**, page 311).

The reactions of Grignard reagents and organolithium compounds are quite simple in that straightforward and irreversible addition usually occurs. The problems arise when we try to define the nature of the attacking nucleophile. The composition of solutions of these reagents is discussed in Chapter 2 (page 144) and the problems of formulating the reaction with a ketone are presented. A possible scheme with a Grignard reagent involves two RMgX units (50). It is easy to see the geometrical advantage of this over transformation of a 1:1 complex (51) when the need for π-attack is remembered. Our knowledge of the geometry of protonated ketones makes it likely that the coordination of magnesium is in the C—C(O)—C plane, as shown. It seems that reduction and enolization, which sometimes compete with addition, are reactions of the 1:1 complex (see 52 and 53), and thus should be favoured in dilute solution.

Rather similar problems arise in discussing the mechanisms of reductions by BH_4^- and AlH_4^-, for here again the formation of Lewis acid–Lewis base complexes may well be important. Again, however, we can be fairly sure that hydride ion delivery occurs by the usual perpendicular π-attack. Related hydride-transfer reactions occur in the aluminium alkoxide catalysed equilibration of ketones and secondary alcohols (Meerwein–Ponndorf–Verley reduction and Oppenauer oxidation)[84] and in the Cannizzaro reaction of aldehydes (see 54 and 55).

One aspect of the addition reactions of RMgX, RLi, AlH_4^-, and BH_4^- which has received considerable attention is their stereochemical course. Since these reactions are normally irreversible, the relative stabilities of the two epimeric products are largely irrelevant. What seems to determine stereo-chemistry is the relative ease of approach of the reagent to each face of the carbonyl group. In other words, the transition-state is 'early' and reactant-like. For example, lithium aluminium hydride reduction of norbornan-2-one gives 89% *endo*-norbornan-2-ol, but camphor gives 92% isoborneol (the *exo*-alcohol).[85] In the first case hydride delivery is from the more open *exo*-face but in camphor this is blocked by the *syn*-7-methyl group and so slower reduction occurs by attack on the *endo*-face.

In conformationally mobile systems the position is more complicated. In acyclic systems the normal result is as shown in **56** (Cram's rule).[86] A typical example is shown in **57** (L = cyclohexyl, M = methyl, S = hydrogen, R = iso-propyl). However in 2-methylcyclohexanone, for example, the major product

$$(56)$$

Major product   Minor product

(S, M, L = small, medium, and large groups; R' = H or alkyl or aryl)

$$(57)$$

Major   Minor

of hydride reduction is *trans*-2-methylcyclohexanol; this corresponds to the minor product in **56** if one equates L = ring, M = methyl, and S = hydrogen. A consistent rationalization of these results has been devised by Felkin.[87] He suggests that the most favoured transition-state is **58**. The aluminium hydride reagent (symbolized by R'⁻) is thought to be relatively bulky and R is considered to be larger than O. Thus **58** minimizes steric repulsions; it also minimizes torsional strain in the (LMS)C—C(O)R bond. Felkin suggests that it is quite important to have R'⁻ staggered with respect to the L, M, S sub-stituents, or in other words that torsional strain in partially formed bonds may be quite important—an interesting idea. The transition-state **58** is the one leading to the major product in the acyclic case. In the cyclohexanone case, *trans*-alcohol results from transition-state **59** which minimizes torsional strain;

(58)   (59)   (60)

torsional strain

cis-alcohol comes from transition-state **60** which minimizes non-bonded steric repulsions. It is not possible in this case to have a transition-state which minimizes both forms of strain simultaneously; apparently the torsional strain effect is the greater in this case.

The reactions just discussed, and many related additions, occur with substrates other than simple aldehydes and ketones, and many of the same considerations apply. Especial interest is attached to α,β-unsaturated carbonyl compounds; these may be viewed as ambident electrophiles (see **61**). Products

(61)

may be formed by direct (path a) or conjugate (Michael) addition (path b). The outcome of reactions of this type frequently depends on whether the addition reaction is irreversible (kinetic control; normal with strong nucleophiles, e.g. RMgBr) or reversible (and thus subject to equilibrium control; normal with weak nucleophiles, e.g. RNH_2, ROH) under the experimental conditions prevailing. Equilibrium control tends to favour conjugate addition because this ultimately leaves the stronger of the two π-bonds (the C=O) intact. Of course, other factors will affect the relative stability of the two products; for example, steric hindrance at one site tends to lead to addition at the other. Kinetic control can favour either pathway in appropriate circumstances. With a powerful nucleophile the transition-state will be 'early' and reactant-like with the new bond long and largely ionic; the greater electron-deficiency at the carbonyl carbon then tends to favour 1,2-addition. Conversely, the more product-like transition-state in addition of a weaker nucleophile favours the often more stable 1,4-addition product. With a given nucleophile the product ratios depend on the substrate structure; PhCH=CHCHO gives nearly 100% 1,2-addition with PhMgBr, whereas PhCH=CHCOBu^t gives nearly 100% 1,4-addition. One case of some interest is that of control of 1,2- vs. 1,4-addition in Grignard reactions by addition of copper salts. When the

(62)

plain Grignard reagent gives 1,2-addition, the presence of CuI always leads to 1,4-addition. It is now thought that the 1,4-addition occurs as shown (62) via organocopper intermediates.[88]

Nucleophilic Addition–Elimination Reactions with Aldehydes and Ketones

When the initial adduct of a nucleophile and a carbonyl compound has an acidic proton a subsequent elimination is possible leading to overall substitution; an example is oxime formation (Figure 38). In this situation it is possible for either the addition or elimination step to be rate-determining and both steps

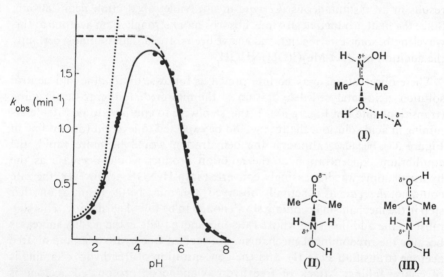

Figure 38. The mechanism of formation of acetoxime. In the plot of the pH–rate profile, the broken line is that calculated for the dehydration step as rate-determining with $k_2 = 1.1 \times 10^8$ M^{-1} min^{-1} (transition-state **I**). The dotted line is for rate-determining hydroxylamine attack with $k_1 = 1.4 \times 10^5$ M^{-1} min^{-1} (transition-state **II**) and $k_1' = 1.0 \times 10^7$ M^{-2} min^{-1} (transition-state **III**). The continuous line is the rate calculated from the steady-state rate equation:

$$\text{Rate} = \frac{k_2(k_1 + k_1'[\text{H}^+])}{(k_{-1}/[\text{H}^+] + k_{-1}' + k_2)(1 + [\text{H}^+]/K)} [\text{Me}_2\text{CO}]\,[\text{NH}_2\text{OH}]_{\text{total}}$$

where $K = [\text{H}^+]\,[\text{NH}_2\text{OH}]/[\text{NH}_3\text{OH}^+]$.

offer opportunities for acid and base catalysis. The mechanistic possibilities are therefore numerous. This reaction has been carefully studied by Jencks[62] who made the following observations.

1. The pH–rate profile (see Chapter 1, page 64) is a bell-shaped curve with a maximum near pH 5 (Figure 38).

2. On the alkaline side of the maximum the rate is given by $k[Me_2CO]$ $[H_2NOH][H^+]$, and there is general acid catalysis by undissociated acids when these are present.

3. On the acid side of the maximum the rate is given by $k_1[Me_2CO]$ $[H_2NOH] + k_1'[Me_2CO][H_2NOH][H^+]$, with the second term making only a minor contribution so that overall the rate falls with increasing acidity (the concentration of free hydroxylamine of course decreases as the acidity rises). In this region there is little general acid catalysis.

4. Near pH 7 (i.e. on the alkaline side of the rate maximum) addition of a concentrated aqueous solution of hydroxylamine to a solution of acetone results in an instantaneous *decrease* in ultraviolet absorption near 280 nm. Since the final product, acetoxime, absorbs more strongly than acetone at this wavelength, some intermediate must have been formed. This is almost certainly the addition product $Me_2C(OH)NHOH$.

These observations may be interpreted as follows. In alkaline and neutral solution, acid-catalysed dehydration of the intermediate is rate-determining (transition-state **I** in Figure 38). If this step were to continue to be rate-determining in acid solution, the rate would be expected to level off (dotted line in Figure 38) because, although the dehydration would be more rapid, the equilibrium concentration of the addition product would decrease as the hydroxylamine was increasingly converted into H_3NOH^+. However, the rate could not *decrease* as is actually observed; this must be the result of another step in the mechanism becoming slow enough to be rate-determining. This step must be the addition step whose rate does indeed fall as the acidity increases because the predominant mechanism is attack of free hydroxylamine on free acetone (transition-state **II**), and the concentration of free hydroxylamine is of course falling. Attack of free hydroxylamine on protonated acetone is independent of acidity, but makes only a minor contribution to the rate (transition-state **III**).

Reactions of other nitrogen nucleophiles (amines and hydrazine derivatives) also fit into this overall scheme (see also Chapter 1, page 39), the actual rate-determining step and the observed catalysis varying in accordance with the conditions and the basicity and nucleophilicity of the attacking reagent. Reactions like the aldol condensation present a very similar mechanistic picture, with the additional complication that creation of the carbon nucleo-

(a)

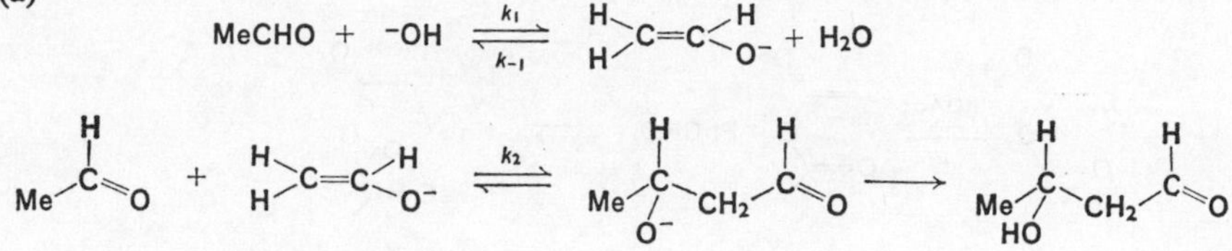

(b) *Condensation in* EtOH *with* NaOEt

Rate = k[PhCHO][PhCOMe][$^-$OEt]

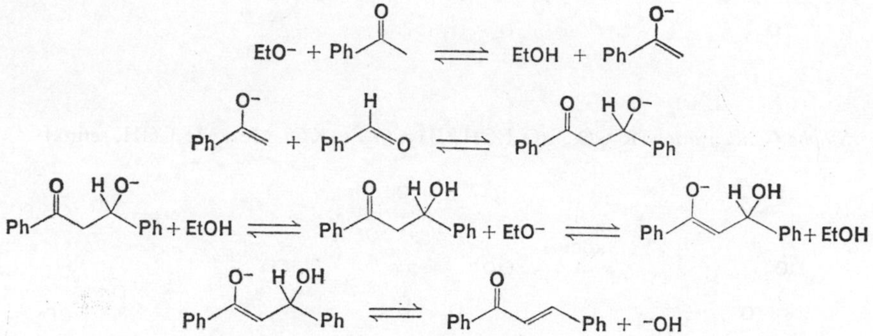

(c) *Condensation in* HOAc–H$_2$O–H$_2$SO$_4$

Rate = k[PhCHO][PhCOMe].h_0, where $h_0 = -$antilog H_0

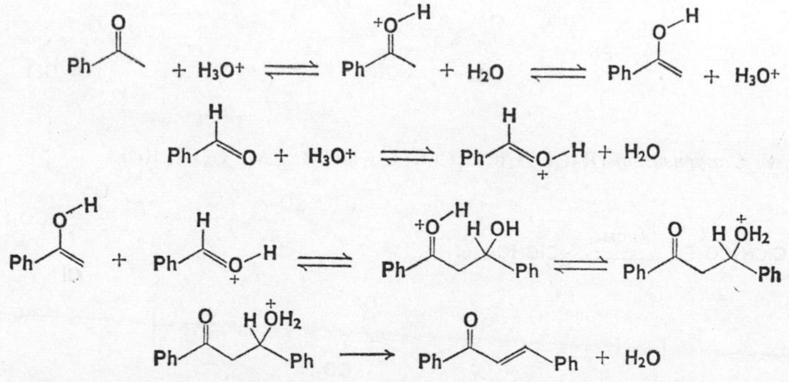

Figure 39. Aldol condensations. (a) Mechanism for the formation of acetaldol. (b) and (c) Rate expressions and mechanisms for condensation of PhCHO with PhCOMe: (b) basic conditions; (c) acid conditions.

Perkin Reaction (ArCHO, Ac$_2$O, KOAc, 150°)

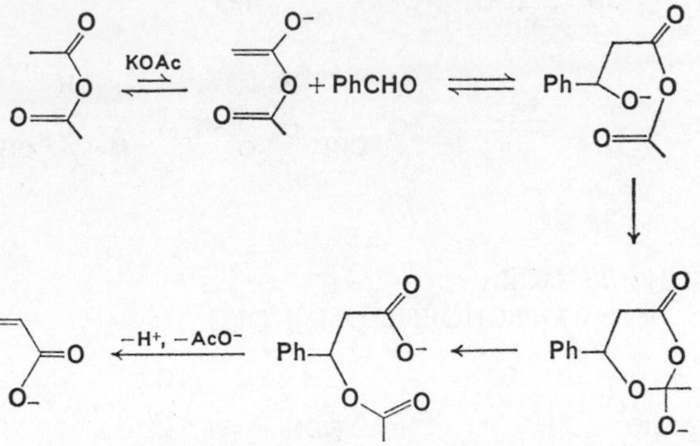

Stobbe Condensation (R$_2$CO, EtO$_2$CCH$_2$CH$_2$CO$_2$Et, KOCMe$_3$, Me$_3$COH, reflux)

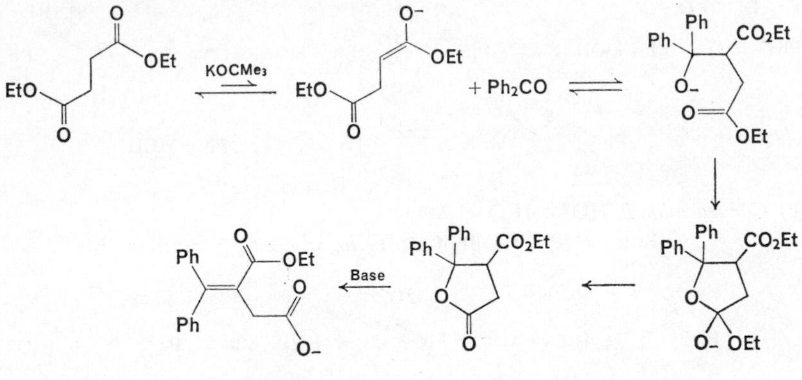

Darzens Condensation (R$_2$CO, ClCH$_2$CO$_2$Et, KOCMe$_3$, Me$_3$COH, 10°)

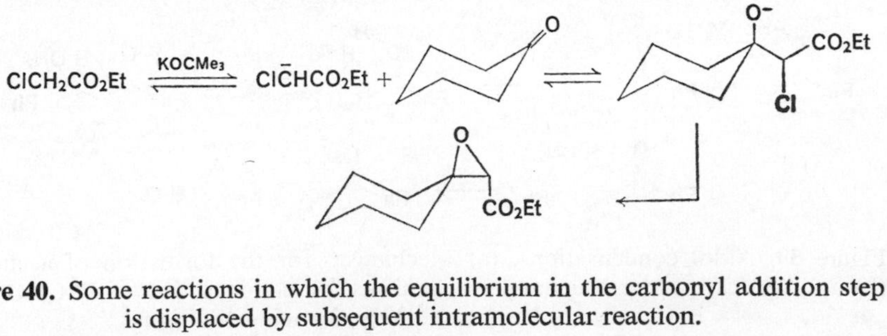

Figure 40. Some reactions in which the equilibrium in the carbonyl addition step is displaced by subsequent intramolecular reaction.

phile must also be considered and may become the rate-determining step. Thus, in the base-catalysed condensation of acetaldehyde the scheme shown in Figure 39 apparently operates.[89] When $[\text{MeCHO}] \approx 1\text{M}$, $k_{-1} \approx k_2[\text{MeCHO}]$, so that neither the first nor the second step is rate-determining and the kinetics are complex; in D_2O, recovered acetaldehyde is deuteriated. When [MeCHO] $\approx 10\text{M}$, the first step is rate-determining, and in D_2O the aldol formed in the early stages of the reaction is almost free from deuterium. Under these conditions and with aliphatic aldehydes there is normally no difficulty in obtaining the addition product (aldol). The reaction is reversible, however, and with most ketones favours the starting materials. Frequently this equilibrium is displaced by a subsequent elimination (dehydration) step, and dehydration almost always occurs in acid-catalysed condensations. The situation is then very similar to that of oxime formation. The condensation of acetophenone with benzaldehyde to give benzylideneacetophenone has been studied under conditions of both base-catalysis (EtO^- in EtOH)[90] and acid-catalysis (H_2SO_4 in H_2O–HOAc).[91] The rate equations and probable mechanistic schemes are shown in Figure 39. Notice how, while the enolate ion can attack the free aldehyde carbonyl group, the less nucleophilic enol requires acid catalysis. In the base-catalysed reaction it has been shown that the β-hydroxy-ketone is converted into benzaldehyde and acetophenone more rapidly than it is dehydrated to benzylideneacetophenone, so dehydration (probably an $E1cB$ reaction via the enolate; see Chapter 3, page 210) is the rate-determining step. In the acid-catalysed reaction either dehydration or the carbon–carbon bond formation step may be rate-determining depending on reaction conditions (and on the substituents on the aryl ring when these are present). In fact the situation is complicated by concurrent conversion of the β-hydroxy-ketone into its acetate.

Space does not permit us to discuss the many other condensations related to the aldol reaction—excellent accounts which describe the main features of the mechanisms are in the books by House and by Gutsche.[70] We would remind readers that in many cases carbon–carbon bond formation [occurring by the π-attack process (a) in Figure 35] is reversible and a subsequent step serves to displace that equilibrium (Figure 40).

The reactions of carbonyl compounds with stabilized ylids (Figure 41) also have features in common with the aldol reaction. In the Wittig reaction[72] the betaine 63 and the cyclic alkoxyphosphorane 64 are likely intermediates, but the real sequence of events is still rather uncertain. Wittig reactions result in retention of configuration at phosphorus. This result is expected if the betaine 63 leads to 64 which then undergoes decomposition to the olefin and the phosphine oxide. This last step may be a two-step process via an intermediate carbanion. Acyclic hydroxyphosphoranes are believed to be

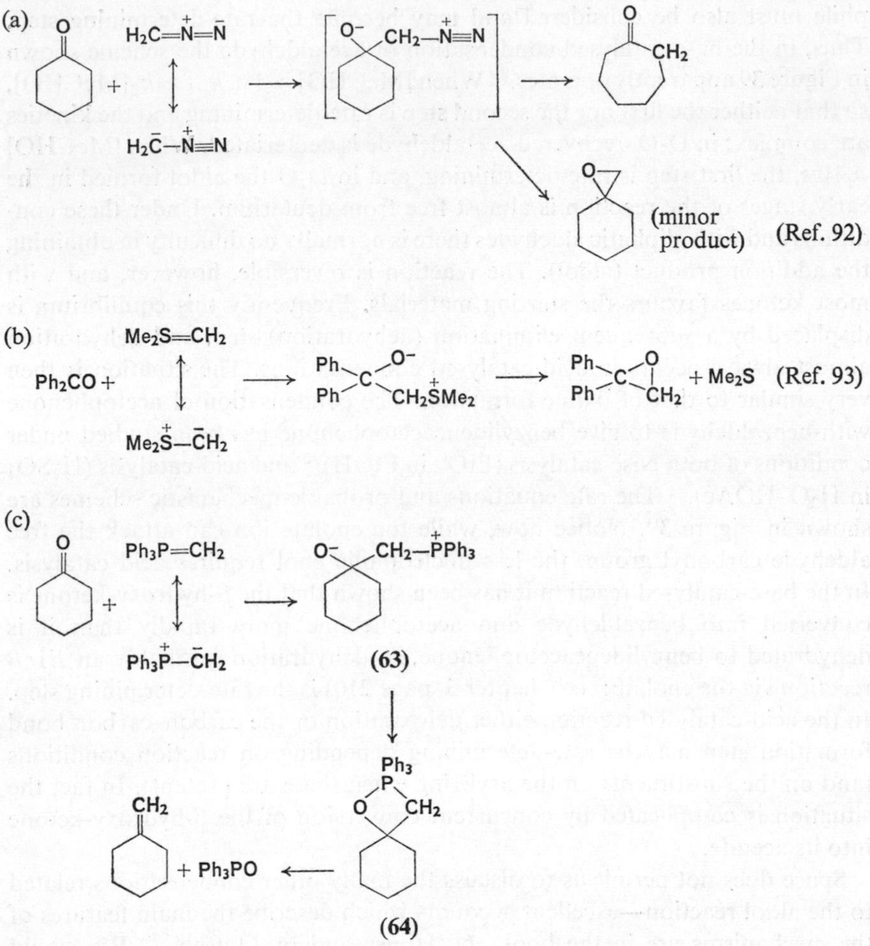

(a)

(minor product)   (Ref. 92)

(b)

(Ref. 93)

(c)

(63)

(64)

Figure 41. Reactions of stabilized ylids with carbonyl compounds. (a) Reaction with diazomethane. (b) Reaction with dimethylsulphonium methylide. (c) The Wittig olefin synthesis.

intermediates in the alkaline decomposition of phosphonium salts and to decompose as shown in **65** with loss of a carbanion from an apical position and overall *inversion* at phosphorus.[73, 94] Obviously in the cyclic case the 4-membered ring cannot span the apical positions but equatorial expulsion of the carbanion (which rapidly fragments to products) is a likely alternative and would result in *retention* at phosphorus. In any case the final step seems to be a *syn*-elimination. The stereochemistry of the olefin is therefore determined by the *erythro*- or *threo*-geometry of the betaine (or the corresponding geometry in **64**) (*erythro*- gives *cis*-olefin).

In practice the stereochemistry of the olefin product has been found to be sensitive to the reaction conditions. In solvents of moderate polarity and under normal conditions, the *trans*-olefin is often the major isomer formed.[70, 95a, b] It is apparently possible to increase the proportion of *cis*-isomer by (a) increasing the concentration of Li^+ or other Lewis acids or by adding protic solvents such as MeOH, or (b) going to non-polar solvents and working in the absence of Li^+. Obviously there must be some changes of mechanism with change of conditions to account for these facts. In polar solvents, betaine formation is probably reversible and, e.g., MeOH stabilizes **67** (the precursor

of *cis*-olefin), at the expense of **66**. In salt-free non-polar solvents at low temperatures, it is likely that betaines such as **63** are not formed, and it has been suggested[95b] that the sequence **68** occurs at least with relatively unstabilized Wittig reagents such as $Ph_3P=CHMe$.

To round off this section and the preceding one, Table 4 shows how catalysis in nucleophilic addition steps varies with the electronic demands of the reagent and substrate.

Table 4. Catalysis in the addition step of nucleophilic attack on carbonyl compounds.

Nucleophile (X)	pK_a of XH$^+$	Catalysis
RC(O$^-$)=CH$_2$	~20	None
RC≡C$^-$	~20	None
OH$^-$	15.7	None
CN$^-$	9.4	None
RNH$_2$	~10	None
SO$_3^{2-}$	7.0	Little or none
NH$_2$OH	6.0	Some acid catalysis
PhNH$_2$	4.6	General acid
NH$_2$NHCONH$_2$	3.7	General acid
H$_2$O, ROH	~−1.5	General acid
RC(OH)=CH$_2$	~−5	Acid catalysis
RMgBr, RLi, AlH$_4^-$, BH$_4^-$	?	Some metal-ion catalysis

Acylation Reactions

Many reactions of acyl derivatives (RCOX) are of the form:

$$RCOX + HY \rightarrow RCOY + HX$$

and may be looked upon as acylations (of HY by RCOX).[68] The great majority of these take place by the now familiar nucleophilic associative–dissociative route involving the usual π-attack. The commonest type of acylation reaction is where HY = H$_2$O (hydrolysis), and the most thoroughly studied example is ester hydrolysis (X = OR). We will now discuss this from several points of view.

Ester Hydrolysis[65, 66, 69]

An ester possesses two possible sites for nucleophilic attack, the carbonyl carbon and the α-carbon of the alkyl (alcohol) group (see **69**). In ester hydrolysis where H$_2$O or OH$^-$ is the nucleophile, it is not immediately apparent from the reaction products where attack has occurred, since both possibilities

(69)

give an alcohol and an acid (with the closely related reaction **70** the outcome does show where attack occurred). Attack at the α-carbon of the alkyl group is an example of substitution at saturated carbon by an S_N2 mechanism, and results in fission of the alkyl–oxygen bond. Nucleophilic attack at the carbonyl

$$
\text{Me—C}\overset{\displaystyle O}{\underset{\displaystyle O-R}{\diagdown}} + \text{R'OH}
\quad
\begin{cases}
\xrightarrow{\text{Attack at } >\!C\!=\!O} & \text{Me—C}\overset{\displaystyle O}{\underset{\displaystyle OR'}{\diagdown}} + \text{ROH} \\
\xrightarrow{\text{Attack at R}} & \text{Me—C}\overset{\displaystyle O}{\underset{\displaystyle OH}{\diagdown}} + \text{ROR'}
\end{cases}
\tag{70}
$$

carbon results in acyl–oxygen fission. We can also imagine comparable unimolecular dissociations, generating an alkyl cation by alkyl–oxygen cleavage of an acylium ion by acyl–oxygen fission. There are therefore four basic mechanisms, which are given the following designations; $_{AC}2$ (acyl oxygen fission, bimolecular), $_{AC}1$, $_{AL}2$ (alkyl–oxygen fission, bimolecular, S_N2), $_{AL}1$. It will be clear by now that there are opportunities for acid and base catalysis with these basic mechanisms. In classifying these reactions, Ingold[96] gave the designation A to acid-catalysed mechanisms and B to uncatalysed and base-catalysed reactions. Thus, the $_{AC}2$ mechanism might be accelerated by acid (attack of water on carbonyl-protonated ester) or by base (attack of OH^- on neutral ester). In fact, these two mechanisms, $A_{AC}2$ and $B_{AC}2$, are found to be much the commonest for ester hydrolysis in dilute acid or base, and will be considered in more detail presently.

Of the other mechanisms, $_{AC}1$ is actually only known in acid-catalysed form, $A_{AC}1$. This requires protonation on the alkyl oxygen of the ester, which is known, however, to be relatively unfavourable from n.m.r. spectroscopic studies of protonated esters which only detect the presence of the carbonyl oxygen protonated form. The rate-determining step in the hydrolysis is then loss of ROH from $R'CO_2HR^+$ to give $R'CO^+$. Several examples of this mechanism are discussed later (page 333). An unusual mechanism (strictly, but misleadingly, a $B_{AC}2$ mechanism) which is related to the $A_{AC}1$ mechanism is shown in **71**.[97] The $B_{AL}2$ mechanism is extremely uncommon but occurs in the uncatalysed hydrolysis of β-lactones in near-neutral solution[98] where alkyl–oxygen fission has the special advantage of relieving almost all the strain in the four-membered ring in the transition-state. Normally however, nucleophilic attack at saturated carbon (S_N2) cannot compete with attack

(71)

at the carbonyl group (see page 314). In concentrated base (OH$^-$ instead of H$_2$O as the nucleophile) even β-lactones hydrolyse by the normal $B_{AC}2$ mechanism.

The $_{AL}1$ mechanism is rather more common[99] especially in its acid-catalysed form $A_{AL}1$ (carboxylic acid rather than carboxylate as leaving group in an S_N1 reaction). Hydrolysis of most esters of tertiary alcohols proceeds by this mechanism in acid solution, and other systems which can form carbonium ions of comparable or greater stability behave similarly (see Chapter 2). The classification scheme is summarized in Table 5.

Table 5. Mechanistic classification for ester hydrolysis.

		Catalysis		
		Acid	None	Base
Acyl–Oxygen fission	Unimolecular	$A_{AC}1$		$B_{AC}1$
RCO-$\{$-O—R	Bimolecular	$A_{AC}2$		$B_{AC}2$
Alkyl–Oxygen fission	Unimolecular	$A_{AL}1$		$B_{AL}1$
RCO—O-$\}$-R	Bimolecular	$A_{AL}2$		$B_{AL}2$

Proof of acyl–oxygen fission in ester hydrolysis is obtained by carrying out the hydrolysis of an ester labelled at the ethereal oxygen. For example, EtCO^{18}OEt yields unlabelled propionic acid and Et^{18}OH on alkaline hydrolysis,[100] and similar experiments have been carried out for the acid- and base-catalysed hydrolysis of several other esters of primary and secondary alcohols.

Base-catalysed hydrolysis of simple esters is first-order in ester and in hydroxide ion so that we would seem to have nucleophilic π-attack by hydroxide ion on the carbonyl group. But how do we know whether the tetrahedral species formed is an intermediate or only a transition-state? The distinction is that the former has a finite lifetime, however short, and so it may have a possible choice of reaction pathways. In the present case the possible choices are shown in **72**. Notice that, provided that proton-transfers are rapid, it

$$
\begin{array}{ccc}
Ph-C{\overset{*O}{\underset{OEt}{\diagdown}}} + {}^-OH & \rightleftharpoons & Ph-C{\overset{*O^-}{\underset{OEt}{\diagup\!OH}}} \\
\end{array}
$$

Ph—C(*O)(OEt) + ⁻OH ⇌ Ph—C(*O⁻)(OH)(OEt) ↘ PhCOO⁻ + EtOH   (72)

Ph—C(O)(OEt) + *⁻OH ⟵ Ph—C(*OH)(O⁻)(OEt) ↗

may return to starting materials but with an exchange of oxygen atoms. In a classic experiment Bender[101] hydrolysed ethyl benzoate labelled in the carbonyl group with ^{18}O under alkaline conditions and recovered the ester after partial hydrolysis. He found that some loss of ^{18}O label had occurred. While it is possible that exchange occurs by an irrelevant side-reaction, it is very hard to imagine what the intermediate involved could be if it is not that also involved in the hydrolysis. Bender measured k_h/k_{ex}, the ratio of the rate constants for hydrolysis and for exchange for a number of substances. It can be shown that $k_h/k_{ex} = 2k_2/k_{-1}$ in the kinetic scheme for ester hydrolysis (**73**), and the value

$$
PhCO_2Et + {}^-OH \underset{k_{-1}}{\overset{k_1}{\rightleftharpoons}} PhC(OH)(OEt)O^- \overset{k_2}{\longrightarrow} PhCOO^- + EtOH \qquad (73)
$$

of this ratio is a measure of the partitioning of the tetrahedral intermediate, which in turn depends on the relative effectiveness of OH^- and OR^- as leaving groups. For some compounds, e.g. phenyl benzoate in alkali, exchange cannot be detected. This probably simply means that $k_2 \gg k_{-1}$, because phenoxide anion is a better leaving group than OH^-, but it is of course possible that a one-step synchronous mechanism has taken over. Similar ^{18}O exchange experiments have been carried out for some acid-catalysed hydrolyses, for example for ethyl benzoate (and in this case exchange is detected).

Ester hydrolysis has been a popular testing-ground for studies of substituent effects, as has been described in Chapter 1 (page 34). In base-catalysed hydrolysis ($B_{Ac}2$) the carbon of the carbonyl group acquires negative charge in the transition-state and accordingly hydrolysis is speeded by electron-withdrawing groups (ρ positive). In acid-catalysed hydrolysis ($A_{Ac}2$) electron-withdrawing groups discourage protonation but encourage attack by water on the protonated species, once it has been formed. The overall result is a

relative insensitivity to polar effects ($\rho \approx 0$). Both $B_{AC}2$ and $A_{AC}2$ hydrolyses
are subject to steric retardation by bulky substituents in either the alcohol or
acid parts of the ester.[102] As might be expected, the steric effect is largest when
the substituent interferes directly with the perpendicular approach of the
nucleophile as in methyl mesitoate (74).

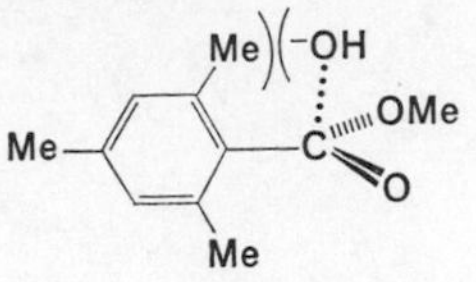

(74)

Ester hydrolysis has been found to be catalysed by a variety of other species
besides H_3O^+ and OH^-. Some of the possibilities have been described in
Chapter 1. One example of intramolecular catalysis was also described there.[103]
Another interesting example is that of methyl o-formylbenzoate[104] which is
hydrolysed in base 10^5 times faster than the p-formyl isomer. We have already
suggested (page 315) that addition to an aldehyde group is energetically more
favourable than addition to an ester group. In this case OH^--addition to the
aldehyde group produces a conveniently positioned nucleophile for the more
difficult attack on the ester group, and the reaction sequence is probably as
shown in 75. The hydrolysis is also catalysed by morpholine, and in this case
an intermediate, probably 76, can be detected spectroscopically.

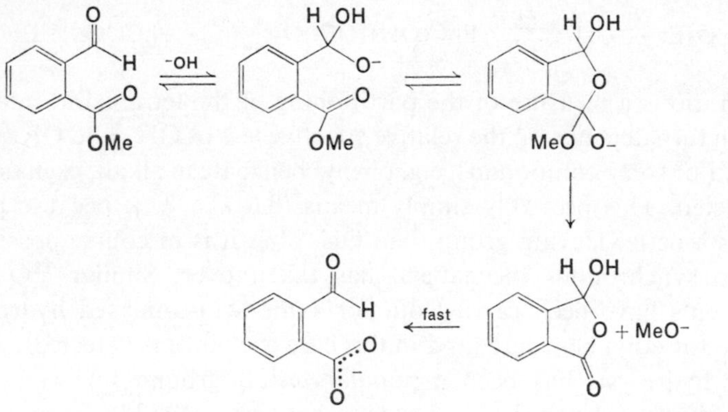

(75)

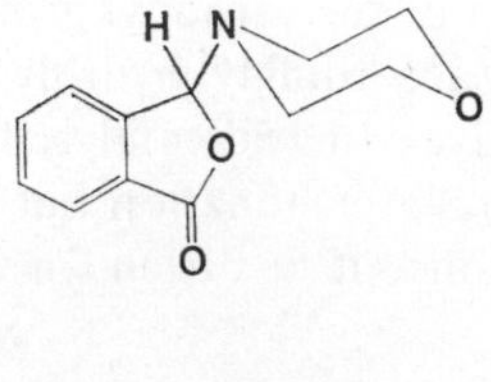

(76)

Hydrolysis of Some Acetates in Concentrated Solutions of Sulphuric Acid

Yates and McClelland[105] have studied the rates of hydrolysis of a series of acetates in sulphuric acid solutions. Their work provides a good example of the use of acidity functions (see Chapter 1, page 58) in mechanistic studies in concentrated solutions, where both acidity and the activity of the nucleophile, water, are varying simultaneously. In addition, interesting changes in the mechanism of hydrolysis occur as the acidity increases.

The activity of water, a_{H_2O}, decreases sharply as the proportion of sulphuric acid increases. In general terms we may expect mechanisms in which water must act as a nucleophile (e.g. the normal $A_{AC}2$ mechanism) to be replaced by unimolecular processes such as $A_{AC}1$ or $A_{AL}1$ (see Table 5, page 330). The reactant ester, its conjugate acid, and the transition-state for hydrolysis are all hydrated by certain numbers of water molecules. If r more water molecules are hydrating or involved in the transition-state than in the protonated ester, water is involved in the equation to the rth power, and in fact we can write:

$$\log k_1 + H = r \log a_{H_2O} + \text{Constant}$$

where k_1 is the pseudo-first-order rate constant (rate $= k_1[\text{Ester}]$), and H is an appropriate acidity function for esters. Yates and McClelland took $H = 0.62 H_0$ (see page 60). The measured rates (see Figure 42) were re-plotted on graphs of $\log k_1 + 0.62 H_0$ versus $\log a_{H_2O}$. It can be seen that these latter graphs typically consist of two linear portions connected by a curved transitional region. What do these graphs mean? The slope is r, which we have said is the number of additional water molecules involved in the transition-state over and above those involved with the protonated ester. A change of r almost certainly indicates a change of mechanism. In fairly dilute acid r is close to 2 for all the esters except Bu^tOAc, and hydrolysis is probably occurring by the normal $A_{AC}2$ mechanism. Apparently water dimer $(H_2O)_2$ is the nucleophile; the second molecule probably helps to disperse the charge or possibly a cyclic mechanism (77) is involved. An ester like Bu^tOAc is known from other work to hydrolyse by an $A_{AL}1$ carbonium ion mechanism. In this case r is about -1.5 as water is released on going from the hydrophilic protonated ester to the relatively hydrophobic carbonium ion. r is also negative (about -0.5) for the secondary acetates and benzyl acetate in concentrated acid solution. These esters can form moderately stable carbonium ions, and it is thought that at high acidities $A_{AL}1$ hydrolysis occurs. With primary and aryl esters $A_{AL}1$ is not likely but clearly a change of mechanism does occur. In strong acid $r \approx -0.2$ and $A_{AC}1$ hydrolysis probably occurs (formation of an acylium ion $MeCO^+$). While the distinction between $A_{AL}1$ and $A_{AC}1$ on the basis of r-values is quite uncertain, it is difficult to see why if they were all

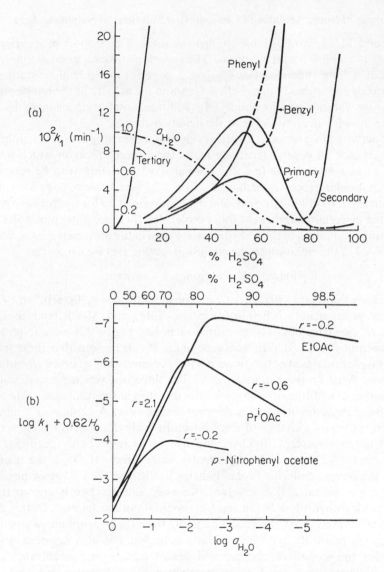

Figure 42. Hydrolysis of acetates in sulphuric acid solutions. (a) Typical rate–acidity plots together with a_{H_2O} at 25°C. (b) The data plotted in the form $\log k_1$ (pseudo-first-order rate constant) + 0.62 H_0 against $\log a_{H_2O}$ for typical primary, secondary, and aryl acetates.

[Part (a) from K. Yates and R.¹A. McClelland, *J. Amer. Chem. Soc.*, **89**, 2686 (1967).]

(77)

$A_{AC}1$ the secondary and benzyl esters should be so much more reactive than the primary ones. The high reactivity of the p-nitrophenyl ester by an $A_{AC}1$ mechanism is quite reasonable because p-nitrophenol should be a better leaving group than, say, methanol.

Yates and McClelland's study was confined to acetates but it is structural changes in the acyl group that are most likely to affect the onset of the $A_{AC}1$ mechanism. The acylium ion intermediate is stabilized by electron-supply but what is perhaps more important is that its formation is subject to steric acceleration; that is, steric strain around the carbonyl group is relieved as the acylium ion forms. Thus, methyl mesitoate, which it is tremendously difficult to hydrolyse by the $A_{AC}2$ mechanism, rapidly yields a stable acylium ion in concentrated H_2SO_4; this is converted into mesitoic acid (78) on quenching the solution in water.[106]

(78)

Mechanism of Other Acyl Transfer Reactions

Having considered ester hydrolysis mechanisms from several points of view, we shall only discuss the general mechanistic trends for other acyl transfer reactions. The possible permutations of nucleophile and leaving group are obviously legion, and with both heteroatom and carbon nucleophiles the reactions are of great biochemical and synthetic importance. Many appear in both acid- and base-catalysed guises. The great majority probably proceed by addition–elimination involving process (a) of Figure 35. The elimination–addition mechanism [process (c) of Figure 35] appears only at the end of the reactivity spectrum when good leaving groups and poor nucleophiles are involved. For example, solvolyses of benzoyl chlorides give Hammett $\sigma\rho$ plots which are strongly solvent-dependent (see Figure 43).[107] It was thought that an acylium ion (S_N1) mechanism takes over in an increasing number of cases (beginning with the p-methoxy-compound) as the solvent polarity rises. However, the situation at the borderline may be more complex than this.

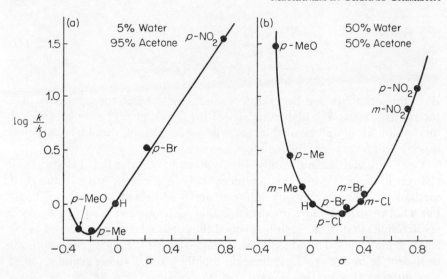

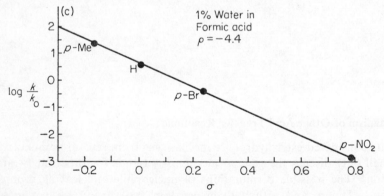

Figure 43. Plots of $\log(k/k_0)$ against σ for hydrolysis of substituted benzoyl chlorides in various solvent mixtures.

[From S. L. Johnson, *Adv. Phys. Org. Chem.*, **5**, 237 (1967).]

Sneen[108] has recently shown how the competition between hydrolysis and reaction with *o*-nitroaniline for benzoyl chloride in 50% H_2O–50% acetone (compare Figure 43b) can be accounted for in terms of his ion-pair mechanism (see Chapter 2, page 91). That is, water and *o*-nitroaniline compete for the $PhCO^+Cl^-$ ion-pair, which is formed in the rate-determining step. In 20% H_2O–80% acetone ($\sigma\rho$ plot probably nearer to Figure 43a) either attack of the nucleophiles on the ion-pair is rate-determining or we have concurrent

bimolecular reactions. These bimolecular reactions might occur via a tetrahedral intermediate or be synchronous (S_N2) reactions. The mechanistic
situation is therefore quite intriguing.

In the hydrolysis of amides, the dissociative (S_N1) mechanism is a possibility
in strong acids ($RCONHR_2^+ \rightarrow RCO^+ + NHR_2$, though this involves the
the less stable tautomer of a protonated amide—amides are predominantly
protonated on oxygen) but in other situations the normal associative mechanism probably predominates. As with esters, ^{18}O exchange has been detected
in recovered starting amides for several alkaline hydrolyses, thus implicating a
tetrahedral intermediate. Thus ^{18}O exchange has been shown to occur[109]
with benzamide and N-methylbenzamide, though it does not occur with
N,N-dimethylbenzamide. The reason for this last result is not clear, but it
must be remembered that the occurrence of ^{18}O exchange demands that the
two oxygens become equivalent during the lifetime of the intermediate, or in
other words, that proton transfer between them is rapid. Possibly this condition is not met in this case. Another interesting facet of amide hydrolysis
is that, since NR_2 is a rather poor leaving group, it is possible for this second
step in the substitution mechanism to become rate-determining; this occurs,
for example, with trifluoroacetanilide.[110] As a final comment on amide
hydrolysis, it is worth pointing out that the opportunities for tautomerism in
the starting amide, in derived ions, and in the tetrahedral intermediates are
particularly rich. The precise pathway through these possibilities is not an
easy one to discover, though different possibilities will certainly give rise to
different substituent and medium effects.

Comparison of the Hydrolysis of Phosphate and Carboxylate Esters[69, 73, 74]

The mechanisms of hydrolysis of many types of esters of non-carboxylic
acids have been studied. In this short section we will compare phosphate esters
with carboxylate esters. Phosphate esters have been chosen because they are
of great biochemical significance and because there is a remarkable similarity
in the mechanistic problems which are encountered with those discussed for
carboxylate esters.

The rate of hydrolysis of trimethyl phosphate is increased by base, and
^{18}O studies show that $(MeO)_2P(O)-\!\!\!\!|-O\!\!-\!\!Me$ cleavage occurs.[111] Hydroxide
ion therefore attacks at phosphorus and the reaction is therefore comparable
with $B_{AC}2$ hydrolysis of RCO_2R. It is not known in this case whether the
reaction is synchronous (5-coordinate transition-state as in an S_N2 reaction)
or two-step via an intermediate. As we shall see shortly, there is now strong
evidence for the formation of well defined 5-coordinate intermediates in some
cases, though in some other not dissimilar examples attempts to detect ^{18}O

incorporation in the starting material (the Bender experiment, page 331) have not been successful. In neutral and acid solution hydrolysis, trimethyl phosphate hydrolyses $(MeO)_2P(O)$—O-$\{$-Me cleavage (counterpart of an $_{AL}2$ hydrolysis).

Hydrolyses of di- and mono-esters of phosphoric acid show fairly complex dependence on pH; this is not surprising since they are themselves acids. Monophenyl phosphate shows a rate maximum at pH 4 where the concentration of $PhOP(O)(OH)O^-$ is at a maximum. If the reaction is carried out in $MeOH$–H_2O, the proportion of $MeOP(O)(OH)O^-$ formed is almost equal to the mole fraction of $MeOH$ in the solvent, which suggests that a very reactive unselective intermediate is formed.[112] This and other evidence suggests that the sequence **79** is followed. This mechanism is a close relative of an $_{AC}1$

$$
\text{PhO—P}\begin{matrix}O^-\\ \diagup \\ \diagdown \\ O\end{matrix}\;\underset{OH}{} \;\rightleftharpoons\; \underset{H}{\overset{Ph}{}}\!\!\text{O—P}\begin{matrix}O^-\\ \diagup \\ \diagdown \\ O\end{matrix}\underset{O^-}{} \;\longrightarrow\; PhOH + \;\text{P}\begin{matrix}O^-\\ \diagup \diagdown \\ -O \quad O\end{matrix}\qquad (79)
$$

$$
\underset{\text{MeOH}}{\swarrow}\qquad \underset{\text{H}_2\text{O}}{\downarrow}
$$

$$
\text{MeO—P}\underset{OH}{\overset{O^-}{}}\!\!\!=\!\!O \qquad\qquad \text{HO—P}\underset{OH}{\overset{O^-}{}}\!\!\!=\!\!O
$$

hydrolysis for RCO_2R. For t-butyl phosphate[113] the rate does not reach a maximum at pH 4, but rises rapidly in acid solution; doubtless C—O cleavage to the carbonium ion occurs ($H_2PO_4^-$ should be a quite good leaving group). Here then is a counterpart of an $A_{AL}1$ hydrolysis.

Some very interesting results have come from the study of some five-membered ring phosphates.[74] Thus, the acid hydrolysis of methyl ethylenephosphate is 10^6 times faster than that for trimethyl phosphate and results in both ring-opening and cleavage of the methoxy group (Figure 44a). The ester in Figure 44b also hydrolyses rapidly but only with ring-opening (<0.2% $MeOH$ is formed). These fast hydrolyses must result from some steric stabilization of the transition-state relative to the strained ground-state. This is possible if the five-membered ring spans the equatorial and apical positions of a trigonal-bipyramid intermediate because the 'natural' angle in this situation is 90° and with the long P—O and P—C bonds this is easily attained. This favourable steric effect explains the high rates but not why cleavage of the methoxy occurs in one case but not in the other. To do this we must take a closer look at these trigonal-bipyramid intermediates. The equatorial and apical positions are bonded in different ways to phosphorus (the equatorial bonds use phosphorus sp^2-orbitals, the apical bond pd-hybrids). It appears to be favourable to place electron-attracting groups in apical positions and electron-donating groups in equatorial positions. Thus, in the proton n.m.r. spectrum of Me_2PF_3 there is only one type of methyl group, but the fluorine

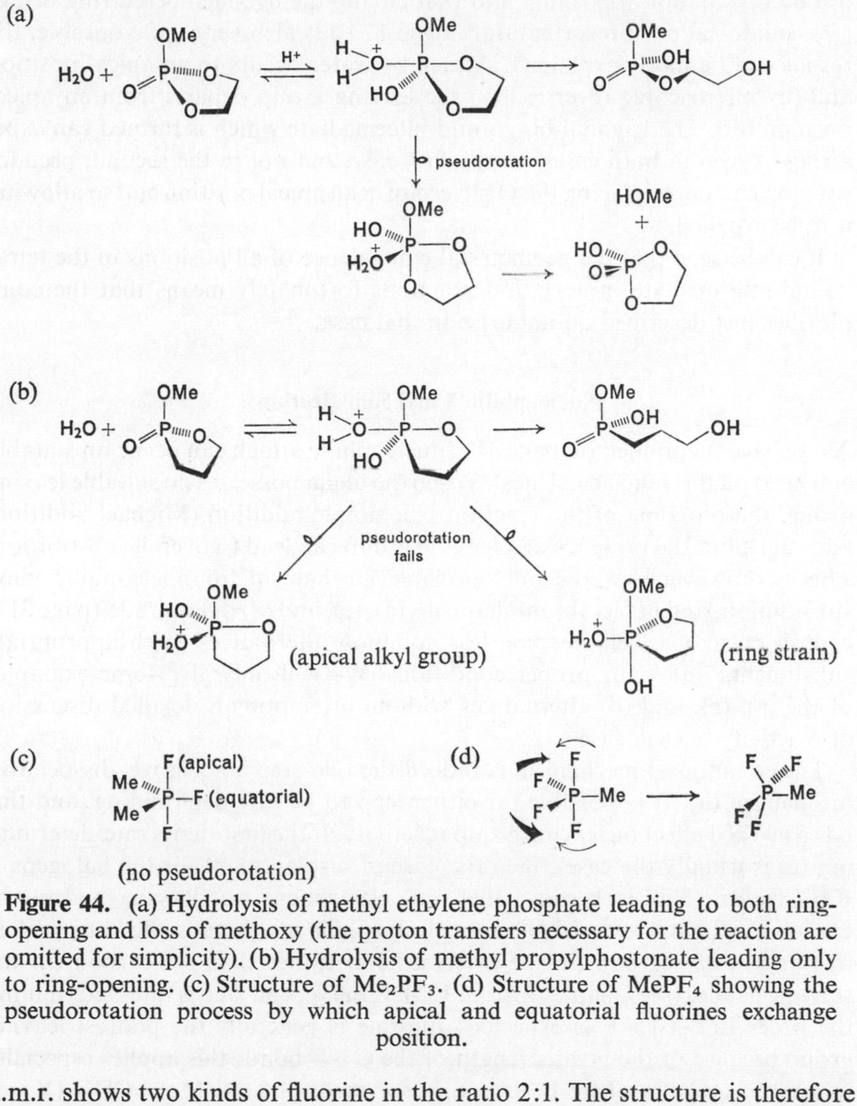

(a)

(b)

(c)

(no pseudorotation)

(d)

Figure 44. (a) Hydrolysis of methyl ethylene phosphate leading to both ring-opening and loss of methoxy (the proton transfers necessary for the reaction are omitted for simplicity). (b) Hydrolysis of methyl propylphostonate leading only to ring-opening. (c) Structure of Me_2PF_3. (d) Structure of $MePF_4$ showing the pseudorotation process by which apical and equatorial fluorines exchange position.

n.m.r. shows two kinds of fluorine in the ratio 2:1. The structure is therefore as shown in Figure 44c. $MePF_4$, however, shows only one kind of fluorine group. This is explained as due to a rapid molecular reorganization called pseudorotation (see Figure 44d), the methyl group remaining in an equatorial position. A pseudorotation in Me_2PF_3, however, puts one methyl into an axial position—apparently this does not occur. If we now return to the ester hydrolyses, and assume that pseudorotation can occur but that carbon cannot occupy an apical position and that having the five-membered ring bridge two equatorial positions ('natural' angle 120°) is also very unfavourable, the results can be nicely explained. Attack by water occurs to an apical position and (by microscopic reversibility) the leaving group departs from an apical position too. The trigonal-bipyramid intermediate which is formed can expel a ring oxygen in both cases. In the first case, but not in the second, pseudorotation can occur placing the OMe group in an apical position and so allowing it to be expelled.

It can be seen that the geometrical equivalence of all positions in the tetrahedral intermediate in acylation reactions fortunately means that the complexities just described do not arise in that case.

Nucleophilic Vinyl Substitution[75]

As we have mentioned on page 311, nucleophilic attack can occur on suitably activated olefins (and acetylenes). When the olefin possesses no suitable leaving group, the outcome of the reaction is a simple addition (Michael addition, page 311), but the presence of a leaving group can lead to overall substitution. This is, however, not the only possible mechanism for nucleophilic vinyl substitution, and in fact the mechanisms (a), (c), and (d) in Figure 35 (page 312) can all occur (as well as some less common alternatives) with appropriate substituents under the proper conditions. We will only give some examples of the (a), (c), and (d) alternatives without attempting a detailed discussion of reactivity in each case.

The commonest mechanism is indeed the two-step associative–dissociative mechanism (a). It is possible for either step to be rate-determining, and this has a marked effect on leaving group reactivity. If the first step is rate-determining (as is usually the case), then the ease of displacement for the halogens is $F \gg Cl \approx Br$. This is because the very electronegative fluorine makes the carbon to which it is attached more electrophilic and because fluorine prefers to bond to sp^3-carbon (in the intermediate) rather than sp^2-carbon (in the starting halide) (see Chapter 2, page 135). If the second step is rate-determining the order $Br > Cl \gg F$ is expected; fluorine is generally the poorest leaving group because of the great strength of the C—F bond; this applies especially in aprotic solvents where F^- is poorly solvated. $PhCOC(Me)=CHX$ (X = F or Cl) reacts readily with piperidine in ethanol at 30°, $k_F/k_{Cl} = 204$. This high ratio strongly suggests a two-step reaction with addition of piperidine rate-determining.[114] The stereochemical outcome of vinyl substitutions depends on the lifetime of, and possible rotations in, the intermediate carbanion. In the example in **80** retention of configuration is the most favoured result; some geometric isomerization of the starting material occurs.[115] In

$$
\begin{array}{c}
\text{(80)}
\end{array}
$$

Ar$_{\prime\prime\prime}\bar{C}$–C$^{\prime\prime\prime}$H with H, Br $\xrightarrow{-Br^-}$ Ar, I / H, H Major kinetically controlled product

60° rotation

Ar$_{\prime\prime\prime}\bar{C}$–C$^{\prime}$Br with H, H $\xrightarrow[\text{rotation}]{120°}$ Ar$_{\prime\prime\prime}\bar{C}$–C$_{\prime\prime\prime}$H with Br, I $\xrightarrow{-Br^-}$ Ar, H / H, I

0° rotation –I⁻

180° rotation

Ar$_{\prime\prime\prime}\bar{C}$–C$^{\prime\prime\prime}$Br with H, I $\xrightarrow{-I^-}$ Ar, H / H, Br

some cases protonation of the carbanionic intermediate occurs and the adduct then undergoes an *E2* or *E1cB* elimination to give overall substitution. In these cases position-isomeric products can arise by alternative elimination routes.

Substitution via a vinyl cation [process (c) of Figure 35] is rather rare. Vinyl cations are relatively high-energy species (see Table 2 in Chapter 2) but stabilized examples can occur in reactions in fairly non-nucleophilic solvents with good leaving groups (see **81**).[116] Elimination–addition mechanisms

$$\text{(81)}$$

Ph, Ph / Ph, OSO$_2$F $\xrightarrow{\text{HOAc}}$ [Ph, Ph / =$\overset{+}{C}$–Ph $\longleftrightarrow$ Ph, Ph / =C–⟨ ⟩+] $\longrightarrow$ Ph, Ph / Ph, OAc

[d in Figure 35] are more common. Thus, reaction of *cis*-1,2-dichloroethylene with sodium toluene-*p*-thiolate is strongly catalysed by EtO⁻ (a stronger base but weaker nucleophile than ArS⁻) and occurs[117] as shown in **82**. Allenic

$$\text{(82)}$$

H, H / Cl, Cl $\xrightarrow[\text{Me–⟨ ⟩–S⁻}]{\text{EtOH, trace EtO⁻}}$ H–C≡C–Cl $\longrightarrow$ H, H / ArS, Cl

↓

H, H / ArS, SAr $\longleftarrow$ ArS–C≡C–H

intermediates can also be formed in suitable cases. In the example shown (83) allene dimerization (see Chapter 4, page 260) is competitive with nucleophilic attack of ethoxide ion.[118]

(83)

Nucleophilic Aromatic Substitution[76]

The same mechanistic alternatives briefly discussed for nucleophilic vinyl substitution have also been recognized for aromatic substitution. The appropriate intermediates are 84, 85, and 86; it is interesting to compare these with

their vinyl counterparts. Although the pentadienyl anion system in 84 is a relatively stable one, its formation requires the loss of aromatic delocalization energy. Thus, other things being equal, 87 should be much more reactive than chlorobenzene. No experimental evidence on this is available but chlorobenzene is certainly very unreactive (see below). Benzyne or dehydrobenzene (85)[77] retains aromaticity but the in-plane non-conjugated π-bond is very weak so we can expect it to be more difficult to form than a normal acetylene intermediate. Finally, phenyl cation (86) is probably also strained relative to an sp-hybridized linear vinyl cation; it is only known as an intermediate in certain reactions of diazonium ions, where the leaving group is molecular nitrogen.

Nucleophilic substitutions of aryl halides possessing electron-withdrawing substituents proceed readily under mild conditions and have been thoroughly studied. The activating group may take several forms (see 88, 89, and 90).

$$\text{(dinitrochlorobenzene)} \xrightarrow{\text{NaN}_3,\ \text{DMF}} \text{(dinitrophenyl azide)} \qquad \textbf{(88)}$$

$$\text{(2-chloro-1-methylbenzimidazole)} + \text{PhSH} \rightleftharpoons \text{(protonated)} + \text{PhS}^- \longrightarrow \text{(2-SPh product)} + \text{Cl}^- \qquad \textbf{(89)}$$

$$\text{(pentafluoronitrobenzene)} \xrightarrow[\text{MeOH}]{\text{NaOMe}} \text{(methoxy product)} \qquad \text{(Ref. 120)} \quad \textbf{(90)}$$

(major isomer produced)

The most thoroughly studied systems are the nitrohalogenobenzenes but many other activating and leaving groups have been studied in the benzene series. Reaction also occurs readily in many π-deficient heteroaromatic systems such as pyridine. The cations derived from such systems are of course still more reactive; in the example shown (89) this effect has the result[119] of making PhSH apparently more reactive than PhS⁻—reaction probably occurs via the cation and PhS⁻.

Polyhalogeno-aromatic systems represent a further type of activation; the different halogen atoms then provide opportunities[120] for the formation of isomeric reaction products, mirroring the situation in electrophilic aromatic substitution. It is probable that all these examples proceed by the two-step associative–dissociative sequence (in this instance called an $S_N\text{Ar}$ reaction), and the activating groups serve further to stabilize the carbanionic intermediate and the related transition-states. Three principal types of evidence have been found for the occurrence of the two-step mechanism as against a one-step synchronous substitution. The first is that in examples closely related to the reactions under study, stable salts of the intermediate anions (Meisenheimer complexes) have been isolated.[121] Thus, 2,4,6-trinitroanisole with KOMe yields the salt 91, whose structure has been established by X-ray diffraction.[122]

$$\textbf{(91)}$$

The second line of evidence concerns the type of leaving group effect discussed under nucleophilic vinyl substitution. The relative rates of reaction[123] of the 1-halogeno-2,4-dinitrobenzenes with piperidine in MeOH at 0° are: F 3300; Cl 4.3; Br 4.3; I 1. As explained on page 340, this is *not* the sort of order expected if the halogen–carbon bond is breaking at the transition-state, for fluorine is the poorest leaving group. This reactivity order (which occurs in many other instances) is, however, the one to be expected if the first, associative, step in the two-step mechanism is rate-determining. Occasionally, the second step may be rate-determining, and this results in a quite different leaving group reactivity order. Thus, for reaction of the 1-halogeno-2,4-

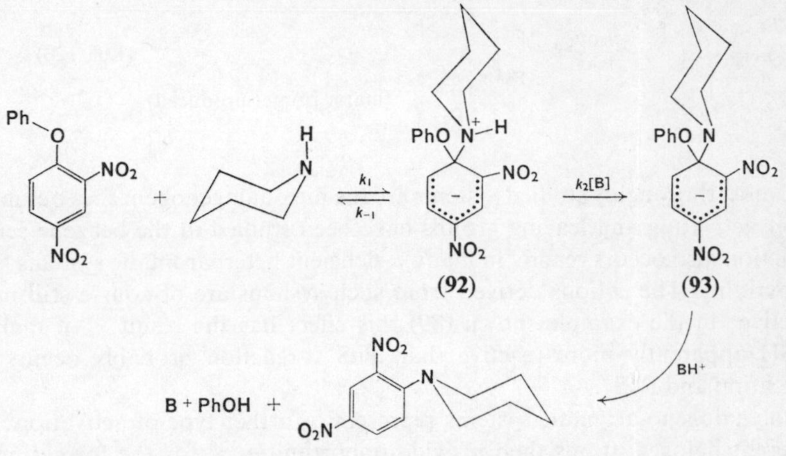

Figure 45. Scheme for the base-catalysed reaction of phenyl 2,4-dinitrophenyl ether with piperidine.

dinitrobenzenes[124] with PhNHMe in nitrobenzene at 120° the relative rates are: F 1; Cl 15; Br 46. In this case the intermediate decomposes to starting materials more frequently than it goes on to products.

The third and clearest line of evidence for the two-step mechanism comes from studies of base catalysis of these reactions. When the slow step is the first one, base catalysis is usually only slight, but when the second step is rate-controlling, base catalysis can be quite powerful. This is the case with the reaction shown in Figure 45, because loss of phenoxide from **93** occurs much more readily than from **92** (the final step may be acid-catalysed as shown). However, the base catalysis will only increase the reaction rate up to the point where the second step has been so accelerated that the first step becomes the slower and thus rate-limiting.

Bunnett and Garst[125] found that the reaction in 60% dioxan–40% water was strongly catalysed by OH⁻, but that above a certain concentration further base was relatively ineffective, just as required by the two-step mechanism. Thus, an intermediate is required and this is probably **92**. This is made more likely by the isotope effect study of Hart and Bourns.[126] They found that at low OH⁻ concentration there was a sizeable $^{16}O/^{18}O$ isotope effect for labelling the ether oxygen (a primary isotope effect due to C—O bond-cleavage in the rate-determining step), but at high base concentrations (first step rate-determining) the isotope effect became negligible.

If the addition–elimination mechanism is well established for activated halogeno-aromatics and related compounds, it is interesting to enquire how far its range extends to unactivated halides. The rates of reaction of a series of *para*-substituted chlorobenzenes ranging from *p*-nitro to *p*-methyl with piperidine in diethylene glycol at 194.5° (see **94**) yield a good plot against σ^-

$$(94)$$

with $\rho = +4.41$.[127] This means that the same mechanism is followed throughout the series (note the strongly positive ρ-value). Since the addition–elimination sequence is well established for the *p*-nitro-compound, it must occur throughout the sequence with this relatively weakly basic nucleophile. When more powerfully basic reagents such as sodamide, lithium piperidide, or phenyl-lithium are used on deactivated halides, different phenomena appear and we enter the area of benzyne chemistry.

Substitution via Benzyne Intermediates[77]

Although Wittig had earlier suggested benzyne as an intermediate, Roberts in 1953 provided the first decisive proof of its formation (see **95**).[128] In reactions such as this, benzyne is formed by a two-step *E1cB* reaction and consumed by the reverse of this process. In the benzyne formation both steps are reversible. *o*-Fluorophenyl anion is much more stable than the other *o*-halogenophenyl anions because of the powerful inductive effect of fluorine. However, for the second step (loss of halide ion) the reactivity order seems to be I⁻ > Br⁻ > Cl⁻ > F⁻. Thus, fluorobenzene in liquid ammonia with KNH_2 does not

12*

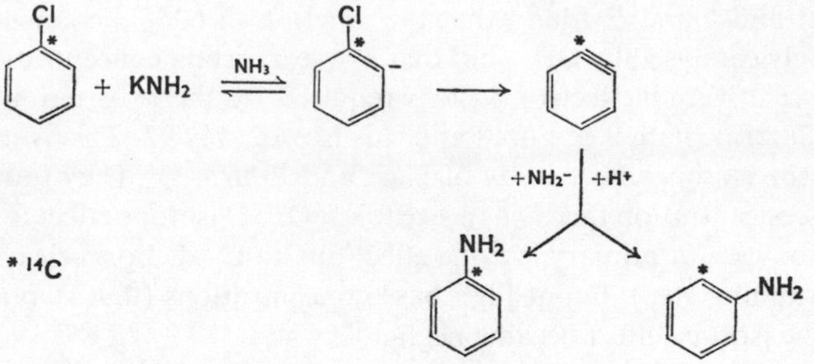

(95)

* 14C

form aniline, therefore no benzyne can be formed, but *o*-deuteriofluorobenzene loses its deuterium, so *o*-fluorophenyl anions are formed. With bromo- and iodo-benzene, however, anion formation is often the slow step.

Numerous other methods have been found to generate benzyne. Precursors such as **96**,[129] **97**,[130] and **98**[131] are particularly useful because the generation of benzyne from them does not require strongly basic conditions and a wider range of reactions of benzynes can be studied.

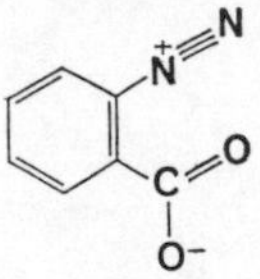

(96) 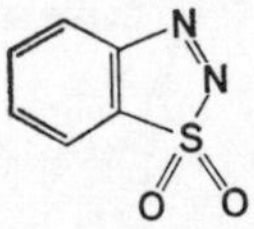   (97)

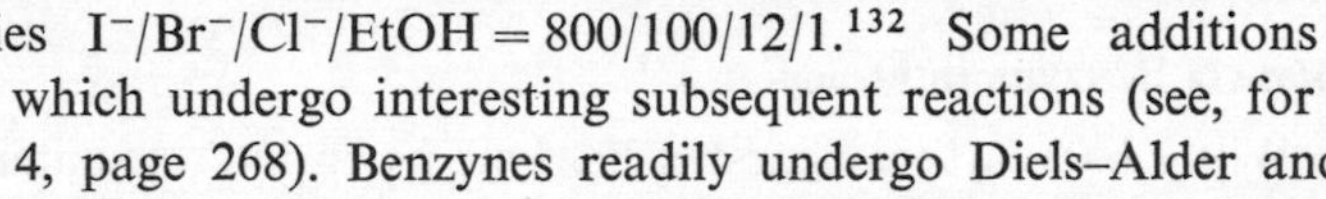

(98)

Benzynes react readily with almost any nucleophile, but they are far from unselective in their reactions. Thus, benzyne from **97** showed the following relative reactivities $I^-/Br^-/Cl^-/EtOH = 800/100/12/1$.[132] Some additions lead to betaines which undergo interesting subsequent reactions (see, for example, Chapter 4, page 268). Benzynes readily undergo Diels–Alder and 'ene'[130] reactions (see **99** and **100**); the former are frequently used to detect benzyne formation using a diene such as tetraphenylcyclopentadienone or furan. 2+2-Additions also occur but are not stereospecific and probably involve diradical intermediates (see **101**).[133]

Benzyne-like intermediates have also been observed in heterocyclic systems. Thus, 3,4-pyridyne is formed from 3-chloropyridine and potassium amide in liquid ammonia.[134, 135]

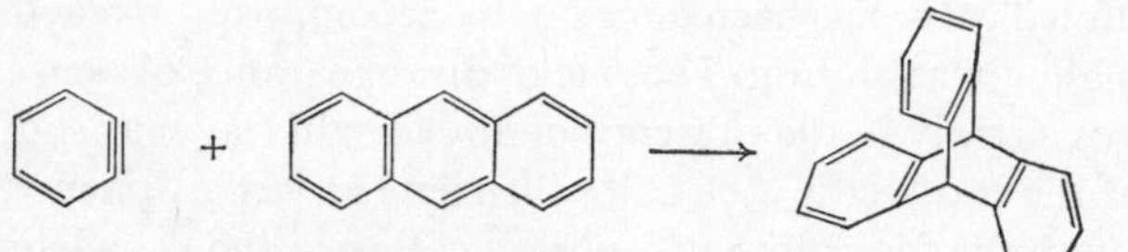

(99)

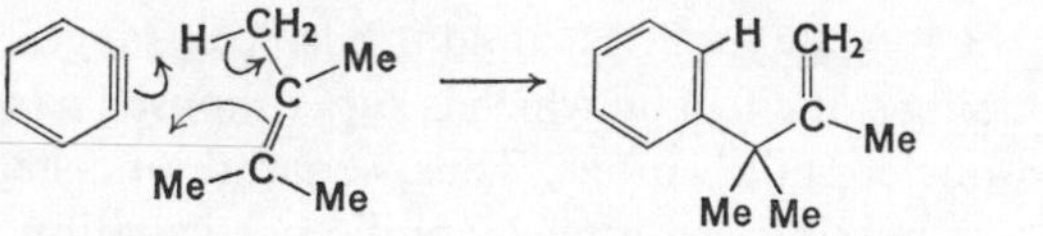

(100)

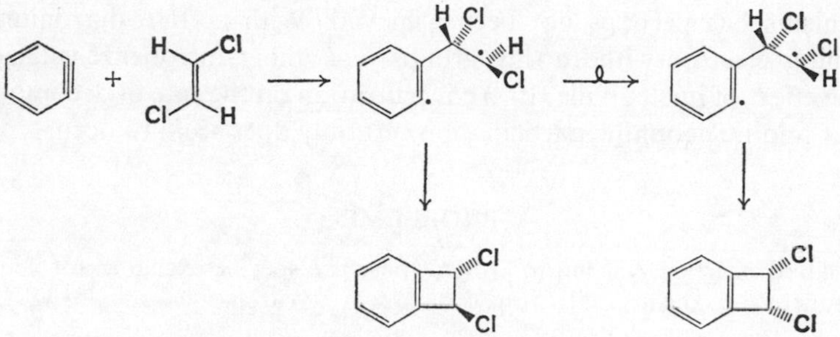

(101)

Reactions via Phenyl Cations

The only reactions for which it has been suggested that phenyl cations are intermediates are ones involving diazonium ions. Many reactions of diazonium ions (for example, those involving copper catalysis such as Sandmeyer reactions) are undoubtedly initiated by one-electron transfer leading to phenylazo and to phenyl radicals. Certain reactions do seem to involve heterolytic mechanisms, however. Decomposition of benzenediazonium tetrafluoroborate in chlorobenzene,[136] diphenyl ether,[136] or tetrahydrothiophen 1,1-dioxide (sulpholan)[137] yields the 'onium salts 102, 103, and 104. A reasonable

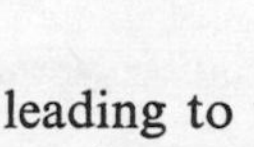

(102) 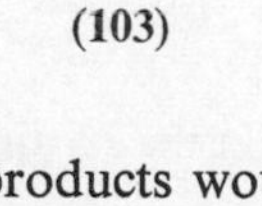   (103) 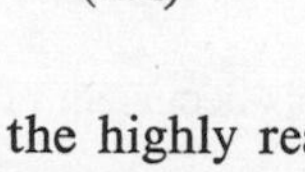   (104)

intermediate leading to these products would be the highly reactive phenyl cation. However, it must be said that reaction occurs in homogeneous solution only in the last example.

More detailed work has been done on the decomposition reaction of diazonium salts in aqueous solution. The rate of disappearance of benzenediazonium ion decreases slightly in the presence of added salts, such as sodium chloride, but some of the product is then chlorobenzene as well as phenol.[138] This is in agreement with the formation of a phenyl cation in the rate-determining step, which, in subsequent fast steps, is captured either by H_2O or by Cl^-. However, a study of a range of salts shows that the rate decreases observed are in the order $HSO_4^- > Cl^- > Br^- > SCN^-$, and it has been suggested that this is due to a rate-depressing salt effect on which is superimposed a (small) rate increase with the more nucleophilic anions. Thus, separate one-step pathways to the two products are possible, with nucleophilic participation by H_2O and, say, Cl^-. However, the effects are so small that a clear distinction between the two mechanisms has perhaps not been achieved. With certain diazonium ions containing electron-withdrawing groups (e.g. the p-nitrobenzenediazonium ion) the effect of nucleophile (Br^-) concentration on the rate of decomposition is larger and nucleophilic participation certainly does seem to occur.

PROBLEMS

1. If the hydrocarbon A is left in D_2SO_4, there is a specific exchange for deuterium of hydrogens β to the aromatic ring.

(A)

Suggest a mechanism by which this might occur. (Psst! consider possible reversible ring-opening routes.)

2. Treatment of $[1,1\text{-}^{14}C_2]$biphenyl **(B)** with aluminium chloride and a trace of water at 100° results in randomization of the label although one radioactive atom is maintained in each ring.

(B)

How does this occur, and how might the starting material be synthesized?

3. Dialkylanilines normally react very rapidly with diazonium salts to produce *para*-substituted azo-coupling products. Explain therefore why, under conditions where dimethylaniline reacts rapidly, **C** reacts very slowly and **D** gives no observable coupling product.

(C)   (D)

4. Cycloocta-1,5-diene reacts with sulphur dichloride in benzene to give one predominant product of formula $C_8H_{12}SCl_2$ which on successive treatment with hydrogen peroxide, sodium thiomethoxide, and hydrogen (in the presence of Raney Nickel) gives a compound $C_8H_{14}SO_2$ with anomalous C—H stretching and bending bands in the infrared spectrum at 2990 and 1490 cm^{-1} also exhibited by bicyclo[3.3.1]-nonane, and additionally a very strong band at 1290 cm^{-1}.

Suggest a likely structure for the addition product and a mechanism for its formation.

[See *J. Org. Chem.*, **31**, 1669 (1966).]

5. In additions to norbornene (**E**) and 7,7-dimethylnorbornene (**F**), the following stereochemical data are available.

[*J. Amer. Chem. Soc.*, **92**, 201 (1970).]

(E)   (F)

Reagent	exo/endo Ratios	
	(E)	(F)
Hydroboration–oxidation	200/1	1/4
Epoxidation	200/1	1/9
Mercuric acetate	>200/1	>200/1
Photoaddition of PhSH	200/1	20/1

Explain these trends by reference to the relevant discussion in this chapter on mechanisms of addition to olefins.

6. (a) Following the discussion in Chapter 5, consider whether the electronic effects revealed by Figure 25 of Chapter 1 for the formation of semicarbazones from substituted benzaldehydes are reasonable. (*Hint*: for those cases where the dehydration step is rate-determining the total effect will be a composite one.)

(b) The rate of reaction of 3×10^{-4}M-pyruvate anion (MeCOCOO$^-$) with hydroxylamine at pH 6.9 rises at first with increasing hydroxylamine concentration but reaches a plateau at [NH$_2$OH] = 0.1M. Explain this.

[See page 72 of ref. 62.]

7. The rates of borohydride reduction of some cyclic ketones relative to di-n-hexyl ketone (=1) are: C_4, 581; C_5, 15.4; C_6, 355; C_8, 0.172; C_{10}, 0.0291; C_{12}, 0.401; C_{17}, 1.31.

Explain the trends in these figures and arrange the following compounds in order of increasing ease of reduction: norbornan-2-one, norbornan-7-one, and camphor.

8. In the ester interchange reaction:

$$RO^- + MeCO_2R' \rightleftharpoons MeC(OR)(OR')O^- \rightarrow MeCO_2R + R'O^-$$

which step is likely to be rate-determining if the pK_a of ROH is greater than that of R'OH? Using 1 mole of RO^-, where is the final equilibrium likely to lie in this case?

9. Hydrolysis of phenyl N-methylacetimidate, $MeC(OPh)NMe$, can yield either phenol and N-methylacetamide or methylamine and phenyl acetate. At pH 7 only phenol (and the amide) is formed, at pH 3—5 ~90% phenol and 10% methylamine result, and below pH 1 only methylamine is formed.

Account for these results by considering a scheme involving tetrahedral intermediates with total charge −1, 0, and +1.

[See W. P. Jencks and M. Gilchrist, *J. Amer. Chem. Soc.*, **90**, 2622 (1968).]

10. (a) In 4N-NaOH at 340°C, [1-^{14}C]chlorobenzene gives 58% 1-^{14}C-labelled and 42% 2-^{14}C-labelled phenol.

Suggest mechanisms to account for this distribution of the label.

[See page 29 of ref. 77.]

(b) In the addition of amide ion to 4-chlorobenzyne, 3- and 4-chloroaniline are formed in the ratio 1/4.2. Explain this.

[See page 141 of ref. 77.]

REFERENCES

Reviews and General Discussions of Aromatic Substitution

1. L. M. Stock, *Aromatic Substitution Reactions*, Prentice-Hall, New York, 1968, pp. 1–81. A general mechanistic discussion of electrophilic substitution.
2. R. O. C. Norman and R. Taylor, *Electrophilic Substitution in Benzenoid Compounds*, Elsevier, Amsterdam, 1965. A clear and systematic exposition of the field.
3. P. B. de la Mare and J. H. Ridd, *Aromatic Substitution, Nitration and Halogenation*, Butterworths, London, 1959. A detailed analysis of earlier work in the field.
4. *Friedel-Crafts and Related Reactions* (ed. G. A. Olah), Wiley, New York, 1963. An exhaustive six-volume review of alkylation, acylation, and related reactions.
5. E. Baciocchi and G. Illuminati, *Progr. Phys. Org. Chem.*, **5**, 1 (1967). Mechanisms of electrophilic substitution in polyalkylbenzenes.

6. H. H. Perkampus, *Adv. Phys. Org. Chem.*, **4**, 195 (1966). A review of physico-chemical studies on σ- and π-complexes.

7. R. Taylor, *Chimia*, **22**, 1 (1968). A review of recent advances in the understanding of electrophilic substitution.

8. L. M. Stock and H. C. Brown, *Adv. Phys. Org. Chem.*, **1**, 35 (1963). Selectivity relationships.

9. J. H. Ridd in *Aromaticity*, Chem. Soc. Special Publ. No. 21, 1967. Electrophilic substitution in deactivated systems.

10. G. A. Olah in *Organic Reaction Mechanisms*, Chem. Soc. Special Publ. No. 19, 1965, p. 22. Olah's earlier views on 'anomalous' substitutions.

11. D. H. Hey, *Adv. Free-Radical Chem.*, **2**, 47 (1967). A review of free-radical arylation.

12. I. C. Calder, P. J. Garratt, H. C. Longuet-Higgins, F. Sondheimer, and R. Wolovsky, *J. Chem. Soc.* (C), **1967**, 1041.

13. H. C. Brown and J. J. Melchiore, *J. Amer. Chem. Soc.*, **87**, 5269 (1965).

14. E. L. Mackor, A. Hofstra, and J. H. van der Waals, *Trans. Faraday Soc.*, **54**, 186 (1958).

15. E. M. Arnett and J. W. Larsen in *Carbonium Ions* (ed. G. A. Olah and P. von R. Schleyer), Interscience, New York, 1968, p. 441 ff.; *idem, J. Amer. Chem. Soc.*, **91**, 1438 (1969).

16. W. von E. Doering, M. Saunders, H. G. Boyton, H. W. Earhart, E. F. Wadley, W. R. Edwards, and G. Laber, *Tetrahedron*, **4**, 178 (1958).

17. D. Bethell, V. Gold, and T. Riley, *J. Chem. Soc.*, **1959**, 3134.

18. C. Perrin and F. H. Westheimer, *J. Amer. Chem. Soc.*, **85**, 2773 (1963).

19. E. Baciocchi, G. Illuminati, G. Sleiter, and F. Stegel, *J. Amer. Chem. Soc.*, **89**, 125 (1967).

20. G. A. Olah, S. J. Kuhn, S. H. Flood, and B. A. Hardie, *J. Amer. Chem. Soc.*, **86**, 2203 (1964).

21. J. P. Colpa, C. MacLean, and E. L. Mackor, *Tetrahedron*, Suppl. 2, 65 (1963).

22. G. A. Olah and N. A. Overchuk, *Can. J. Chem.*, **43**, 3279 (1965), and references therein.

23. R. G. Coombes, R. B. Moodie, and K. Schofield, *J. Chem. Soc.* (B), **1968**, 800; J. G. Hoggett, R. B. Moodie, and K. Schofield, *ibid.*, **1969**, 1.

24. See C. K. Ingold, *Structure and Mechanism in Organic Chemistry*, 2nd Edn., Bell, London, 1969, p. 320 ff.

25. See A. Streitwieser, Jr., *Molecular Orbital Theory for Organic Chemists*, Ch. 12, Wiley, New York, 1961.

26. C. Eaborn, P. Golborn, R. E. Spillett, and R. Taylor, *J. Chem. Soc.* (B), **1968**, 1112.

27. C. Rüchardt, *Angew. Chem. Internat. Edn. Engl.*, **4**, 964 (1965).

28. R. Ito, T. Migita, N. Morikawa, and O. Simamura, *Tetrahedron*, **21**, 955 (1965).

Reviews and General Discussions on Additions to Unsaturated Systems

29. *The Chemistry of Alkenes* (ed. S. Patai), Wiley-Interscience, London, 1964. Contains chapters on many aspects of olefin chemistry but unfortunately none on electrophilic reactions. See particularly the chapter on free-radical chemistry by J. I. G. Cadogan and M. J. Perkins, p. 585.

30. P. B. de la Mare and R. Bolton, *Electrophilic Additions to Unsaturated Systems*, Elsevier, Amsterdam, 1966. A systematic review.
31. H. C. Brown, *Hydroboration*, Benjamin, New York, 1962. A lucid account, mostly of the author's own work.
32. R. C. Fahey, *Topics Stereochem.*, **3**, 237 (1968). A very valuable review on stereochemical aspects of olefin addition reactions.
33. W. R. Dolbier, *J. Chem. Educ.*, **46**, 342 (1969). Mechanisms of olefin addition reactions.
34. M. J. S. Dewar and R. C. Fahey, *Angew. Chem. Internat. Edn. Engl.*, **3**, 245 (1964). Stereochemistry and mechanism in HBr additions.
35. C. S. Foote, *Accounts Chem. Res.*, **1**, 104 (1968). A review of singlet oxygen reactions, including those with olefins. See also W. Fenical, D. R. Kearns, and P. Radlick, *J. Amer. Chem. Soc.*, **91**, 7771 (1969).
36. R. W. Murray, *Accounts Chem. Res.*, **1**, 313 (1968). Mechanisms of ozonolysis.
37. T. G. Traylor, *Accounts Chem. Res.* **2**, 152 (1969). Additions to strained olefins.
38. J. I. G. Cadogan, *Pure Appl. Chem.*, **15**, 153 (1967). Radical additions to olefins. This topic is also covered in C. H. Walling, *Free Radicals in Solution*, Wiley, New York, 1957.
39. Data obtained from enthalpies gathered by S. W. Benson *et al.*, *Chem. Rev.*, **69**, 279 (1969).
40. P. D. Bartlett and G. D. Sargent, *J. Amer. Chem. Soc.*, **87**, 1297 (1965).
41. I. Roberts and G. E. Kimball, *J. Amer. Chem. Soc.*, **59**, 947 (1937).
42. J. Strating, J. H. Wieringa, and H. Wynberg, *Chem. Comm.*, **1969**, 907.
43. G. A. Olah, J. M. Bollinger, and J. Brinich, *J. Amer. Chem. Soc.*, **90**, 2587 (1968).
44. J. E. Dubois and G. Mouvier, *Bull. Soc. Chim. France*, **1968**, 1426; G. Mouvier and J. E. Dubois, *ibid.*, p. 1441.
45. Hydroxymercuration: J. Halpern and H. B. Tinker, *J. Amer. Chem. Soc.*, **89**, 6427 (1967); Thallium(III) oxidation: P. M. Henry, *ibid.*, **87**, 4423 (1965); Acid-catalysed hydration: private communication by R. W. Taft cited in ref. 40.
46. R. T. Dillon, W. G. Young, and H. J. Lucas, *J. Amer. Chem. Soc.*, **52**, 1953 (1930); **51**, 2528 (1929).
47. R. Huisgen and G. Boche, *Tetrahedron Letters*, **1965**, 1769.
48. H. Kwart, L. Kaplan, and P. von R. Schleyer, *J. Amer. Chem. Soc.*, **82**, 2341 (1960).
49. M. J. S. Dewar and R. C. Fahey, *J. Amer. Chem. Soc.*, **85**, 3645 (1963).
50. Ref. 34, and M. J. S. Dewar and R. C. Fahey, *J. Amer. Chem. Soc.*, **85**, 2245 (1963).
51. S. Wolfe, private communication quoted in ref. 32, p. 248.
52. H. C. Brown and A. W. Moerikofer, *J. Amer. Chem. Soc.*, **85**, 2063 (1963).
53. E. W. Garbisch, Jr., S. M. Schildcrout, D. B. Patterson, and C. M. Sprecher, *J. Amer. Chem. Soc.*, **87**, 2932 (1965).
54. L. Ghosez, P. Laroche, and L. Bastens, *Tetrahedron Letters*, **1964**, 3745.
55. Data taken from: K. D. Bingham, G. D. Meakins, and G. H. Whitham, *Chem. Comm.*, **1966**, 445; A. S. Bailey and J. E. White, *J. Chem. Soc.* (B), **1966**, 819; R. E. Erickson and R. L. Clark, *Tetrahedron Letters*, **1969**, 3997.
56. W. E. Parham and E. E. Schweizer, *Org. Reactions*, **13**, 55 (1963).
57. M. S. Kharasch and F. R. Mayo, *J. Amer. Chem. Soc.*, **55**, 2468 (1933).
58. Intermolecular examples from refs. 29 and 38; intramolecular example: M. Julia and M. Maumy, *Bull. Soc. Chim. France*, **1969**, 2415.

59. C. Moussebois and J. Dale, *J. Chem. Soc.* (C), **1966**, 260.
60. P. D. Readio and P. S. Skell, *J. Org. Chem.*, **31**, 753 (1966).
61. P. S. Skell and R. G. Allen, *J. Amer. Chem. Soc.*, **81**, 5383 (1959).

Reviews and General References for Nucleophilic Associative Reactions

62. W. P. Jencks, *Progr. Phys. Org. Chem.*, **2**, 63 (1964). An excellent discussion of mechanism and catalysis in simple carbonyl group reactions. W. P. Jencks, *Catalysis in Chemistry and Enzymology*, McGraw-Hill, New York, 1969.
63. R. P. Bell, *Adv. Phys. Org. Chem.*, **4**, 1 (1966). Reversible hydration of carbonyl compounds.
64. E. H. Cordes, *Progr. Phys. Org. Chem.*, **4**, 1 (1967). Mechanism and catalysis in the hydrolysis of acetals and orthoesters.
65. M. L. Bender, *Chem. Rev.*, **60**, 53 (1960). Mechanisms of catalysis of the nucleophilic reactions of carboxylic acid derivatives.
66. S. L. Johnson, *Adv. Phys. Org. Chem.*, **5**, 237 (1967). General base and nucleophilic catalysis of ester hydrolysis and related reactions.
67. S. I. Miller, *Adv. Phys. Org. Chem.*, **6**, 185 (1968). An interesting discussion of the intermediates in nucleophilic substitution at unsaturated carbon is on page 265.
68. D. P. N. Satchell, *Quart. Rev.*, **17**, 160 (1963). A brief survey of acylation reactions.
69. T. C. Bruice and S. J. Benkovic, *Bio-organic Mechanisms*, Benjamin, New York, 1966. Discussion of acyl transfer reactions in Vol. 1; phosphate esters in Vol. 2.
70. C. D. Gutsche, *The Chemistry of Carbonyl Compounds*, Prentice-Hall, Englewood Cliffs, New Jersey, 1967. H. O. House, *Modern Synthetic Reactions*, Benjamin, New York, 1965. These two books contain much relevant information, especially on synthetically important carbonyl reactions.
71. A. T. Neilsen and W. J. Houlihan, *Org. Reactions*, **16**, 1 (1968). Contains a section on the mechanism of aldol condensations.
72. A. Maercker, *Org. Reactions*, **14**, 270 (1965); S. Trippett, *Quart. Rev.*, **17**, 406 (1963). Two reviews on the Wittig reaction.
73. A. J. Kirby and S. G. Warren, *The Organic Chemistry of Phosphorus*, Elsevier, Amsterdam, 1967. Discusses the hydrolysis of phosphate esters.
74. F. H. Westheimer, *Accounts Chem. Res.*, **1**, 70 (1968). Pseudorotation in phosphate ester intermediates.
75. Z. Rappoport, *Adv. Phys. Org. Chem.*, **7**, 1 (1969). Nucleophilic vinyl substitution.
76. J. Miller, *Aromatic Nucleophilic Substitution*, Elsevier, Amsterdam, 1968; F. Pietra, *Quart. Rev.*, **23**, 504 (1969). A book and a review on this subject.
77. R. W. Hoffmann, *Dehydrobenzene and Cycloalkynes*, Academic Press, New York, 1967. A good book on benzynes.
78. A. Lapworth, *J. Chem. Soc.*, **83**, 995 (1903); **85**, 1206 (1904).
79. J. O. Edwards and R. G. Pearson, *J. Amer. Chem. Soc.*, **84**, 16 (1962).
80. J. F. Bunnett, *J. Amer. Chem. Soc.*, **79**, 5969 (1957).
81. A. Albert and W. L. F. Armarego, *Adv. Heterocyclic Chem.*, **4**, 1 (1965).
82. D. Samuel and B. L. Silver, *Adv. Phys. Org. Chem.*, **3**, 123 (1965).
83. T. H. Fife and L. K. Jao, *J. Org. Chem.*, **30**, 1492 (1965).

84. V. J. Shiner and D. Whittaker, *J. Amer. Chem. Soc.*, **91**, 394 (1969).
85. H. C. Brown and H. R. Deck, *J. Amer. Chem. Soc.*, **87**, 5620 (1965).
86. D. J. Cram and F. A. Abd Elhafez, *J. Amer. Chem. Soc.*, **74**, 5828 (1952); D. J. Cram and F. D. Greene, *ibid.*, **75**, 6005 (1953).
87. M. Chérest, H. Felkin, and N. Prudent, *Tetrahedron Letters*, **1968**, 2199, 2205.
88. H. O. House and W. F. Fischer, *J. Org. Chem.*, **33**, 949 (1968).
89. R. P. Bell and M. J. Smith, *J. Chem. Soc.*, **1958**, 1691.
90. E. Coombs and D. P. Evans, *J. Chem. Soc.*, **1940**, 1295; D. S. Noyce, W. A. Pryor, and A. Bottini, *J. Amer. Chem. Soc.*, **77**, 1402 (1955).
91. D. S. Noyce and W. A. Pryor, *J. Amer. Chem. Soc.*, **77**, 1397 (1955).
92. C. D. Gutsche and D. Redmore, *Carbocyclic Ring Expansion Reactions*, Academic Press, New York, 1968, Ch. 4.
93. E. J. Corey and M. Chaykovsky, *J. Amer. Chem. Soc.*, **87**, 1353 (1965).
94. R. F. Hudson, *Structure and Mechanism in Organophosphorus Chemistry*, Academic Press, London, 1965, Ch. 7.
95a. M. Schlosser and K. F. Christmann, *Ann. Chem.*, **708**, 1 (1967); E. J. Corey and H. Yamamoto, *J. Amer. Chem. Soc.*, **92**, 226, 3523 (1970).
95b. W. P. Schneider, *Chem. Comm.*, **1969**, 785.
96. C. K. Ingold, *Structure and Mechanism in Organic Chemistry*, Cornell University Press, Ithaca, N.Y., 1953, p. 754.
97. B. Holmquist and T. C. Bruice, *J. Amer. Chem. Soc.*, **91**, 2993 (1969).
98. A. R. Olson and R. J. Miller, *J. Amer. Chem. Soc.*, **60**, 2687 (1968); Ch. 1, p. 21.
99. A. G. Davies and J. Kenyon, *Quart. Rev.*, **9**, 203 (1955).
100. D. N. Kursanov and R. V. Kudryavtsev, *J. Gen. Chem.* (U.S.S.R.), (Engl. Transl.), **26**, 1183 (1956).
101. M. L. Bender., *J. Amer. Chem. Soc.*, **73**, 1626 (1951).
102. R. W. Taft in *Steric Effects in Organic Chemistry* (ed. M. S. Newman), Wiley, New York, 1956.
103. B. Capon, *Quart. Rev.*, **18**, 45 (1964).
104. M. L. Bender and M. S. Silver, *J. Amer. Chem. Soc.*, **84**, 4589 (1962).
105. K. Yates and R. A. McClelland, *J. Amer. Chem. Soc.*, **89**, 2686 (1967).
106. M. S. Newman, *J. Amer. Chem. Soc.*, **63**, 2431 (1941).
107. D. A. Brown and R. F. Hudson, *J. Chem. Soc.*, **1953**, 883; R. F. Hudson and J. E. Wardill, *ibid.*, **1950**, 1729; E. W. Crunden and R. F. Hudson, *ibid.*, **1956**, 501.
108. R. A. Sneen and J. W. Larsen, *J. Amer. Chem. Soc.*, **91**, 6031 (1969).
109. C. A. Bunton, B. Nayak, and C. O'Connor, *J. Org. Chem.*, **33**, 572 (1968); C. O'Conner, *Quart. Rev.*, **24**, 553 (1970).
110. S. O. Eriksson and C. Holst, *Acta Chem. Scand.*, **20**, 1892 (1966).
111. P. W. C. Barnard, C. A. Bunton, D. R. Llewellyn, K. G. Oldham, B. L. Silver, and C. A. Vernon, *Chem. and Ind.*, **1955**, 760; C. A. Bunton, *Accts. Chem. Res.*, **3**, 257 (1970).
112. J. D. Chanley and E. Feageson, *J. Amer. Chem. Soc.*, **85**, 1181 (1963).
113. A. Lapidot, D. Samuel and M. Weiss-Broday, *J. Chem. Soc.*, **1864**, 637.
114. P. Beltrame, G. Favini, M. G. Cattania, and F. Guella, *Gazzetta*, **98**, 380 (1968).
115. S. I. Miller and P. K. Yonan, *J. Amer. Chem. Soc.*, **79**, 5931 (1957).
116. W. M. Jones and D. D. Maness, *J. Amer. Chem. Soc.*, **91**, 4314 (1969); see also M. Hanack, *Accounts Chem. Res.*, **3**, 209 (1970).
117. W. E. Truce, M. M. Boudakian, R. F. Heine, and R. J. McManimie, *J. Amer. Chem. Soc.*, **78**, 2743 (1956).

118. P. Beltrame, D. Pitea, A. Marzo, and M. Simonetta, *J. Chem. Soc.* (B), **1967**, 71.
119. A. Ricci and P. Vivarelli, *J. Chem. Soc.* (B), **1968**, 1280.
120. J. G. Allen, J. Burdon, and J. C. Tatlow, *J. Chem. Soc.*, **1965**, 1045.
121. M. R. Crampton, *Adv. Phys. Org. Chem.*, **7**, 211 (1969).
122. H. Ueda, N. Sakabe, J. Tanaka, and A. Furusaki, *Bull. Chem. Soc. Japan*, **41**, 2866 (1968).
123. J. F. Bunnett, E. W. Garbisch, and K. M. Pruitt, *J. Amer. Chem. Soc.*, **79**, 385 (1957).
124. G. S. Hammond and L. R. Parks, *J. Amer. Chem. Soc.*, **77**, 340 (1955).
125. J. F. Bunnett and R. H. Garst, *J. Amer. Chem. Soc.*, **87**, 3879 (1965).
126. C. R. Hart and A. N. Bourns, *Tetrahedron Letters*, **1966**, 2995.
127. C. L. Liotta and D. F. Pinholster, *Chem. Comm.*, **1969**, 1245.
128. J. D. Roberts, H. E. Simmons, L. A. Carlsmith, and C. W. Vaughan, *J. Amer. Chem. Soc.*, **75**, 3290 (1953).
129. M. Stiles, R. G. Miller, and U. Burckhardt, *J. Amer. Chem. Soc.*, **85**, 1792 (1963); L. Friedman and F. M. Logullo, *ibid.*, p. 1549.
130. G. Wittig and R. W. Hoffmann, *Chem. Ber.*, **95**, 2718 (1962).
131. C. D. Campbell and C. W. Rees, *Proc. Chem. Soc.*, **1964**, 296.
132. G. Wittig and R. W. Hoffmann, *Chem. Ber.*, **95**, 2729 (1962).
133. M. Jones and R. H. Levin, *J. Amer. Chem. Soc.*, **91**, 6411 (1969).
134. J. A. Zoltewicz and C. L. Smith, *Tetrahedron*, **25**, 4331 (1969).
135. Th. Kauffmann, *Angew. Chem. Internat. Edn. Engl.*, **4**, 543 (1965).
136. A. N. Nesmayanov, L. G. Makarova, and T. P. Tolstaya, *Tetrahedron*, **1**, 145 (1957).
137. G. R. Chalkley, D. J. Snodin, G. Stevens, and M. C. Whiting, *J. Chem. Soc.* (C), **1970**, 682.
138. E. S. Lewis, L. D. Hartung, and B. H. McKay, *J. Amer. Chem. Soc.*, **91**, 419 (1969).

Author Index

Abd Elhafez, F. A., 319, 222
Abrahamson, E. W., 245
Abraitays, V. Y., 257
Adams, R., 105
Albert, A., 315
Albery, W. J., 120
Alden, R. A., 126
Alder, K., 240
Alexander, R., 49
Allen, J. G., 343
Allen, L. C., 84
Allen, R. G., 310
Allred, E., 98
Almy, J., 184
Andersen, K. K., 34
Angus, J. C., 211
Appel, B., 86, 87
Armarego, W. L. F., 315
Arnett, E. M., 40, 47, 78, 81, 82, 283
Arnold, D. R., 257
Ashby, E. C., 120, 144
Asperger, S., 214
Atkinson, J. G., 129
Ayrey, G., 214
Ayscough, P. B., 155

Baciocchi, E., 280, 287
Baddeley, G., 93
Bailey, A. S., 307
Baird, R., 111
Baker, E. B., 128
Baker, R., 86, 87, 90, 101
Baldwin, J. E., 242, 261, 274
Bank, S., 136
Banthorpe, D. V., 209, 213, 273
Bargon, J., 155
Barnard, P. W. C., 337
Bart, J. C. J., 131
Bartlett, P. D., 9, 78, 107, 159, 160, 202,
    238, 256, 298
Bartsch, R. A., 217, 218, 220
Bastens, L., 306

Bateman, L. C., 82, 83
Bates, R. B., 135
Bathgate, R. H., 203
Battiste, M. A., 106
Bean, C. M., 194
Bear, J. L., 49
Becker, W. E., 144
Belanger, P., 129
Bell, J. A., 145
Bell, R. P., 51, 67, 310, 315, 316, 325
Beltrame, P., 340, 342
Bemis, A., 164
Bender, M. L., 53, 310, 328, 331
Benjamin, B. M., 97
Benkovic, S. J., 51, 69, 310, 328, 337
Benson, S. W., 6, 145, 270, 297
Bentrude, W. G., 47
Bergson, G., 184
Bernstein, R. B., 211
Berson, J. A., 268
Bethell, D., 78, 145, 286
Bickel, A. F., 185, 186
Biggs, A. I., 33
Bingham, K. D., 307
Birch, A. J., 139
Black, D. St. C., 52
Blackburn, G. M., 274
Blanchard, E. P., 153
Boche, G., 30
Bohlmann, F., 45
Boikess, R. S., 238, 269
Bollinger, J. M., 299
Bolton, R., 295
Bordwell, F. G., 208, 232
Bottini, A. T., 39, 325
Boudakian, M. M., 341
Bourns, A. N., 214, 345
Boyd, R. H., 51
Boyton, H. G., 283
Brauman, J. I., 251, 255
Breslow, R., 129, 211
Brettschnieder, H., 156

357

Brinich, J., 299
Broaddus, C. D., 120
Brook, P. R., 261
Brouwer, D. M., 187
Brown, F., 113
Brown, D. A., 335
Brown, H. C., 22, 26, 32, 93, 97, 100, 106, 115, 116, 189, 204, 216, 220, 280, 281, 295, 305, 319
Brown, J. F., 69
Brown, J. K., 138
Brown, J. M., 256
Brown, L. M., 280
Brown, P., 19
Brown, T. L., 190
Broxton, T. J., 49
Bruice, T. C., 51, 53, 69, 310, 328, 329, 337
Brunner, H., 156
Buckles, R. E., 97
Bumgardner, C. L., 232
Bunnett, J. F., 34, 60, 61, 194, 202, 209, 217, 218, 219, 220, 313, 344, 345
Bunton, C. A., 78, 194, 337
Burckhardt, U., 346
Burdon, J. G., 343
Burke, J. J., 47
Burnham, D. R., 138
Burnett, W. A., 120
Bushby, R. J., 136
Buttner, H., 157
Byram, S. K., 191

Cadogan, J. I. G., 295
Calder, I. C., 280
Caldwell, R. A., 122, 126
Campbell, C. D., 346
Campbell, N. C. G., 95, 101
Candlin, J. P., 51
Caple, G., 240
Capon, B., 78, 97, 98, 99, 105, 106, 111, 332
Carlsmith, L. A., 345
Carlson, M., 143
Carlsson, D. J., 158
Carnighan, R. H., 135
Casanova, J., Jr., 164
Caserio, M. C., 266
Cassar, L., 56

Cattania, M. G., 340
Chadwick, J., 93
Chalkley, G. R., 347
Chanley, J. D., 338
Charman, H. B., 192
Chaykovsky, M., 326
Cheema, Z. K., 97
Chen, A., 121
Chérest, M., 24, 319
Chiang, Y., 66
Childs, R. F., 269
Chitwood, J. L., 112
Chivers, T., 121
Chloupek, F. J., 115
Cholod, M. S., 155
Christmann, K. F., 327
Ciuffarin, E., 120
Clark, H. R., 53
Clark, R. L., 307
Clayman, L., 48
Cleveland, J. D., 93
Clippinger, E., 89
Clopton, J. C., 151
Closs, G. L., 127, 145, 154, 168
Closson, W. D., 107
Cocivera, M., 101
Cockerill, A. F., 214
Coetzee, J. F., 40
Coke, J. L., 108
Collman, J. P., 51
Collins, C. J., 97, 187
Colpa, J. P., 289
Comisarow, M. B., 111
Commeyras, A., 120
Cookson, R. C., 19, 257
Coombes, R. G., 290
Coombs, E., 325
Cooper, J. S., 49
Cooper, K. A., 230
Cordes, E. H., 51, 57, 65, 310, 316
Corey, E. J., 132, 164, 326, 327
Cossee, P., 54
Coulson, C. A., 239
Coyle, J. J., 154
Craig, D. P., 130
Cram, D. J., 43, 106, 113, 120, 123, 182, 183, 184, 222, 232, 319
Cramer, R., 51
Crampton, M. R., 343

Criegee, R., 251
Cristol, S. J., 151, 223, 224, 226, 232
Crosby, J., 212
Crunden, E. W., 335
Cunningham, L., 51
Curtin, D. Y., 228, 230
Csizmadia, I. G., 133

Dahmen, A., 253
Dale, J., 308
Dance, J., 257
Dauben, W. G., 112
Davidson, R. S., 157
Davies, A. G., 330
Davis, G. T., 34
Day, A. C., 238
De Boer, C. D., 24
De la Mare, P. B. D., 200, 208, 280, 295
De Member, J. R., 120
De More, W. B., 145
De Puy, C. H., 37, 214, 223, 230, 254
De Salas, E., 113
De Wolfe, R. H., 208
Deck, H. R., 319
Degrup, C. L., 106
Deno, N. C., 78
Dessy, R. E., 121, 189
Dewar, M. J. S., 6, 125, 245, 295, 303
Deyrup, A. J., 58
Dhar, M. L., 230
Diaz, A., 45, 86, 87
Dillon, R. L., 129, 302
Dimroth, K., 45
Dix, D. T., 143
Doering, W. von E., 152, 242, 270, 271, 283
Dolbier, W. R., 295
Dolman, D., 61
Donoghue, E. M., 245, 248
Donaldson, M. M., 97
Doorakian, G. A., 252
Doran, M. A., 128
Douek, M., 129
Draper, F., Jr., 34
Drenth, W., 208
Dubois, J. E., 299
Duggleby, P. M., 47
Duke, A. J., 261

Duncan, J. H., 151
Dunlap, R. B., 51, 57
Dyckes, D., 111
Dytham, R. A., 136

Eaborn, C., 292
Earhart, H. W., 283
Eberson, L., 111
Edwards, J. O., 194, 202, 313
Edwards, W. R., 283
Ehrensen, S., 26
Ehrenson, S. J., 145, 199
Eigen, M., 51, 120, 123
Eiter, K., 229
Eldred, N. R., 204
Eliason, R., 66
Eliel, E. L., 201
Elliott, C. S., 149
England, B. D., 208
Erickson, R. E., 307
Eriksson, S. O., 337
Evans, D. P., 325
Exner, J. H., 188

Fahey, R. C., 295, 303
Fainberg, A. H., 47
Favini, G., 340
Fawcett, J. K., 191
Feageson, E., 338
Felkin, H., 24, 319
Fendler, E. J., 51
Fendler, J. H., 51
Fenical, W., 295
Feuer, J., 263
Fierens, P. J. C., 201
Fife, T. H., 316
Fife, W. K., 37
Fischer, H., 120, 155, 187
Fischer, W. F., 321
Fletcher, R. S., 94, 100
Flood, S. H., 288
Fonken, G. J., 253
Foote, C. S., 116, 295
Ford, W. T., 183
Forrester, A. R., 155
Fort, R. C., Jr., 78, 84
Fowden, L., 200
Fox, I. R., 34
Fraenkel, G., 143

Frankhauser, R., 104
Franzen, V., 147
Fraser, G. M., 112, 229
Fráter, G., 272, 273
Freedman, H. H., 252
Frey, H. M., 4, 145, 149, 238
Friedman, L., 151, 346
Friedrich, A., 268
Frisone, G. J., 196
Frost, A. A., 162
Fuchs, R., 49
Fujimoto, H., 220, 245
Fukui, K., 220, 245
Funahashi, T., 159
Furusaki, A., 343

Gaasbeek, G. J., 185, 186
Gac, N. A., 270
Gall, J. S., 203
Gaoni, Y., 136
Garbisch, E. W., Jr., 21, 305, 344
Garner, A. Y., 153
Garratt, P. J., 280
Garst, R. H., 345
Gerstl, R., 154
Ghirardelli, R. G., 212
Ghosez, L., 306
Gilbert, J. C., 155, 270
Gilchrist, T. L., 145
Gill, G. B., 238
Girtis, D., 204
Glass, D. S., 238, 269
Gleave, J. L., 78, 194
Glick, R. 98
Goering, H. L., 85, 87, 90, 117
Golborn, P., 292
Gold, V., 67, 78, 286
Golden, D. M., 251, 270
Goldstein, M. J., 17
Gommper, R., 139
Gosser, L., 183
Graham, E. J., 183
Grashey, R., 238, 241, 256, 266
Green, S. I. E., 189
Greene, F. D., 202, 319
Grimm, H. G., 43
Grimsrud, E. P., 196
Grob, C. A., 102, 103, 105
Grovenstein, E., 137

Grunewald, G. L., 272
Grunwald, E., 3, 45, 97, 120
Guella, F., 340
Gunsher, J., 209
Gutfreund, H., 51
Guthrie, R. D., 156
Gutsche, C. D., 310, 325, 326

Haber, R. G., 201
Haberfield, P., 43, 49, 141
Habich, A., 272
Hackler, R. E., 274
Hagan, E. L., 80, 266
Halevi, E., 18
Hall, L. H., 263, 265
Halpern, J., 56, 301
Hammett, L. P., 28, 58
Hammond, G. S., 13, 24, 344
Hammons, J. H., 120
Hanack, M., 341
Hansen, H.-J., 238, 272
Hanson, C., 97
Hardie, B. A., 288
Harper, J. J., 97, 108, 109
Harrington, J. K., 151, 232
Harrison, A. G., 37
Harrison, J. M., 261
Hart, C. R., 345
Hartung, L. D., 348
Hause, W. L., 223, 224, 226
Hauser, C. R., 228
Havinga, E., 271
Hay, J. M., 155
Haynes, P., 242
Haywood-Farmer, J., 106
Heck, R. F., 51, 98
Heine, R. F., 341
Hendrickson, J. B., 23
Hennion, G. S., 267
Henry, P. M., 301
Herndon, W. C., 263, 265
Hesse, R. H., 165
Hey, D. H., 280
Hiatt, R. R., 160
Higginson, W. C. E., 67
Hill, R. R., 95, 101
Hine, J., 134, 145, 199, 212
Hinshelwood, C. N., 44
Hoegerle, R. M., 103

Hoffmann, A. K., 151
Hoffmann, H. M. R., 112, 196, 229, 238, 243, 245
Hoffmann, R., 146, 239, 242, 245, 249, 265, 272
Hoffmann, R. W., 310, 345, 346
Hofmann, J. E., 122
Hofstra, A., 281
Hogen-Esch, T. E., 141
Hogeveen, H., 56, 185, 186
Hoggett, J. G., 290
Hollaway, M. R., 51
Holmquist, B., 329
Holness, N. J., 95
Holst, C., 337
Horauf, W., 251
Houk, K. N., 265
Houlihan, W. J., 310, 317
House, H. O., 310, 321, 325
Houser, T. J., 211
Huber, H., 253
Hückel, W., 222
Hudec, J., 101, 257
Hudson, B. E., 187
Hudson, R. F., 326, 335
Huebner, C. F., 245, 248
Hughes, E. D., 78, 82, 83, 113, 192, 194, 195, 200, 208, 230
Huisgen, R., 238, 240, 241, 253, 256, 260, 266, 302
Hyde, J. L., 22

Ichikawa, K., 22
Illuminati, G., 287
Impastato, F. J., 130
Ingold, C. K., 40, 43, 78, 82, 83, 113, 192, 194, 200, 216, 230, 291, 329
Ingold, K. U., 158
Irie, T., 106
Itô, R., 294

Jacobsen, P., 156
Jaffe, H. H., 26
Jao, L. K., 316
Jefford, C. W., 209
Jeffrey, E. A., 190
Jencks, W. P., 51, 310, 322
Jensen, F. R., 190
Jewett, J. C., 96

Johnson, A. W., 120
Johnson, S. L., 310, 328
Jones, M., Jr., 346
Jones, M. G., 108
Jones, M. M., 53
Jones, W. M., 341
Joy, D. R., 245, 248
Julia, M., 309
Juliusburger, F., 195

Kabbe, H. J., 229
Kapecki, J. E., 261
Kaplan, L., 303
Kauffmann, Th., 346
Kaufmann, H., 46
Kauranen, P., 43
Kearns, D. R., 295
Kebarle, P., 37
Keller, C. E., 255
Kellom, D. B., 230
Kelly, D. P., 274
Kemp, D. S., 54
Kenyon, J., 194, 330
Kepner, R. E., 208
Kerr, J. A., 155
Kettle, S. F. A., 27
Kharasch, M. S., 12, 308
Kiefer, E. F., 260
Kim, C. J., 106
Kimball, G. E., 298, 302
King, R. W., 230
Kingsbury, C. A., 43
Kirby, A. J., 310, 326, 337
Kirkwood, J. G., 26
Kirmse, W., 145, 151, 274
Kirner, W. R., 198
Kitching, W., 121
Klasinc, L., 214
Klimisch, R. L., 216, 220
Klinedinst, P. E., Jr., 89
Klopman, G., 6
Knowles, J. R., 51, 56
Knox, L. H., 202
Ko, E. C. F., 49
Kobrich, G., 145, 157
Kochi, J. K., 164, 168
Koenig, T., 159
Kollmar, H., 187
Kollmeyer, W. D., 123

Kornblum, N., 141
Kosower, E. M., 40, 45, 105
Kramer, G. M., 187
Krasnobajev, V., 104
Kraut, J., 126
Kreevoy, M. M., 66
Kreiter, C. G., 255
Kresge, A. J., 66
Krusic, P. J., 168
Kudryavetsev, R. V., 330
Kuhn, S. J., 288
Kuivila, H. G., 155
Kursanov, D. N., 330
Kwart, H., 224, 303

Laber, G., 283
Lancamp, H., 156
Lancelot, C. J., 108, 109, 110
Lapidot, A., 338
Lapworth, A., 310
Laroche, P., 306
Larrabee, R. B., 127
Larsen, J. W., 78, 81, 82, 91, 283, 336
Lawton, R. G., 107
Lazdins, I., 45
Ledwith, A., 153
Lee, W. G., 211
Leffler, J. E., 3
Legutke, G., 222
Leisten, J. A., 38
Lepley, A. R., 267
Leung, C., 27
Levin, R. H., 346
Lewis, I. C., 34
Lewis, E. S., 348
Liotta, C. L., 134, 345
Liu, R. S. H., 253
Llewellyn, D. R., 337
Logullo, F. M., 346
Long, F. A., 51
Longuet-Higgins, H. C., 245, 280
Lorenz, P., 161
Lossing, F. P., 37, 78
Lowry, T. H., 132
Lucas, H. J., 98, 302
Lui, C. Y., 120
Lukas, E., 20
Lukas, J., 20, 266
Lupton, E. C., Jr., 33, 251

MacClelland, R. A., 333
Maccoll, A., 130, 230
Mackay, B. H., 348
Mackie, J. D. H., 200
Mackinley, S. V., 142
Mackor, E. L., 281, 289
Maclean, C., 156, 289
Maclean, S., 242
Maerecker, A., 310, 317, 325
Mahajan, S. N., 209
Makarova, L. G., 347
Malhotra, S. K., 13, 14
Malone, L. G., 134
Maness, D. D., 341
Mango, F. D., 51
Mansoor, A. M., 151
Marchand, A. P., 134
Mares, F., 134
Margolis, E. T., 12
Martin, J. C., 163
Marvell, E. N., 238, 240
Marzo, A., 342
Masterman, S., 195
Matt, J. W., 165
Maumy, M., 309
Maya, W., 228
Mayo, F. R., 12, 308
Mclennan, D. J., 209, 210
Mcmanimie, R. J., 341
Mcmichael, K. D., 90
McNulty, B. J., 230
Meakins, G. D., 307
Meek, J. S., 223
Melander, L., 14
Melchior, M. T., 187
Melchiore, J. J., 281
Merenyi, R., 135, 238
Merriman, P. C., 208
Miekka, R. G., 211
Migita, T., 294
Miller, J., 310, 342
Miller, K., 187
Miller, R. G., 346
Miller, R. J., 329
Miller, S. I., 194, 211, 238, 310, 341
Mobius, L., 240
Moelwyn-Hughes, E. A., 203
Moerikofer, A. W., 305
Mole, T., 190

Moller, F., 229
Montgomery, L. K., 165
Moodie, R. B., 290
Moreland, W. T., Jr., 26
More O'Ferrall, R. A., 78
Morikawa, N., 294
Morris, G. F., 37, 214, 223
Moss, R. A., 154
Motes, J. M., 129
Mouvier, G., 299
Moussebois, C., 308
Muir, D. M., 95, 101
Muller, G., 240
Murray, R. W., 295
Murrell, J. N., 27

Nakagawa, M., 100
Namanworth, E., 111
Nauta, W. T., 156
Nayak, B., 337
Nebe, W. J., 253
Neilscn, A. T., 310, 317
Nelson, G. L., 268
Nesmayanov, A. N., 347
Nevell, T. P., 113
Newman, M. S., 335
Nickols, R. E., 122
Nickon, A., 231
Noble, P., Jr., 40
Nordlander, J. E., 142
Norman, R. O. C., 155, 280
Norton, C., 106
Noyce, D. S., 39, 325, 326
Nyburg, S. C., 191
Nye, J. L., 224
Nyholm, R. S., 130

Oakenfull, D. G., 67
O'Brien, R. J., 191
Occolowitz, J. L., 256
O'Connor, C., 337
Oediger, H., 229
Oelderik, J. M., 187
Ohlsson, L., 184
Ohta, M., 103
Okamoto, Y., 32
Okamura, M. Y., 260

Olah, G. A., 20, 51, 78, 111, 119, 120, 128, 185, 266, 280, 288, 289, 299
Oldham, K. G., 337
Ollis, W. D., 274
Olsen, A. R., 22, 329
Olsen, F. P., 61
Orgel, L., 130
Ostermayer, F., 102
Oth, J. F. M., 135, 238
Otto, P., 260
Overchuk, N. A., 289

Paquette, L. A., 257
Parham, W. E., 307
Parish, J. H., 95, 101
Parker, A. J., 40, 43, 49, 194, 196, 203
Parks, L. R., 344
Parsons, C. A., 56
Patterson, D. B., 305
Paul, M. A., 51
Pearson, R. C., 163
Pearson, R. G., 129, 194, 202, 205, 313
Perkampus, H. H., 280
Perkins, M. J., 295
Perrin, C., 287
Petrovich, J. P., 111
Pettit, R., 255
Phillips, H., 194
Phillips, L., 153
Pichat, L., 208
Pickles, N. J. T., 44
Pietra, F., 310, 342
Pincock, R. E., 106
Pinholster, D. F., 345
Pitea, D., 342
Pittman, C. U., 78
Pocker, Y., 188
Podall, H., 204
Pombo, M. M., 90
Pot, J., 270
Powell, J. W., 151
Price, C. C., 136
Price, E., 34
Prudent, N., 319
Pruitt, K. M., 344
Pryor, W. A., 155, 325
Psarras, R., 121
Pudjaatmaka, A. H., 134

Purmont, J. I., 142
Puterbaugh, W. H., 228

Rabone, K. L., 101
Radlick, P., 295
Ramsey, B., 111
Rappe, C., 134
Rappoport, Z., 310, 340
Raudenbusch, W., 102
Rauk, A., 133
Readio, P. D., 310
Redmore, D., 326
Rees, C. W., 145, 346
Rei, M.-H., 115
Reichardt, C., 40, 45, 161
Reutov, O. A., 193
Rewicki, D., 120
Ricci, A., 343
Rickborn, B., 43, 190
Ridd, J. H., 280
Riley, T., 67, 286
Ringold, H. J., 13, 14
Ritchie, C. D., 26, 40, 51, 63, 121
Roberts, I., 298, 302
Roberts, J. D., 26, 142, 181, 266, 345
Robinson, G. C., 89
Robinson, R. A., 33
Rochester, C. H., 51
Rodewald, R. F., 49
Rogan, J. B., 107
Rogers, N. A. J., 138
Roman, S. A., 107
Rony, P. R., 69
Rose, J. B., 113
Rosenfeld, J. C., 254, 266
Rothberg, I., 97
Roth, W. R., 242, 268, 271
Rowe, C. A., Jr., 136
Ruchardt, C., 293
Ruf, H., 43
Ryason, P. R., 34

Sager, W. F., 26
Sakabe, N., 343
Salem, L., 238, 265
Salinger, R., 121
Samuel, D., 315, 338

Sargent, G. D., 78, 107, 111, 298
Satchell, D. P. N., 310, 328
Sauer, J., 238, 241, 256, 261, 266
Saunders, M., 80, 119, 254, 266, 283
Saunders, W. H., Jr., 214, 217, 218
Savedoff, L. G., 203
Schacht, E., 161
Schadt, F. L., 108
Schaefer, H., 137
Schatz, B., 240
Schellenberg, W. D., 251
Schechter, H., 151
Schewene, C. B., 117
Schildcrout, S. M., 305
Schlatmann, J. L. M. A., 270
Schleyer, P. von R., 8, 78, 84, 96, 97,
    108, 109, 110, 116, 119, 254, 303
Schlosberg, R. H., 185
Schlosser, M., 120, 327
Schmid, H., 238, 272, 273
Schmir, G. L., 53
Schnieder, W. P., 327
Schofield, K., 290
Schollkopf, U., 137, 268
Schreiber, K. C., 108
Schreiber, M. R., 217, 218
Schriesheim, A., 122, 136
Schroeder, G., 238, 272
Schweizer, E. E., 307
Scott, C. B., 202, 203
Seltzer, R., 141
Seltzer, S., 238, 256, 263
Senders, J. R., 135
Shapiro, R. H., 151
Shatavsky, M., 106
Shatenshtein, A. I., 120
Shine, H. J., 273
Shiner, V. J., 96, 318
Shoemaker, M. J., 267
Shoppee, C. W., 105
Sicher, J., 24
Siepmann, T., 45
Silver, B. L., 337
Silver, M. S., 332
Silversmith, E. F., 87
Simamura, O., 294
Simmons, H. E., 153, 345
Simonetta, M., 342
Singer, M. S., 232

Sipos, F., 24
Skell, P. S., 153, 155, 310
Sleiter, G., 287
Slomp, G., 257
Smat, R. J., 37, 214, 223
Smid, J., 141
Smith, C., 274
Smith, C. L., 346
Smith, G. G., 37
Smith, H., 139
Smith, H. O., 187
Smith, J. A., 151
Smith, J. F., 113
Smith, J. S., 37, 214, 223
Smith, L. I., 238, 241, 256, 266
Smith, M. B., 144
Smith, M. J., 325
Smith, R. D., 153
Smith, S. G., 39, 90, 203
Sneen, R. A., 91, 92, 336
Snodin, D. J., 347
Snyder, W. H., 136
Sondheimer, F., 136, 280
Songstad, J., 205
Southam, R. M., 95, 101
Spillett, R. E., 292
Sprecher, R. F., 21
Sprecker, C. M., 305
Srivastava, R. C., 52
Staab, H. A., 156
Stang, P. J., 8
Stangl, H., 240, 266
Staples, C. E., 135
Stefanovic, D., 214
Stegel, F., 287
Stein, G., 240
Stevens, G., 347
Stevens, I. D. R., 151, 203
Stevens, T. S., 120
Stewart, R., 61
Stiles, M., 346
Stirling, C. J. M., 212
Stock, L. M., 26, 280
Stolow, R. D., 228
Stork, G., 208
Strating, J., 298
Streitwieser, A., Jr., 24, 78, 91, 92, 94,
    120, 122, 126, 128, 134, 194, 200,
    205, 291

Stuart, R. S., 129
Sturm, H. J., 266
Su, T. M., 254
Sustmann, R., 84
Sutherland, I. O., 274
Sutton, L. E., 130
Swain, C. G., 33, 69, 202, 203
Symons, M. C. R., 155
Szeimies, G., 240
Szwarc, M., 86, 87, 120

Tadanier, J., 113
Taft, R. W., Jr., 26, 34, 301, 332
Takeshita, T., 224
Tanaka, J., 343
Tanida, H., 106
Tappe, W., 222
Tatlow, J. C., 343
Taylor, H. T., 93
Taylor, J. W., 196
Taylor, K., 51
Taylor, R., 37, 231, 280, 292
Tedder, J. M., 27, 163
Thomas, C. H., 199
Thompson, D. T., 51
Thompson, J. A., 106
Thomson, R. H., 155
Thornton, E. R., 78, 196
Tichý, M., 24
Tinker, H. B., 301
Tolstaya, T. P., 347
Tommila, E., 43
Tonkyn, R. G., 34
Topley, B., 195
Toporcer, L. H., 189
Toscana, V., 270
Traylor, T. G., 126, 295
Trifan, D., 113
Trotman-Dickenson, A. F., 155
Trozzolo, A. M., 148
Truce, W. E., 341
Tsuji, J., 51, 106
Tsuno, Y., 33
Turner, D. W., 78
Turnquest, B. W., 53
Turro, N. J., 11

Ueda, H., 343
Uschold, R. E., 121

Valange, P., 135
Van Dine, G. W., 146
Van der Walls, J. H., 281
Vaska, L., 51
Vaughan, C. W., 345
Vernon, C. A., 208, 337
Vernon, J. M., 240
Verschelden, P., 201
Viehe, H. G., 135
Vivarelli, P., 343
Volger, H. C., 56, 192
Vyas, V. A., 214

Waack, R., 128
Wadley, E. F., 283
Wagenhofer, H., 266
Wagner, E., 157
Walborsky, H. M., 129, 130
Walker, F. W., 144
Walling, C., 155, 295
Walsh, R., 4, 238
Walsh, T. D., 91
Wardill, J. E., 335
Warren, S. G., 310, 326, 337
Wassermann, A., 46
Weedon, B. C. L., 136
Weisman, G. R., 156
Weiss, J., 195
Weiss-Broday, M., 338
Wells, P. R., 26
Wentworth, G., 137
Werstiuk, N. H., 231
Werth, R. G., 97
West, P., 142
Westheimer, F. H., 8, 26, 286, 310, 337, 338
Wheeler, O. H., 216
White, A. M., 119, 120
White, J. E., 307
White, W. N., 208

White, W. W., 37
Whitesides, G. M., 142
Whitham, G. H., 307
Whiting, M. C., 95, 101, 151, 347
Whittaker, D., 318
Wieringa, J. H., 298
Wiesboeck, R., 212
Wilcox, C. F., 27
Williams, J. E., Jr., 8, 84, 232
Williams, K. C., 190
Wilson, C. L., 113
Winstein, S., 45, 47, 85, 86, 87, 89, 90, 95, 97, 98, 101, 105, 106, 108, 111, 113, 203, 208, 238, 255, 269
Winter, R. E. K., 240
Wittig, G., 346
Wold, S., 184
Wolf, R., 159
Wolfe, S., 133, 304
Wolff, H., 43
Wolovsky, R., 280
Woodward, R. B., 106, 239, 242, 245, 249, 265, 272
Woodworth, R. C., 153
Woolf, L. I., 230
Wynberg, H., 298

Yamamoto, H., 327
Yates, K., 333
Yonan, P. K., 341
Young, A. E., 130
Young, W. G., 85, 208, 302
Young, W. R., 126, 128
Youssef, A. A., 129
Yukawa, Y., 33

Zahler, R. E., 34
Zeiss, G. D., 146
Ziegler, G. R., 24
Zimmerman, H. E., 135, 245, 272
Zoltewicz, J. A., 346

Subject Index

Acenaphthylene, addition of DBr, 303
Acetaldehyde
  dehydration of hydrate, 67
  hydration, kinetics, 316
Acetals, hydrolysis, 316
Acetamidate, phenyl N-methyl, 350
Acetates
  1-arylethyl, thermolysis, 37
  hydrolysis in concentrated acids, 333
  thermolysis, 230, 238
Acetic acids, ionization of substituted,
  table, 26
Acetone
  cyanohydrin, 310
  enolization, 66
Acetoxime, 321
α-Acetoxyacrylonitrile, 259
2-Acetoxycyclohexyl tosylate, 97
Acetylenes,
  additions to, 307
  as intermediates in nucleophilic sub-
    stitution, 311, 341
  base-catalysed isomerization, 136
Acid–base reactions
  table of rates of, 62
  rates of, 61
Acidity functions, 58, 121, 280, 333
Acidity
  of cycloalkanes, 126
  of hydrocarbons, table, 124
  scales of, 120, 124
Acylation reactions, 285, 287, 328, 335
Acylium ion, 312, 333
Adamantyl derivatives, 84, 171
Adamantylideneadamantane, 298
Addition reactions, electrophilic, see
  Electrophilic addition reactions,
  295
Addition,
  anti and syn, 302
  cis and trans, 302
Aldehydes and ketones, hydration, 315

Aldehydes, radical addition to olefins,
  309
Aldol condensation, 322
Alkylation, C vs O, 140
Alkyl halides, reactivity in S_N2 reac-
  tion, tables, 198, 200
Alkyl lithiums, structure, 142
Allenes
  as intermediates in nucleophilic
    substitution, 311, 341
  cycloadditions, 260
Allene, 1,1-dimethyl, 260
Allyl anion, 27
Allyl anions, stabilities of cis and trans
  substituted, 136
Allyl cation, 27
Allyl cations, n.m.r., 80
Allyl lithium, 142
Allyl radical, conjugation in, 27
Allyl vinyl ether, Claisen rearrange-
  ment, 2
Aluminium, trimethyl, 191
Ambident anions, reactions of, 139
Amides,
  hydrolysis, 38, 337
  protonation, 337
Amine oxides, thermolysis, 230
1-Aminobenzotriazole, 346
2-Amino-5-t-butylcyclohexanol,
  deamination, 24
Ammonia, liquid, and dissolving
  metals, 138
Anchimeric assistance, 97
Angle strain, 8, 10
Anilines, diazo-coupling, 349
Anions,
  α-sulphone, 131
    calculations on, 133
  α-sulphoxide, 131
  solvation in dipolar aprotic solvents,
    43, 197
Anisole, reduction, 138

3-*p*-Anisyl-2-butyl brosylate, 87, 89
[18]-Annulene, nitration, 280
Antarafacial, 248
Anthracene, 19
   reaction with benzyne, 347
   11,12-dichloro-9,10,-dihydro-9,10-
     ethano, 224
Anti and *syn* addition, 247, 302
Anti-elimination, 221
Anti-Markownikov addition, 308
Aprotic solvents, 43
Aromatic substitution,
   electrophilic, *see* Electrophilic aro-
     matic subs., 280
   nucleophilic, *see* Nucleophilic aro-
     matic subs., 311, 342
3-Aryl-2-butyl tosylates, 110
1-Aryl-2-propyl tosylates, 109, 110
Aryl radicals, aromatic substitution
   by, 294
Association-prefaced catalysis, 56
Associative mechanism for nucleo-
   philic substitution, 311
Autoxidation, 157
Axial and equatorial derivatives, solvo-
   lysis, 95
Aziridines, ring opening, 275
Azo-compounds, thermolysis, 158

Bamford–Stevens reaction, 151
Barton reaction, 165
Benzaldehyde acetals, hydrolysis, 316
Benzenediazonium ion, and nucleo-
   philes, 348
Benzenediazonium-2-carboxylate, 346
Benzenes, substituted, bromination, 33
Benzhydryl chloride, solvolysis, 36, 83
Benzidine rearrangement, 273
Benzoate esters, hydrolysis, 31
Benzoic acids, ionization, 29
Benzothiadiazole-1,1-dioxide, 346
Benzothiazolium salt, carbene from,
   147
Benzoyl chlorides, solvolysis, 335
2-Benzoyl-1-halopropenes, 340
Benzyl anion, conjugation in, 27, 128
Benzyl chlorides, solvolysis, 36, 93
Benzyl radicals, ionization potentials,
   37

Benzylideneacetophenone, 325
Benzyne,
   precursors, 346
   reaction with anthracene, 347
   reaction with dichloroethylene, 347
   reaction with tetramethylethylene,
     347
Benzynes, 342, 345, 350
Bicyclo[X,2,0]alkenes, 251
Bicyclobutane, ring opening, 274
Bicyclo[3,2,0]heptenes, rearrange-
   ment, 268
Bicyclo[3,1,0]hexenes, rearrangement,
   268
Bicyclo[3,1,0]hexenyl cation, hepta-
   methyl, 268
Bicyclo[6,1,0]nona-3,5-diene, 270
Bicyclo[3,2,1]octadienyl anion, 255
Bicyclo[2,2,2]octane-1-carboxylic
   acids, 4-substituted, 26
Bond enthalpies, table, 7
Bond-length strain, 8
Boranes, cleavage of, 189
Borderline solvolysis, 85, 91
Bridged radicals in addition reactions,
   310
Bridgehead halides, solvolysis, 84 lack
   of S_N2 reactivity, 202
Bromination, 285
   of olefins
     in methanol, 297
     substituent effects on, 298
1-Bromoadamantane, 84
Bromobicyclo[2,2,1]heptane, 84
Bromobicyclo[2,2,2]octane, 84
3-Bromobutan-2-ols, reaction with
   hydrobromic acid, 98
ω-Bromonitrostyrene, substitution,
   341
Bromonium ion, 298
2-Bromooctane, solvolysis, 85
Brønsted equation, 67, 122
Bullvalene, 272
Butadiene
   addition of HBr, 12
   1-methyl-4-($^2H_3$)methyl-1,2,3,4-
     tetraphenyl, 252
2-Butene, addition of $BrCCl_3$, 308
1-t-Butylallyl chloride, 208

t-Butyl bromide hydrolysis, 83
2-Butyl cation, 266
Butyl cations, 80
t-Butyl chloride
    ionization, 40
    solvolysis, solvent effects, 45, 46
3-t-Butylcyclopentadienone, 20
t-Butyl derivatives, solvolysis, 101
t-Butyl p-nitrobenzoate, 9

Cage effect, 159
Camphor, reduction by BH_4^-, 319
Camphor tosylhydrazone, 151
Cannizzaro reaction, 318
Carbanions, 120
    fluoro-substituted, 134, 172
    ion-pairing in, 141
    isomerizations through, 135
    proton exchange via, 122, 182
    rearrangements, radicals in, 137, 266
    stability, 125
        and $d\pi$ conjugation, 130
        and $p\pi$ conjugation, 127
        and hybridization, 125
        and inductive effects, 133
    stable, 141
Carbenes, 145
    acyl, 150
    alkyl, 150
    and carbenoids, 153
    halogeno, 152
    nucleophilic, 147
    singlet and triplet, 146
    structure, 146
Carbenoids and carbenes, 153
Carbonium ions, 78
    bridged, 105
    bridgehead, 84
    elimination from, 100
    formation by associative reactions, 279
    fragmentation from, 100, 102
    from ionization of radicals, 79
    heats of formation, 81
    n.m.r. studies, 80
    non-classical, 105
    reactions, 99
    rearrangements in, 100, 105

Carbonium ions—cont.
    stability, 78
    stability of primary, secondary and tertiary, 80
    stable salts, 80
    strained, 84
    structural factors affecting formation, 92
    substitution from, 100
    [1,2]-shifts, 266
Catalysis, 51
    acid–base, 57
    bifunctional, 69
    by association, 56
    by transition metals, 54
    electrophilic, 52
    enzymic, 70
    general acid, 64
    in Friedel–Crafts reaction, 52
    intramolecular, 69
    micellar, 56
    nucleophilic, 52
    specific acid, 65
Chain reactions, 156
Cheletropic reaction, 242
Chloral, 315
Chlorination, 285
2-Chloro-1-methylbenzimidazole, 343
3-Chloro-5-methylcyclohexene, 86
Chlorobenzene, reaction with KNH_2, 345
Cholesteryl tosylate, solvolysis, 101, 105
CIDNP, 168, 267
Cinnamic acids, isomerization, 73
Cinnamyl p-tolyl ethers, rearrangement, 37
cis-Ligand migration reaction, 54
Claisen rearrangement, 2, 4, 242, 270
    substituent effects, 37
Cleavage
    of carbon–metal bonds, 188
    of organoboranes, 189
    of organomercurials, 190
Common-ion effect, 82, 85
Condensations, see Aldol condensation
Conjugate addition, 320
    of lithium dimethyl copper, 320

Conjugative effect, 26
Conrotation, 243
Cope rearrangement, 4, 242, 270, 275
  stereoelectronic effect, 24
Correlation diagram, 244
Cram's rule, 319
Cumyl chlorides, solvolysis, 32, 93
Cuprous iodide and 1,4-addition, 320
Cyanohydrin formation, 316
Cyclic olefins, strained, additions to,
  306
Cyclic transition state, 238
Cycloadditions,
  mechanisms of, 256
  via dipolar ions, 259
  via diradicals, 256
  definition, 240
  orbital symmetry rules, 245
  1,3-dipolar, 240, 248, 266
Cycloalkyl bromides, reactivity in S_N2
  reactions, 201
Cycloalkyl derivatives, solvolysis rates,
  94
Cycloalkyldimethylamine N-oxides,
  225
Cycloalkyl radicals, stability of, 161
Cycloalkyl trimethylammonium ions,
  225
Cyclobutane, cleavage to ethylene, 6
Cyclobutene,
  reaction with methylene, 149
  ring opening, 240, 243, 251
Cyclohexadienyl anions, 135
Cyclohexadienyl cation, 276, 281, 282
Cyclohexanol, oxidation by CrO_3, 11
Cyclohexene, tetradeuterio, addition of
  DBr, 304
Cyclohexenyl esters in S_N2' reactions,
  208
Cyclohexyl halides in S_N2 reactions,
  201
Cyclohexylamine, caesium salt, 122,
  126
Cyclohexyl tosylates, solvolysis, 72, 94,
  113
Cyclonona-1,4,7-triene, 270
Cycloocta-1,5-diene, 349
Cyclooctatetraene, bromination, 302
Cyclooctatrienes, cyclization, 253, 275

Cyclopentadiene, 239, 240, 261, 262
Cyclopentadiene, hydrogen shifts, 241
Cyclopentadienone dimer, 242
Cyclopentadienones, dimerization of,
  21
Cyclopent-3-enylethyl derivatives, 107,
  298
Cyclopropane,
  bonding and hybridization, 125
  protonated, 187
Cyclopropanone, 315
Cyclopropyl anion, configurational
  stability, 129
Cyclopropylcarbinyl radicals, 166
Cyclopropyl cation, ring opening, 254
Cyclopropyl derivatives, ring opening in
  solvolysis, 254

Darzens condensation, 324
Deamination, stereoelectronic effect,
  24
Decalyl tosylates, 101
Detritiation, 285
Deuteration and proton exchange, 182
Diazoalkanes,
  decomposition, 151
  effect of copper, 152
  protonation, 151
Diazomethane,
  diphenyl, 265
  reaction with carbonyl compounds,
  326
Diazonium ions, decomposition, 347
2,4-Di-t-butylcyclopentadienone, 21
Dichlorocarbene, 145
1,1-Dichloro-2,2-difluoroethylene,
  241, 256
cis-1,2-Dichloroethylene, substitution,
  341
2,3-Dichloro-1,1,1,4,4,4-hexafluoro-
  but-2-ene, 311
Dichloromethyl-lithium, 154
Dielectric constant and reaction rates,
  43
Diels–Alder reaction, 19, 239, 247, 261
  isotope effects in, 264
  solvent effect, 46
  stereochemistry, 262, 264

Diffusion control,
  in nitration, 290
  in proton transfer reactions, 61
Difluorocarbene, 147
Diimide, 305
1,1-Dimethylallyl chloride, 85, 208
3,3-Dimethylallyl chloride, 86
1,3-Dimethylallyl p-nitrobenzoates, 90
4,4'-Dimethyl benzhydryl chloride,
  hydrolysis, 83
Dimethyl-t-butylcarbonium ion, 266
2,2-Dimethylcyclopropanone, cleavage
  by base, 10
3,4-Dimethylhexa-1,5-diene, 242, 271
Dimethylsulphonium methylide, 326
Dimethylsulphoxide,
  anions in, 43
  carbanions in, 121
2,4-Dinitrohalobenzenes, 314, 343
2,4-Dinitrophenyl phenyl ether, 344
1,1-Diphenyl-2-bromopropene, 342
Diphenylchloronium ion, 347
Dipolar aprotic solvents, 43
  effect on S_N2 reactions, 196, 197
Dipolar cycloadditions, 240, 248, 266
Dipolar intermediates in cycloadditions, 239, 259, 266
Diradicals in cycloadditions, 239, 256,
  263
Dispersion forces in solvation, 42
Disrotation, 243
1,2-Divinylcyclobutane, 24
Divinyl cyclopropanes, rearrangement,
  272

E1 mechanism, 209
E1cB mechanism, 210
  isotope effects, 212
E2 eliminations, stereochemistry, 220
E2 mechanism, 209
  isotope effects, 213
  orientation, 214
  substituent effects, 214
E_s, 35
E_T-scale, 45
Electrocyclic reactions,
  definition, 240
  mechanisms of, 251
  orbital symmetry rules, 243

Electronic effects, 26
Electron spin resonance spectra, 166
Electron transfer to radicals, 163
Electrophilic addition reactions, 295
  charge-controlled, 298
  orientation in, 295
  overlap controlled, 305
  stereochemistry, 302
  to acetylenes, 307
  to strained olefins 306,
Electrophilic aromatic substitution,
  280
  acylation, 285
  benzylation, 286
  bromination, 285
  by benzhydryl cation, 286
  catalysis in, 52
  chlorination, 285
  detritiation, 285
  ethylation, 285
  in fused and heteroatomic systems,
    291
  isotope effect, 286
  mercuration, 285
  nitration, see Nitration
  ortho substitution, 288
  selectivity and reactivity, 285
  substituent effects, 283
Electrophilic attack on π-bond, 279
Electrophilic catalysis, 52
Electrophilic substitution,
  aliphatic, 181
    by metals, 190
    by protons, 182
  aromatic, 280
  bimolecular, 181
Electrostatic forces in solvation, 42
Electrostatic model of solvent effects,
  43
Eliminations,
  α, 145
  anti- and syn-, 221
  β, see Eliminations, 209
  bimolecular, 209
  competition with substitution, 227
  cyclic, 230, 238
  from carbonium ions, 100
  γ, 231
  orientation and leaving group, 214

Eliminations—*cont.*
  spectrum of transition states, 210,
    213
    and orientation, 219
    thermal *syn-*, 230
Ene reaction, 242
Enolate anions,
  conjugation in, 129
  protonation, 13, 140
  alkylation of, 140
Enolates, protonation of, 13
Entropy of activation, 5
Enzymic catalysis, 70
Epoxidation, 298
Ester hydrolysis, 311, 328
  acyl–oxygen fission, 328, 330
  alkyl–oxygen fission, 328
  catalysis by imidazole, 53
  in concentrated acid, 333
  intramolecular catalysis, 69, 332
  mechanisms, 329
  substituent effects, 331
Ester interchange, 329, 350
Esters,
  ^{18}O exchange, 331
  protonation, 329
Ethane,
  hexamethyl, 186
  1,1,1-trifluoro-2,2-dihalogeno, 211
Ethylation, 285
Ethylene, dimerization, 245
Ethyl *p*-nitrophenyl malonate, 329
Ethyl tosylate, 45
Ethyl vinyl ether, 66
External return, 86

Favorski reaction, 232
Feist's ester, 172
Field effect, 26
Fluorenone, cleavage by base, 10
Fluorenyl carbanions, 141
  racemization and exchange, 183
Fluorophenyl anion, 345
Foote–Schleyer equation, 116
Fragmentation, synchronous, 102
Friedel–Crafts acylation, *see* Acylation,
  steric effects, 20
Friedel–Crafts alkylation, *see also*
  Ethylation

Gallium, trimethyl, 191
Galvinoxyl, 159
General acid catalysis, 64
Grignard reagents,
  additions to carbonyl compounds,
    317
  enolization by, 318
  reduction by, 318
  structure, 144
Grunwald–Winstein equation, 45

H_0, 58
Halide ions, nucleophilicity, 202
Haloforms, hydrolysis, 145
Halogenocarbenes, 152
Hammett equation, 28
Hammett's ρ, table of values, 36
Hammond postulate, 13, 68
Hard and soft acids and bases, 52, 205,
  233, 313
  solvent effects on, 207
Heptafulvalene, 257
Heptamethylbenzenonium ion, 283
Hexa-1,5-diene, 270
Hexa-2,4-diene, 241, 256
Hexafluoroacetone, 315
Hexatrienes, cyclization, 240
2-Hexyl halides, eliminations, 217
Hofmann rule, 214
Homoallyl radicals, 165
Homolytic aromatic substitution, 293
Homotropylium ion, 255
Hybridization,
  and acidity, 125
  and $J_{13_{C-H}}$, 126
  of carbanions, 127
  of radicals, 160
Hydration equilibria for carbonyl
  compounds, 315
Hydration,
  of heterocyclic compounds, 315
  of olefins, 301
Hydroboration, 296, 297, 305
Hydrocarbons, acidities, 120
Hydrogen atom, abstraction, 161
Hydrogen-bonding in solvation, 42
Hydrolysis
  of acetals, 65, 316
  of acetates in concentrated acids, 333

Hydrolysis—*cont.*
  amides, 38, 337
  of aryl acetates, catalysis of, 67, 69
  of benzoate esters, 31
  of esters, *see* Ester hydrolysis
  of ethyl vinyl ether, 66
  β-lactones, 21, 329
  of *p*-nitrophenyl acetate, catalysis by
    imidazole, 53
  of orthoesters, 66
Hyperfine splitting, 167

I-strain, 93, 201
Imidazoles, catalysis by, 53, 56
Indanetrione, 315
Indenyl carbanions, in racemization
    and exchange, 185
Indicators, for acidity functions, 58
Indium, trimethyl, 191
Inductive effect, 26
Insertion reaction, 54
Inversion and retention in S_E2 and S_N2
    reactions, 180
Inversion in [1,3]-shifts, 268
Iodomethylenezinc iodide, 153
Ion pair return, stereochemistry, 90
Ion pairs, 85
  and carbanions, 141
  and racemization, 86
  in borderline solvolyses, 91
  in solvolysis, 83
  internal return, 86
  intimate, 88
  solvent effects on formation, 90
  solvent separated, 88
  special salt effect, 87
Ion-pairing,
  and nucleophilicity, 203
  in eliminations, 227
Isomerizations of olefins by base, 135
Isopropyl cation, 80
Isotope effects,
  in carbanion formation, 122
  in carbon–metal bond cleavage, 188
  in retro-Diels–Alder reaction, 18
  in Diels–Alder reactions, 264
  in $E1cB$ reactions, 212
  in $E2$ reactions, 213

Isotope effects—*cont.*
  in electrophilic aromatic substitu-
    tion, 282, 286
  in nucleophilic aromatic substitu-
    tion, 345
  on solvolysis of cyclohexyl tosylates,
    72, 96
  primary, 12, 14
  secondary, 18

Ketene,
  chloro, 261
  dimethyl, 260
  diphenyl, 261
Ketenes,
  cycloadditions, 260
  in ester hydrolysis, 329
Ketones,
  cyclic, reduction by borohydride,
    22, 350
  ^{18}O exchange, 315
Kinetic control, 12
Kinetic isotope effect, *see* Isotope
    effects

β-Lactones,
  hydrolysis of, 21, 329
Lead tetraacetate, 164
Linear free energy relationships, 28
  in eliminations, 218
  in olefin bromination, 300
Lithium alkyls and aryls, addition to
    carbonyl compounds, 317
Lithium alkyls, structure, 142
Lithium dimethyl copper, conjugate
    addition, 320
London forces in solvation, 42

m-Value, 45
Markownikov rule, 301
Meerwein–Ponndorf–Verley reduc-
    tion, 318
Meisenheimer complexes, 343
Menthyl chloride solvolysis, 113
Menthyl chlorides in eliminations, 222
Mercuration, 285, 287
  of olefins, 296, 301

Mercurials,
  cleavage of, 190
  exchange, 192
Mesitoic acid and methyl ester, 332, 335
Mesitylene, protonated, 283
Metal ion oxidations, radicals in, 164
Metal–metal exchange, 190
  and carbanion stability, 121
Methane, protonated, 185
Methoxypentyl brosylates, 98
4-Methoxystyrene, 259
2-Methylbutyl-lithium, 143
2-Methylcyclohexanone reduction, 319
Methylene,
  reaction with cyclobutene, 149
  singlet and triplet, 146, 148
Methylethylcarbonium ion, 266
Methyl ethylene phosphate, 338
2-Methylfuran-maleic anhydride adduct, 263
Methyl 2-formyl benzoate, 70, 332
Methyl iodide,
  substitution by azide, solvent effect, 43
  substitution by thiocyanate, solvent effect, 49
4-Methylpent-3-enyl tosylate, 107
2-Methyl-3-phenylpropionitrile, racemization, 43
2-Methyl-2-phenylpropyl tosylate, 45
Michael addition, 320
Microscopic reversibility, 3
Molecular orbital calculations,
  and aromatic substitution, 291
  and multicentre reactions, 239
  CNDO method, 6
  on α-sulphone anions, 133
Multicentre reactions, 238

Naphthalenes, electrophilic substitution, 292
α-Naphthyl carbene, 147
Neighbouring group participation, 69, 97
  by acetoxy group, 97
  by π-bonds, 105
  by σ-bonds, 111
  by hydrogen, 112

Neighbouring group participation—cont.
  by methoxy group, 98
  by phenyl, 106
  ring size and, 99
Neopentyl chloride, 112
Neopentyl halides in S_N2 reactions, 200
Nitration, 280, 285
  diffusion control in, 290
  with nitronium tetrafluoroborate in sulpholan, 289
4-Nitroanilinium ion, acidity, 32
2-Nitrobromobenzenes, substituted, S_NAr of, 37
2-Nitrochlorobenzenes, 4-substituted, methanolysis, 33
Nitroethane, deprotonation, 72
4-Nitroiodobenzene, reaction with azide ion, solvent effect, 49
Nitronium ion, evidence for, 280
4-Nitrophenol acidity, 32
4-Nitrophenyl acetate,
  hydrolysis catalysed
    by imidazole, 53
    by trimethylamine, 69
4-Nitrophenyl γ-(N,N-dimethyl-amino)butyrate, 69
Nitrosyl chloride, attack on cyclo-hexane, 157
Non-bonded interactions, 8
Norbornadiene, 257
Norbornane, 2,3-dihalogeno, 224
Norbornanone, reduction, 319
Norbornene, 241, 349
  bromination, 303
Norborn-5-ene-2,3-dicarboxylic anhydride, 239, 262
Norborn-2-en-7-yl tosylate, 106
2-Norbornyl acetates, equilibration, 117
2-Norbornyl brosylate, 113
Norbornyl cation, n.m.r., 119
Norbornyl derivatives, 170
  classical ions, 115
  non-classical ions, 115
  solvolysis and σ-bond participation, 113
  torsional effects, 96
7-Norbornyl tosylate, 106

Nucleophilic addition reactions, 310, 315

Nucleophilic addition, stereochemistry, 318

Nucleophilic addition-elimination reactions with aldehydes and ketones, 321

Nucleophilic aromatic substitution, 311, 342
base catalysis, 344
leaving group effects, 344

Nucleophilic associative reactions, 310
catalysis, 314, 328

Nucleophilic catalysis, 52

Nucleophilic constants, table, 203

Nucleophilic substitution,
aliphatic, S_N1, 78
aliphatic, S_N2, 194
associative mechanism, 311
bimolecular, 194
in carbonyl compounds, 321

Nucleophilic vinyl addition, 311

Nucleophilic vinyl substitution, 311, 340
leaving group effects, 340

Nucleophilicity, 202
and polarizability, 204
and steric effects, 204
of halide ions, factors affecting, 202
towards carbonyl carbon, 312

Octatetraenes, cyclization, 253, 275

2-Octyl iodide, racemization and exchange, 195

2-Octyl tosylate, 91

Olefins,
addition of radicals, 308
heats of hydrogenation, 298
homogeneous hydrogenation, 55
isomerization by phenylthiyl radicals, 309
Ziegler polymerization, 55

Oppenauer oxidation, 318

Orbital symmetry rules, 239
and photochemical reactions, 245
cycloadditions, 245
electrocyclic reactions, 243
pericyclic reactions, 249
sigmatropic reactions, 248

Oximes, formation, 321

Oxygen, singlet, reaction with olefins, 296

Ozonolysis, 296

pH–rate profiles, 64

π-complex, 281, 290

Partial rate factor, 284

Participation, see Neighbouring group participation, 97

Pentafluoronitrobenzene, 343

t-Pentyl cation, 81

2-Pentyl derivatives, orientation in eliminations, 216

Peresters, thermolysis, 156, 159

Pericyclic reactions, orbital symmetry rules for, 249

Perkin reaction, 324

Peroxides, thermolysis, 158

Phenacyl chloride, 199

Phenols, ionization, 33

Phenonium ions, 106

2-Phenoxyethyl derivatives, elimination, 212

Phenoxytetramethylenesulphoxonium ion, 347

Phenylacetic acids, ionization, 29

Phenyl azide, 241

3-Phenyl-2-butyl tosylate, 87, 106

Phenyl cations, 312, 342, 347

2-Phenylcyclopentyl tosylate, 223

2-Phenylethyl derivatives, elimination, 213

6-Phenylhex-5-ynyl brosylate, 107

Phenyl phosphate, 338

1-Phenylpropene, stereochemistry of additions to 303

2-Phenyl-2-propyl tosylate, 194

Phenyl radicals, 293

Phosphate ester hydrolysis, 337

Phosphonium salts, alkaline decomposition, 326

Phosphorus ylids, 131, 172, 325

Photochemical reactions and orbital symmetry rules, 245

Photolysis of cycloocta-1,3-diene, 252

Piperidines, 1-alkyl-4-chloro, 104

Polar effects, 26
on radical reactions, 162

Polarizability and nucleophilicity, 204
Precalciferol, 270
Protic solvents, 43
Protolysis of carbon–metal bonds, 188
Protonated cyclopropanes, 187
Proton abstraction, rates of, and acidity, 121
Proton exchange,
  and deuteration, 182
  in arenes, 281
  mechanisms, 182
  racemization, 182
  via carbanions, 122, 182
Proton-transfer rates, table, 62
Pseudorotation, 339
Pyridil rearrangement, 52
2,2′-Pyridine, reaction with benzyl chloride, solvent effect, 49
2-Pyridone,
  in bifunctional catalysis, 69
  in cleavage of organoboranes, 189
2-Pyridone, N-methyl, 257

Quadricyclene, conversion to norbornadiene, 56
Quaternization of amines, solvent effect on rate, 44
Quinuclidine, 10
  4-bromo, 104
  2-tosyloxylmethyl, 103
2-Quinuclidone, 73

ρ, 29
Racemization
  in proton exchange, 182
  via ion pairs, 86
Radical anion of anisole, 138
Radicals, 156
  additions to olefins, 308
  and CIDNP, 168
  aromatic substitution by, 293
  e.s.r. spectra, 166
  electron transfer to, 163
  generation, 158
  ionization potentials, 79
  intramolecular additions, 309
  intramolecular reactions, 165
  rates of combination, 158

relative rates of hydrogen abstraction by, 162
stability, 160
Ramberg–Backlund reaction, 232
Reductions by BH_4^- and AlH_4^-, 318
  by electron transfer, 138
  with diimide, 305
Retention and inversion in S_N2 and S_E2 reactions, 180

σ, 29
σ^*, 34
σ^-, 33
σ^+, 32
σ^+ in electrophilic aromatic substitution, 284
σ_R, 34
σ-complex, 281
S_E2 reaction, 181
S_NAr reaction, 343
S_N1 reactions,
  common ion effect, 82
  kinetic evidence, 82
  salt effects, 83
  stereochemistry, 84
S_N2' reactions, 194
  electronic effects, 195, 199
  I-strain effects, 201
  loose and tight transition states, 196
  solvent effects, 195
  stereochemistry, 194
  steric effects, 195, 200
S_N2' reactions, 208
Salt effects, 82, 85
Saytzeff rule, 214
Selectivity relationship, 285
Semibullvalene, 272
Semicarbazones, formation, 39, 311, 349
[1,j] shifts, 248
[1,2] shifts, 266
[1,3] shifts and [1,4] shifts, 268
[1,5] shifts, 269, 275
[1,7] shifts, 270
Sigmatropic reaction,
  definition, 241
  mechanisms of, 266
  orbital symmetry rules, 248

Simmons–Smith reagent, 153
Solvation,
   by dipolar aprotic solvents, dissection of effects, 48
   effects on ΔG^+, ΔH^+ and ΔS^+, 48
   of anions, 43, 197
   of ground states, 48
   of transition states, 48
Solvent activity coefficients, 49, 196
Solvent effects, 40
   on S_N2 reactions, 196
Solvent parameters, table, 41
Solvolysis,
   B-strain effects, 93
   conformational effects, 94
   I-strain effects, 93
   of axial and equatorial derivatives, 95
   solvent effects on, 45
   stereochemistry, 84
   steric inhibition of ionization, 97, 116
   structural effects on rate, 92
   torsional effects, 96
Special salt effect, 87
Specific acid catalysis, 65
Stereochemistry
   of nucleophilic addition, 318
   of S_N2 and S_E2 reactions, 180
   of S_N2 reactions, 194
Stereoelectronic effects, 23
Steric assistance to ionization, 8
Steric inhibition of conjugation, 24
Steric substituent constant, E_S, 35
Stilbene dibromide, 221
Stobbe condensation, 324
Strain energy, 8
   see also Angle strain, torsional strain, non-bonded interactions, bond-length strain
Structure breaking, 40
Styrene, addition of HBr, 308
Styrenes, substituted, polymerization, 37
Substituent constants, table, 30
Substitution, competition with elimination, 227
Sulphones, anions from, 131
Sulphonium ylid, rearrangement, 274
Sulphoxides, anions from, 131

Super-acid media, 80, 111, 119, 185, 281
Suprafacial, 248
Swain–Scott equation, 202
Syn-elimination, 221

Taft equation, 34
Tetracyanoethylene, 19, 257, 259
2,3,4,6-Tetramethyl glucose, bifunctional catalysis of mutarotation, 69
2,3,3,4-Tetramethylpentane-2,4-diol, 102
Thermodynamic control, 12
Toluene, partial rate factors for substitution in, 284
Toluenes, radical bromination, 37
Torsional strain, 8
Tosylate/bromide rate ratio, 196
Transition state theory, 3
1,3,5-Tri-t-butylbenzene, 287
Tri-t-butylmethyl p-nitrobenzoate, 9
Trichloromethyl radicals, additions to olefins, 309
Trigonal-bipyramidal intermediates, 338
Triisopropylmethyl p-nitrobenzoate, 9
Trimethyl phosphate, 337
Trimethyl-4-(2,2-dimethylnonyl)-ammonium ions, 226
Trinitrobenzene, addition of cyanide ion, 73
2,2,2-Triphenylethyl lithium, 137
Triphenylmethyl anion, 23, 129
Triphenylmethyl cation, 80
Triphenylmethyl radical, 156
Triphenyloxonium ion, 347
2,2,3-Triphenylpropyl lithium, 137
Triphenyltin hydride, 157
Triphenylvinyl fluorosulphonate, 341
Tropanyl chlorides, 104
Tropone, 257
Tropylium ion, 80
Tryptycene, bridgehead anion, 23, 124

Van der Waals forces in solvation, 42
Varrentrapp reaction, 136

Vinyl cations, 311, 341
Vitamin D$_2$, 270

Wacker process, 55
Winstein–Holness equation, 95
Wittig reaction, 325
    stereochemistry, 327
Wittig rearrangement, 137
Woodward–Hoffmann rules, *see*
    Orbital symmetry rules, 239

Xanthates, thermolysis, 230
Xanthyl cation, 80
o-Xylylene, 257

Y-value, 45
Ylids, 131, 325
Yukawa–Tsuno equation, 33

Z-scale, 45